THE LYSENKO AFFAIR

The Lysenko Affair

David Joravsky

The University of Chicago Press
Chicago & London

For Esther

1923–1962

The University of Chicago Press, Chicago 60637
The University of Chicago Press, Ltd., London

University of Chicago Press edition 1986
Printed in the United States of America

95 94 93 92 91 90 89 88 87 86 5 4 3 2 1

Originally published by Harvard University Press
in their *Russian Research Center Studies* series.

Library of Congress Cataloging in Publication Data

Joravsky, David.
The Lysenko affair.

Reprint. Originally published: Cambridge, Mass.: Harvard University Press, 1970 (Russian Research Center studies; 61)
Bibliography: p.
Includes index.
1. Agriculture and state—Soviet Union. 2. Science and state—Soviet Union. 3. Lysenko, Trofim Denisovich, 1898–1976. I. Title.
HD1993.J67 1986 338.1′847 86-11303
ISBN 0-226-41031-5 (pbk.)

Contents

Preface

Mean motives will probably be ascribed to me, for I am presenting a history of scandal. I am even dispelling the romantic myth of the scandal, which has been common in the West ever since T. D. Lysenko won the highest Communist support for an effort to abolish the science of genetics. The Marxist theoretical heritage has usually been pictured as the cause. According to this myth genetics was denounced because it subverted faith in the malleability of human nature, and the dying Lamarckist tradition in biology was revived to prove that new conditions will create new men. Thus the Lysenko affair has been pictured as a latter-day version of Galileo versus the Church, or Darwin versus the churches: new science denounced to save old theology. The historical reality was far less high-minded, far more serious. Lysenko's school did not derive from a moribund tradition in science; it rebelled against science altogether. Farming was the basic problem, not theoretical ideology. Not only genetics but all the sciences that impinge on agriculture were tyranically abused by quacks and time-servers for about thirty-five years. The basic motivation was not a dream of human perfectibility but a self-deceiving arrogance among political bosses, a conviction that they knew better than scientists how to increase farm yields. The Lysenko affair, in short, was thirty-five years of brutal irrationality in the campaign for improved farming, with severe convulsions resulting in the academic disciplines that touch on agriculture.

A Soviet historian of this protracted disorder may reasonably expect widespread understanding of his motives: he is bravely trying to set his own house in order. An American historian will be suspected of *Schadenfreude,* and no doubt people who derive satisfaction from tales of Soviet woe will take up this book with expectation of perverse pleasure. At the least I will be suspected of trying to bolster the American self-esteem that rests on disdain for the Soviet Union. My conscious motives have been quite different. I began with the assumption that Lysenko's school must have boosted farm yields—why else would commissars of agriculture repeatedly say so? And I was inclined to make the corollary assumption: however crude, the Lysenkoites must have grasped some truth invisible to academic science.

Even now that I have discovered both assumptions to be wrong, I feel embarrassed sympathy with enthusiasm for a willful leap out of agricultural backwardness, the enthusiasm that Bertolt Brecht captured in his paean to Lysenkoism:[1]

> Lasst uns so mit immer neuen Künsten
> Ändern dieser Erde Wirkung and Gestalt
> Frölich messend tausend jährige Weisheit
> An der neuen Weisheit, ein Jahr alt.
> Träume! Goldnes Wenn!
> Lass die schöne Flut der Ähren steigen!
> Säer, nenn
> Was du morgen schaffst, schon heut dein Eigen! *

Of course my chief sympathies have been with the scientists who suffered persecution because their analysis of the real world subverted the dream of a great leap forward. But the instinctive contempt that their persecutors arouse in me is diluted by the realization that most of the persecutors were ignorant brutes; they only dimly understood the world they were trying to beat into modernity. Indeed, the reader who is quite innocent of biological and agricultural knowledge may share the Soviet leaders' incomprehension as he follows them from faith in autonomous science to authoritarian pseudoscience and back again to faith in science. (If such a reader wishes to replace faith by understanding, he will find in Chapters

* In pedestrian, literal English:

> Let us thus with ever newer arts
> Change this earth's form and operation,
> Gladly measure thousand-year-old wisdom
> By new wisdom one year old.
> Dreams! Golden If!
> Let the lovely flood of grain rise higher!
> Sower, what
> You will create tomorrow, call it yours today!

7 and 9 simplified explanations of the scientific and technical issues that the Soviet leaders stumbled over.)

I hope the reader will move with me beyond idle sympathies to serious engagement with two difficult problems. By what mechanisms of institutional development was the Soviet campaign for modern farming plunged into tyrannical irrationality and then extricated from it? Can this process be objectively analyzed by a foreigner, or is he merely capable of pitting one ideology against another, contrasting "their" wicked blundering with "our" good sense? A sensible first response is to imagine oneself discussing these problems with Soviet scholars, hoping to provoke illuminating disagreement, anxious to avoid rancorous quarrels. In fact imagination can sometimes produce reality. One of the encouragements that kept me at my task was the achievement of such discussions, in 1962 and again in 1965, with a number of Soviet scholars who are gratefully named at the end of this Preface. They differed strongly among themselves, and none of them agreed with all of my interpretations, but we could and did discuss acutely sensitive issues without strife. We argued our cases as scholars should, by constant reference to an abundant stock of factual material.

Many American scholars are surprised to hear that there is abundant factual material presently available for the study of Soviet history. They know that high politics has been effectively screened from public view, except for the first five or at most ten years of the Soviet regime. Since then the locked archives seem to preclude serious inquiry. That is a mistake. I had no access to Soviet archives, and I have concentrated on the years from 1929 to 1965, which include the most secretive period of Stalinist politics and twelve dimly lit years of the spasmodic "return to Leninist norms." I have nevertheless found much factual material bearing on important questions, because I have postponed the conventional first question of historical inquiry: Exactly which high-placed men got together with which others to effect this and that policy? That traditional method of beginning historical inquiry must wait for the opening of the archives, which may not happen until the state begins to wither away. In the meantime we can learn a great deal about the history of the Soviet Union, if we will concern ourselves with other processes than maneuverings at the apex of the political hierarchy. High politics is hidden, but the public record contains much evidence of changing policies, and of the realities they have been designed to master and shape. From the interaction one can draw inferences about the changing patterns of thought and behavior at various levels of Soviet society. Even "they," the bosses, who take great pains to conceal their individuality, reveal their changing group mentality to the historian as they have to the diverse groups of Soviet citizens that they have been trying to manipulate.

I have analyzed the evolving interaction of agriculture, natural science, ideology, and political power. Analogous studies can be made of many other aspects of Soviet society. The prime rule is to determine the political unknown — to the extent that it can be determined — by its interaction with strategically selected knowns, as they are evidenced in the public record. I claim no originality in this approach to Soviet history. It has long been used by economists, literary scholars, and in general by those who appreciate the decisive importance of other institutions and social processes than high politics. I have simply applied their basic viewpoint to a study of the voluminous record left by Soviet scientists, philosophers, and administrators of science, concentrating this time on those involved in the campaign for scientific agriculture. The reader should be constantly aware that Soviet agricultural difficulties have been exceptionally acute. In many respects, therefore, this case study is not typical. Exactly which respects I do not know, for there are no analogous studies of, say, the interaction between physics and engineering within the context of forced industrialization.

In an earlier book I analyzed the philosophical arguments that were provoked by Bolshevik efforts, during their first fifteen years of rule, to win the allegiance of natural scientists. This sequel shifts to drastically different matters, as the Bolsheviks shifted, during the crisis of forced industrialization and collectivization of agriculture. Philosophical debate was petrified, and an intense insistence on practicality moved the focus of discussion from the philosophy of nature to such topics as cropping systems and breeding methods. One might expect ideological influences to abate as such matters came to the fore; in fact, they were greatly intensified. The effort to explain that paradoxical turn of events and the extended trouble it occasioned led me to the intractable problem of ideology. How can we distinguish it from other types of thought? And how can we be precise and factual in analyzing ideological influences in Soviet history? The basic purpose of this study is to help answer those two questions.

Many generous people shared such diverse opinions with me that

I could not possibly have written a book to satisfy them all. Those who provoked me to disagreement were perhaps even more helpful than those who will find their ideas repeated in the following pages. In any case I wish to express my profound gratitude to them all. I owe special thanks to Theodosius Dobzhansky, L. C. Dunn, and Walter Landauer for allowing me to examine their personal papers and to pick their brains. I wish H. J. Muller were still here, so he could judge whether I have made good use of his fascinating reminiscences, and the one item of his Soviet correspondence that accidentally survived the McCarthy years. (He burned the rest when threatened by investigation.) I feel a special debt to C. O. Erlanson, Chief of New Crops Research in the U. S. Department of Agriculture, who not only directed me to relevant archival material, but did the much greater service of patiently explaining elementary matters to an interloper. Many thanks also to E. S. Schultz, G. F. Sprague, William Coleman, and Richard Jeffrey, who had the generosity and patience to criticize sections of a manuscript that has only minor relevance to the topics on which they are profound authorities. Hubert Dyer, Walter Kenworthy, Walter Quevedo, and Sergei Utechin probably do not realize that I derived great benefit from their friendly and learned interest in my topic. I have already explained the crucial value I derived from discussion with friendly Soviet scholars: L. Ia. Bliakher, V. V. Bunak, S. M. Gershenzon, D. V. Lebedev, S. Iu. Lipshits, M. E. Lobashev, Zh. A. Medvedev, I. I. Prezent, E. S. Smirnov, B. P. Tokin, and P. I. Valeskaln. I also wish to thank F. G. Kirichenko, the first disciple of Lysenko I ever met; his passionate apologia helped decide me to undertake this study.

For material support that enabled me to work on this book, I wish to thank the National Science Foundation, the American Council of Learned Societies, the Guggenheim Foundation, and Northwestern University. The Russian Research Center of Harvard University helped me both materially and intellectually. The most important support came, as usual, from my wife Doris.

Two chapters have appeared previously in somewhat different form: Chapter 1 in *Soviet Studies,* July 1966 and the three parts of Chapter 9 in *Scientific American,* November 1962; in *Survey,* July 1964; and in Jerzy Karcz, ed., *Soviet and East European Agriculture* (Berkeley, 1967).

1

Soviet Ideology as a Problem

WHAT ARE IDEOLOGICAL BELIEFS?

When other efforts fail, tired minds invoke the magic name of "ideology" to explain the behavior of Soviet Communists. Sometimes this is a shrug of bewilderment, sometimes a shibboleth, calling on us to distinguish the crackbrained fanaticism of Communists from our own humane good sense. It is rarely an explanation; the key term is too vague. What do we mean by "Soviet ideology?" According to an American political scientist, it "is easily defined. I mean by it the body of doctrine which the Communist Party teaches all Soviet citizens." [1] But why should that body of doctrine be called ideology? Why not call it science, philosophy, theology, *Weltanschauung*, poetry, common sense, or nonsense? The obvious answer is that these types of thought may be found here and there in the body of doctrine but are not the chief distinguishing feature of the body as a whole.

Is there then a distinctive type of thought that deserves the special name of ideology? Some sociologists and philosophers believe that there is. They define it in so many different ways — and use it in still more senses — that one is tempted to agree with Arne Naess and associates, who found ideology a vague pejorative, of little or no use to social science.[2] But the fact that the term is used pejoratively more often than not suggests that one can define it precisely and hope to be understood, if one is willing to define and use it pejoratively. Naess and associates, in common with many scholars, resist such a choice. They wish to be dispassionate students of conflict rather than participants in it, to avoid "classification of out-groups in relation to [in]-groups of social scientists." [3] The trouble with the concept of ideology is that it makes such scholarly neutrality impossible.

A charge of unwitting but self-serving falsehood is the persistent, depreciatory element in the concept of ideology.[4] We can define ideology most briefly as the supposedly rational beliefs shared by a group only because they are members of the group. That is not only a definition; when applied to a group that believes its beliefs to be rational, it is an accusation. It is a charge that certain beliefs and modes of thinking are characteristic of the group not because they satisfy the criteria of truth recognized by any rational mind but because they satisfy, consciously or subconsciously, this group's interests, real or imaginary. It might be pleasant to avoid such pejorative statements, but the decision to subject thought to social analysis rules out the peace of the academy. The philosopher or scientist can aspire to irenic rationality, if he accepts the basic rule that another man's thought should be criticized as such, without regard to its emotional and social functions. But the social functions of beliefs are just what the sociologist of knowledge wishes to discover. If he wants to explain why a group that claims a rational basis for its beliefs actually holds those beliefs, he cannot disregard its claim to rationality. He must ask whether the group's beliefs do have a rational basis; whenever he finds that they do not, he must ask why they are nevertheless accepted as true. At this point the concept of ideology becomes unavoidable.

I will attempt a precise definition. When we call a belief "ideological," we are saying at least three things about it: although it is unverified or unverifiable, it is accepted as verified by a particular group, because it performs social functions for that group. "Group" is used loosely to indicate such aggregations as parties, professions, classes, or nations. "Because" is also used loosely, to indicate a functional correlation rather than a strictly causal connection between acceptance of a belief and other social processes.[5]

The philosophical problems of distinguishing between unverified and verified beliefs become critical in analyses of groups working at genuine knowledge. Thomas Kuhn, for example, argues that natural scientists do not ordinarily set themselves the task of discovering unacknowledged dogma in their own beliefs. When they are forced to such a discovery by unforeseen difficulties, a transformation of dogmatic "paradigms," a "scientific revolution," occurs.[6] That process has a distant resemblance to the development of ideologies, but the basic difference can be sensed by comparing the resistance of scientists and ideologists to the discovery of unacknowledged dogma. In groups of natural scientists the resistance is comparatively slight and brief; in ideological groups such as political parties it is usually intense and prolonged. I believe one can draw a general conclusion from this contrast: The rationality of the scientific enterprise is not absolute, neither is the irrationality of the political enterprise. But the group's commitment to rational examination of its own beliefs is strong enough in the case of natural scientists, and weak enough in the case of politicians, to give the sociologist drastically different problems in analysing the beliefs of one or the other. As a group, natural scientists have little or no interest in clinging to dogma once its existence is discovered. Groups of politicians often have a great interest in doing so; it is just that — their reasons for being unreasonable — that the analyst of ideology seeks to discover. Professional groups of social thinkers — such as economists, jurists, or historians — present the analyst with problems on both levels: the social process of systematic verification, and the social functions of persistent mistaken belief in verification.

Ideology, then, is unacknowledged dogma that serves social functions. The main purpose of scholarly analysis of ideology is to discover the functions. The concept of ideology is the analogue in sociology of the concept of rationalization in psychology, only it is much more disruptive of social harmony, for analysis of ideology challenges the coherence of groups rather than individual personalities. Ideology is one of those nasty terms, like exploitation or aggression, that threaten to poison the well of rational discourse yet cannot be avoided, for there are nasty aspects of human affairs that must be named. It should be used only when rigorous investigation has shown that a group accepts dogmas as rationally verified knowledge. Until then, one should use neutral terms like "beliefs," "social psychology," or "*Weltanschauung*."

Ideology is a special type of *Weltanschauung*, a constituent part of social psychology. Ideology is no more hypocritical or mendacious than religion. If subjects feign belief out of fear of their rulers, the subjects are hypocrites or liars and the rulers are ideologists — unless it can be shown that the rulers themselves do not believe what they are forcing their subjects to profess. (One sometimes gets the feeling that this nightmarish situation was reached in the last years of

Stalin's reign.) Widespread subtle hypocrisy presents a special problem for the student of ideology; it is the hazy border between ideology and mendacity.

It may be said that those are unnecessary scruples for the student of Soviet affairs. Communist writers use the term "ideology" to describe their own thought, and we may simply take them at their word. But the matter cannot be settled so simply. We do not take them at their word when they call Marxism-Leninism a science or when they call their government a representative democracy. Why should we insist on the proper meaning of words when the Communists appropriate eulogistic terms but readily concede their appropriation of pejoratives? [7] We certainly do not accept the special pleading by which they have tried to convince themselves that ideology ceases to be a pejorative when applied to Communist doctrine. They keep telling themselves that theirs is the only ideology to escape the distortions of self-serving group belief, because theirs is the only group whose interests are served by true philosophy and genuine science. Indeed, the Communists themselves have been finding—at great expense—that the equation of their ideology with science and philosophy is an intolerable circular argument. In 1931, Stalin suppressed the efforts of Communist thinkers to disentangle those three types of thought,[8] only to find himself obliged, in 1950, to resurrect the problem.[9] Communist thinkers have been struggling with it ever since.[10]

The heart of their difficulty is simply this. If one does not distinguish between service of group interest and objective verification as essentially different bases for belief, then the cognitive enterprise becomes wretchedly inefficient, including that part of cognition that distinguishes between real and imaginary group interests. How can the leaders tell if their beliefs and policies are really in the group interest? They can do it in Stalin's "practical" way, by adhering to a belief—and a policy based on it—until its presumed service to the group is overwhelmingly disproved by sad experience. Or they can do it the "theoretical" way, by permitting autonomous thinkers to subject beliefs and policies to rational discussion. Either way is painfully awkward for political chiefs in the Soviet Union, as it is in varying degree for chiefs in any system. The way out of the dilemma, it often seems, is to distinguish between a sacred class of absolutely true beliefs and policies, which may not be subjected to discussion, and a class that may. Some people distinguish ideology from other types of thought in this way.[11]

But clearly that is not a genuine escape from the dilemma. At any given time there are issues in the nonsacred class that are causing the leaders to worry: Does it endanger group interests to have such matters subjected to public discussion? Shouldn't we settle this matter, rally the masses around our solution, and get on with the business of building communism? At the same time, there are some beliefs in the sacred class that are causing the leaders to worry: Are these beliefs really in the group interest or do we only imagine them to be so? Will not the group interest be better served by allowing autonomous thinkers to scrutinize these beliefs? The conflict between the two styles of Communist thought is not only logically possible, it is historically observable. The first, the boldly intuitive authoritarian style, was increasingly dominant from the appearance of Lenin's *What Is To Be Done?* in 1902 until 1950, when Stalin excluded language from the "ideological superstructure" and suggested that the exclusion of science was a debatable issue. The second style has been increasingly evident since then, but both have coexisted and conflicted all along. Seen in historical context, Stalin's denunciation of an Arakcheev regime in science, his solemn observation that "no science can develop and flourish without a battle of opinions, without freedom of criticism," [12] was not only outrageously hypocritical. It was also an unwitting confession of self-defeat. The conviction of the leaders that their views were scientifically true and therefore *not* subject to discussion had been derided by a great many "revisionists" and "deviants"; now it was placed in doubt by Stalin himself.

To say that one is simply following Communist usage in calling their thought ideology is to assume their tormenting obligation to believe that their group is unique in its partisan apprehension of objective truth, in its exemption from any conflict between self-serving belief and verified knowledge. It is far more sensible to note that Communists are like the rest of us. Reasoned analysis can discover both ideological and nonideological elements, both among their sacred beliefs and among their nonsacred. Indeed, I hope that the preceding discussion has established one of the elements of verified truth in the sacred doctrine of *partiinost'* (partyness or partisanship): In analysing group beliefs, the sociologist cannot avoid taking sides — or making them — for he cannot avoid accusations of ideology. Even if he is purely rational, rational discourse between the sociologist and the group he is studying will be as unlikely as it is between men of opposed ideologies.

In the analysis of ideology the distinction between unverified and unverifiable belief is crucial. Beliefs that have had the most powerful emotional appeal — such as "all men are equal" or "the land belongs to the people" — are so vague as to be unverifiable, unless they are translated into one of their many possible meanings. In that respect, the grand ideological slogans resemble poetry, which often affects the emotions by its ambiguous reverberations in the intellect. Whatever the psychological explanation of their appeal may be, the social consequence of the ambiguity of ideological slogans is that efforts to translate them into specific beliefs and policies are unavoidably arbitrary and fiercely disputed. Let us call such beliefs, and the arguments they arouse among their adherents, "grand ideology." Some people consider this to be the only kind of ideology. In such a view, thinking can be called "ideological" only to the extent that it is a serious effort to draw specific conclusions from the vague precepts of grand ideology.[13] But this restricted definition inhibits the sociological analysis of thought, for it blinds one to the importance of specific, verifiable beliefs that cluster about the grand ideologies.

Consider, for example, an argument that was fairly common in the seventeenth to nineteenth centuries among the growing number of men who were attracted to the proposition that all men are equal. They could not agree whether all men should have equal voting rights, in large part because many assumed that universal suffrage would allow the poor to control or even to take the property of the rich.[14] To call this contention over voting rights ideological merely because the contenders were divided on the meaning of the grand, vague principle of equality, is to overlook the ideological nature of the specific, verifiable proposition that played a major role in that quarrel: Universal suffrage would allow the poor to control or even to take the property of the rich. Many supporters of private property took that for granted without anything like adequate evidence because they were afraid to put it to the test. Hope rather than fear warped the thinking both of the socialists, who took it for granted, and of the nonsocialist radicals who tried to brush aside the potential conflict between the equal right to vote and the equal right to hold property (in unequal amounts). In any case, the historian is dealing with thinking that had to be warped by unverified belief, even though it concerned a fairly clear and verifiable proposition. A purely rational view would have been that the relationship between universal suffrage and property systems is not known, for universal suffrage has not been tested in enough countries for a long enough period. But this purely rational view could not have justified the choice between testing universal suffrage and refraining from testing it. Only ideological hope or fear could do that — or political "realism," as radical and socialist ideologists forced enough testing to reveal that universal suffrage is compatible with a variety of property systems.

Ideology, then, includes verifiable but unverified beliefs as well as unverifiable or grand beliefs. We may as well call the former "petty ideology."

Social functions are, of course, performed by other types of thought than theoretical ideology and genuine knowledge, which have been contrasted so far. For present purposes, in order to isolate and identify ideological elements in thought, it is necessary to consider what politicians call "common sense" or "realism." Many political scientists give this concept the philosophical name of "pragmatism" or "empiricism" and consider it to be the opposite of ideology. Here it will be called political "realism" for it can be quite unrealistic in any meaningful sense of the term, and philosophical dignity it does not deserve.

Political "realism" is a constantly shifting jumble of commitments to particular judgments and persons, but basic rules of thought can be discovered in it. Perhaps the most basic, the golden rule of the "realistic" politician, is his practical way of recognizing that politics is the business of arranging people in hierarchies of power: One must avoid statements that might lose the support of important people, that is, people on whom one's office and career depend. The rule can also be put in positive form, which is characteristic of the more energetic and ambitious "realist": One must seek out statements that would gain the support of important people. A person who takes chances with that rule is considered courageous and farsighted, if he wins more influence than he loses. He is brushed off as a fool, a trouble-maker, or an "ideologist," if he persistently loses. (In the vocabulary of many "realists" — Napoleon, for example — ideology gets its special meaning of impractical theorizing.[15]) To the extent that a politician is committed to that basic rule, he is implicitly committed to an equation between what is politically useful and what is genuinely right, without distinction between the factual and moral meanings of "right." His basic principle, though rarely stated, is evident in his behavior: If a belief reduces one's influence in one's group, it is wrong; if it increases one's influence, it is right. This is an ideological principle. It satisfies not only the three conditions of my definition of ideology; it also exhibits the confusion of and factual and moral judgment that some writers consider another essential characteristic of ideology.[16]

It may be objected that I have been too hasty calling that principle "ideological." Careful definition of the terms and elaborate specifications as to the organization of the political system might ensure a coincidence of moral rightness and political usefulness. Moreover, few politicians are so committed to unadulterated "realism" as to forget other standards of right and wrong. Those objections are beside the point. Political "realism" is being treated as a type of thought. Of course it may be mixed up with other types of thought in the mind of an actual person; and, of course, the definition of political usefulness is conditioned by the particular system within which a particular "realist" operates. (But I know of no political system that ensures a coincidence of moral rightness and political usefulness.)

As for factual rightness, the political "realist" may be aware of the need for genuine knowledge to solve technical problems, but the technical specialist — to quote another rule of political "realism" — must be on tap, not on top. (That is also an ideological belief. It assumes that political and technical issues can be separated, while in fact they are deeply and increasingly interdependent.) The point is that the complete political "realist" — the ideal type — is concerned above all with getting and keeping power. To the extent that a flesh-and-blood politician is "realistic," he is unwilling to consider other criteria of rightness than political usefulness; to that extent, he is extremely susceptible to unverified and unverifiable beliefs. If the influential people in his group are going mad, the political "realist" goes mad with them.

WHAT ARE SOVIET IDEOLOGICAL BELIEFS?

It is clear to most Western students of Soviet affairs that grand and petty ideology have been giving way to political "realism" in the thinking of the Soviet elite. That process has been called the routinization, erosion, exhaustion, or even the end of ideology. Such phrases are accurate, if one specifies that it is the old theoretical ideology, grand and petty, that is being routinized, eroded, or exhausted. They are misleading in their suggestion that political "realism" is devoid of ideology, simply because it is averse to serious theorizing.

Consider, for example, Stalin's favorite ideological slogan during the last twenty-four years of his reign: In the dialectical unity of theory and practice, theory guides practice, but practice is the criterion of theoretical truth. This sounds like theoretical ideology; it invokes the intellectual heritage of Marx (and Hegel too, though Stalin was apparently unaware of that). With precise definition of terms, it could even be turned into genuine philosophy. In context, however, Stalin's slogan functioned most often as a decorative flourish on one of the maddest versions of political "realism" imaginable. The implicit rule of the "realist"—important people are usually right—was turned into the rule that the most important people are always right. They read the lessons of practice, and their reading ceases to be right when they cease to be important. This was ideology at its fiercest, even though—or because—theorizing was reduced to such impotence that it could serve neither as a guide to action nor as an effective rationalization of action. The triumph of such extreme political "realism" in 1929–30 explains the paradox of Stalin's wildly ideological revolution, which petrified theoretical ideology. "The body of doctrine taught to all Soviet citizens" became, after 1929, so leaden-spirited and pointless as to defy belief, except where it seethed with hatred for those who doubted the transient but absolute wisdom of transient but absolute chiefs. After 1929, the ideology actually at work in the minds of the chiefs is to be found much more in their intuitive judgments of practical matters than in the largely irrelevant texts of theoretical ideology.[17] To a considerable extent, that is still the case. We ought to recognize what is overwhelmingly apparent to any student yawning over *Osnovy marksistskoi filosofii* (*The Fundamentals of Marxist Philosophy*, a recent Soviet textbook). Much of what passes for theoretical ideology in the Soviet Union is a traditional survival. It performs the same function as Corinthian columns on modern public buildings or the invocation of Jesus' name at the launching of a nuclear submarine.

Of course, obviousness is not proof. The only way to prove which ideological beliefs have performed what functions in the social process is to study the beliefs and the social process from the vantage point of genuine knowledge. Consider, for example, this belief, which was mandatory in the 1930s: The land belongs to the people, and therefore collective farmers hold their land rent free.[18] This seems to be a characteristic conjunction of grand and petty ideology; it presents a specific, verifiable statement as a logical consequence of a vague but stirring principle. (It is comparable to the belief that all men are equal, and therefore Soviet schoolchildren are graded only by achievement, without reference to native intelligence.)[19] An outsider need not assume the group member's obligation to establish a logical connection between statements that have only an emotional correlation. Neither should time be wasted on elaborate investigation of the supposed connection. The logical filiation of ideas, which is the first, most obvious concern of the historian of science or philosophy, is the last, most subtle problem for the historian of ideology. His model should be, not Voltaire's brilliant mocking of religious illogic, but the anthropologist's strenuous effort to discover the social functions of various types of thought. As the student of primitive religion begins his analysis of rainmaking ceremonies with the quiet assumption that they do not affect the weather, the student of Soviet ideology should begin his analysis with the observation that rent has existed in the Soviet Union, whether or not Soviet leaders have been aware of it.

That initial step is worth pondering. The outside observer has easily identified an illogical argument and an unverified belief by reference to his own genuine knowledge of logic and economics. Very serious difficulty is encountered when one chooses to study beliefs about things like mass terror or political factions. One cannot start with well established principles concerning such things. Political theory cannot serve as the vantage point of genuine knowledge. For example, a careful student of the Soviet political process remarks: "Apparently, it is in popular insecurity that the regime's security has been thought to lie."[20] Ignore for a moment the telltale qualification "apparently"; assume that this has been an implicit belief of Soviet leaders more often than not. Is it an ideological belief? There are two ways of discovering whether it is. One is by reference to a mass of data that might contradict either the Soviet leaders' reading of their subjects' feelings or their reading of the connection between popular feelings and the security of the regime, or both. But one does not have access to the Soviet archives that would make possible such an important contribution to the analysis of ideology in the Soviet political process. The other method is to refer to a general principle already established by study of many other political systems. One might say: Whether or not the Soviet leaders have been aware of this truth, regimes that rest on popular insecurity are themselves insecure.[21] But such a statement is itself ideological, for it puts the seal of truth on an hypothesis, a proposition that requires verification precisely by such studies as, in the crucial Soviet case, cannot be made.

I think that example can be generalized. The meagerness of data on the Soviet political process and the deficiency of science in political science are mutually reinforcing limitations, which cripple the analysis of Soviet ideology in purely political matters, or turn the analyst into an ideologist himself, contesting beliefs he is not sure the Soviet leaders have with the unacknowledged dogmas of his own group. The only significant exception to this rule is the first five or at most ten years of the Soviet regime, when Soviet leaders made serious efforts to state their operative political beliefs in explicit theoretical form, and a significant part of the political process was publicly documented. After that brief period, the increasing replacement of theoretical ideology by closemouthed "realism," and the growing passion for closed politics, limit the Western student to very gross inferences about the interaction of political beliefs and political processes.[22]

But political leaders must think about other things than politics — about economic and educational processes, for example — and many of these thoughts must be publicized, along with data on the processes they concern. Here the Western student of Soviet ideology gets a chance to do thorough studies of the interaction between beliefs and social processes. Consider further the example of rent. Serious analysis begins when one asks how the systems of agricultural procurement have been distributing rent from the 1920s to the present, and how beliefs and systems have been interacting and changing each other. Those are difficult questions, but they can be answered by examination of material available in the public record. Indeed, the economists who study the Soviet Union have already gone a long way toward answering them. They have analysed the changing systems of agricultural procurement in considerable detail and have discovered that Soviet economists and political leaders have been growing increasingly aware of the troubles they cause themselves by their reluctance to recognize the problems of measuring and distributing rent. Since the analysis of ideology is not the primary interest of economists, they have limited themselves to noting the economic dysfunctions resulting from the denial of rent.[23] They have not asked whether the denial has also served useful functions.

Of course, it may not. Stalin may have been simply mistaken or dysfunctionally mad — not crazy like a fox — in repressing the discussion of rent. But one should note that he did so in 1929–30, as he got deeply involved in collectivization.[24] The discussions of the six years prior to 1929 ought to be studied to discover what considerations and proposals Stalin was repressing. Then one could ask whether, in the conditions attending forcible collectivization, rent could have been rationally measured and distributed by any of the procedures suggested by the economists of the 1920s. The sudden increase in the state's extraction of produce from the peasants may have been so great as to make sport of any effort to promote long-run efficiency by a discriminating distribution of the proceeds of basic ground rent, land improvement, capital intensification, and well-motivated labor.[25] If so, then a latent function of the denial of rent, like other drastic repressions of theoretical enquiry in the same period, was to reassure "realistic" leaders that an insoluble problem, the result of their own wild action, did not exist. Their recurrent tendency to evade the knotty problems of increasing productivity, to take the deceptively simple expedient of increasing acreage under cultivation, enables one to study the self-perpetuation of this ideological pattern of thought. (Even now there are Soviet economists who say, "Under socialism there is no poor land. There is poor work, and there is good work." [26])

Of course, one can test other hypotheses, not only by comparing the theoretical discussions of the 1920s with the violent "realism" of the 1930s, but also by examining the revival of theoretical discussion in the last years of Stalin's reign and especially in the period since his death, as the "realistic" chiefs came to acknowledge many problems they once denied.[27] By detailed study of the process of acknowledgment before and since Stalin died, one can even measure the relative weight of Stalin's personal whims, the group mentality of Soviet "realists," and the thinking of genuine economists. The main point here is simply that one can frame and test hypotheses about the interaction of beliefs and economic processes. There is enough science in economics and enough evidence of beliefs and processes in the public record.

The Western student of Soviet ideology faces a choice. He can limit himself to areas of thought where Soviet ideology can be identified from the vantage point of genuine knowledge and its social functions discovered by rigorous empirical scholarship. Or he can turn boldly to the political process, where ideology is most important — and nearly impossible to study in a rigorously empirical scholarly fashion. He can aspire to a scientific analysis of Soviet ideology or to an ideological critique of it. The choice can hardly be thoroughly rational; it is unavoidably influenced by ideological hopes and fears concerning the relationship between politics and scholarship. I think Western students of Soviet ideology should limit themselves to areas where rigorous empirical scholarship is possible; but I also think that boldly ideological critiques of Soviet political ideology are a vital necessity. I cannot justify this apparent inconsistency without indulging in a little ideological speculation of my own.

The chief function of ideology — whether theoretical, "realistic," or a mixture of the two — is to rationalize a group's readiness to act, or to refuse to act.[28] I am revising Sorel's observation that *myth* is an *affirmation* of a group's readiness to act. Few important groups in the modern world — the age of science — are willing to cast their justifications of policy in plainly mythical and willful forms. Even the fascists were not consistent in their boasts of irrationalism and voluntarism. We may be as brutish as our ancestors, but the Enlightenment has taught us to justify our brutishness by appeals to reason. That slight restraint on irrational action is one of the most beneficial influences of science; but it makes ideology, the apparently rational justification of group policies, unavoidable. The argument that ideology is avoidable comes down either to the absurd belief that political leaders can have enough genuine knowledge to justify all their policies, or to Burke's belief that, since reason cannot solve political problems, a commitment to muddling through is rational. The first belief is absurd because there is no prospect of a comprehensive science of society in the near future. As for Burke's view, it may turn out to be true — muddling through is being tested all the time; all we need is an adequate control. But two features of it are already indisputable: It is presented as a verified truth, though it is either unverified or unverifiable, and it serves the interests of groups resisting change. In short it is also ideology, conservative ideology.

Recognition of the inevitability of ideology in the thought that accompanies political action (or inaction) does not entail recognition of its inevitability in the detached analysis of such thought. That point can be made by philosophical argument.[29] It can also be made by social observation. The outside observer of a group's thought can demand rigorous proof of the beliefs that seem quite obvious or sacred within the group, for he does not aspire to leadership within it. A less flattering way to say the same thing is that the outside critic of a group's ideology is an intellectually arrogant, politically impotent critic of men who must humble their minds in order to accept responsibility. But his political impotence is not absolute. When the functions served by ideological beliefs are increasingly outweighed by dysfunctions, the beliefs decay, giving way either to genuine knowledge or to new ideological beliefs. In the process, the outside critic can have an influence within the group.

We have seen this happen even in the relationship between Western scholars and Soviet ideologists, where the sense of in-group and out-group has been extremely exacerbated. The knowledge cultivated by Western linguists, physicists, chemists, cyberneticists, and biologists has been corroding the beliefs of Soviet ideologists. Of course, those disciplines have had native Soviet practitioners, which has reduced the resistance of Soviet ideologists. (Though Soviet scientists are an out-group with respect to Soviet ideologists, they are not as far out as Western scientists.) But even Western economists and philosophers, who seemed for a long time to have no Soviet colleagues, have been exercising a corrosive influence on Soviet ideology.[30] Soviet politicians have been accommodating themselves to that process by narrowing the extension of their "realistic" ideology — granting autonomy to more and more fields of knowledge — while still insisting that bourgeois and revisionist devils have nothing useful to say within the shrinking field of ideology itself. Nevertheless the narrowing of extension and the change of intention it necessitates, that is, the redefinition of theoretical ideology by a slightly autonomous group of philosophers, are impressive evidence of the power of genuine knowledge to corrode ideology.[31]

How far can it go? That is really two questions. One concerns

Soviet institutions. How far can Soviet rulers divest themselves of prophetic and sacerdotal pretensions, how much autonomy can they grant to intellectuals, including the philosophers working at theoretical ideology? Soviet rulers show signs of sensing the dilemma that tormented their monarchic predecessors (as it did the rulers of seventeenth-century England and eighteenth-century France). If Soviet leaders grant more intellectual autonomy, they may undermine the ideological reverence they expect from their subjects. If they do not, they may undermine it anyway, by creating an ever widening chasm between themselves and the intellectuals. But there is a broader question here than the relations between Soviet rulers and intellectuals. How far can rulers dispense with ideology in their own thinking? Can genuine knowledge of human society ever reach such a level that ideology will become unnecessary in political action?

This is the kind of question that drives a person to ideology. If the answer is no, the concept of ideology becomes a grim joke, as far as political action is concerned. Reason is called upon to discover the social functions of beliefs that cannot but be irrational. The critic of ideology becomes, at worst, a cynical partisan, turning against the beliefs of groups he dislikes the rational criticism he refrains from using against his own. At best, he can hope to be a tragic isolate, whose social function it is to help corrode existing ideologies so that new ones may take their place. If the answer is affirmative or tentative — we cannot know whether genuine knowledge will ever suffice for political action until we have tried in every way to make it so — the prospect is only a bit less bleak. The major political problems will not wait. While the patient scholar toils at his dream of political science, ideologies must rule the world, or ruin it. If criticism of ideology is the scholar's part of the great enterprise, he must constantly end his exposure of the unacknowledged dogma in a group's beliefs with the admission that he has no verified belief to take its place. That is most painfully clear in the field of international politics. Machiavelli's basic rule — defense of the national state is worth all sacrifices — has become so clearly dysfunctional — "all sacrifices" now include destruction of the nation — that many "realistic" politicians are deeply troubled by their adherence to it. Yet they can see no alternative, and the political scientist cannot offer them one without becoming, to some degree, a proponent of an anti-Machiavellian ideology.[32] The scholar's natural impulse in such circumstances is to mix ideology with genuine knowledge, and

I approve the mixture, if it suits my ideological beliefs and if it is done with a strong show of rationality. This is not said cynically; a strenuous effort to think rationally about our irrational beliefs is one of the best services we can do them.

That is why I think that we Western students of Soviet ideology should restrain our natural impulse to deal with the large political issues where our own ideology must come into play. In such a context, our own ideology is not pressed toward rational doubt; it is too easily mistaken for the vantage point of genuine knowledge. Consider again Stalin's mad plunge into breakneck industrialization and forcible collectivization. Our natural impulse is to accuse him of criminal blunder. But what shall we do with his conviction that industrialization could not indefinitely depend on the peasants' requirement of reasonable terms of trade? That was as much a political as an economic conviction. Before we can judge it, we must understand what led Stalin to it. All the meaningful hypotheses point vainly to the locked archives of 1928–29, when a contrary conviction, previously shared by Stalin himself, was suddenly rejected. As long as our ideological fancies are not confined in those archives, they soar too easily. Some lecture Stalin on the wisdom of bargaining with peasants rather than bludgeoning them. Others condemn or pity the poor brute because Russia — or Bolshevism, or the universal devils of the twentieth century — had trained him to bludgeon rather than bargain. Others shrug off the problem with the "realistic" tautology, "the past cannot be undone." And how shall we deal with Stalin's conviction that Russia had to catch up with Western military potential in ten years or be crushed? We can only counter with ideological convictions of our own, beginning with a position on Machiavelli's basic rule that defense of the national state is worth all sacrifices. If we do not recognize that Machiavelli's rule is as much our problem as Stalin's, we fall into smug chatter about the totalitarian disregard for human life. If we do, we become either utopian dreamers or retrospective Herman Kahns, calculating the relative cost in megadeaths of this policy or that, finding Stalin's "revolution from above" a successful though inefficient feat of engineering.

My own conviction is that it will be a great day when Soviet intellectuals begin serious ideological controversy over such issues. But I do not think that we Western specialists on Soviet affairs can do the job for them. We cannot forge in the smithies of our academic souls the uncreated conscience of Communist society. We lack two vital conditions. The ideological theorist uses intuition to leap over gaps in genuine knowledge, and one must live within a society to have a fair chance of doing that well. In the second place, a strong revulsion against the outworn stereotypes one was trained to accept is as vital to ideological creativity as to artistic. Our revulsion comes from without; and it comes too easily, too often degenerating into a self-serving judgment of Soviet ideology by reference to the stereotypes we were trained to accept.

Do we know the actual beliefs of Communist leaders and the social functions of those beliefs? When asked with dogged persistence in Western societies, that question produces crawling empiricism. When asked with dogged persistence in Communist societies, it produces rebellious ideology ("revisionism"). Social thought is socially conditioned. But it does not follow that social thought must be futile. Our crawling empiricism and their rebellious ideology corrode the stereotyped beliefs of the dominant "realists" on both sides.

2

A Crisis of Faith in Science

In most of the countries we call advanced or highly developed, modern farming emerged before scientists could explain it or help it along, yet faith in agricultural science grew widespread and strong. In Russia, the sciences that bear on agriculture ran far ahead of actual farming methods, pointing so long to unattainable improvements that a crisis of faith occurred — a disruption of belief in the utility and truth of those sciences. The utility of science depends more on the readiness of practical people to accept scientific advice than on the ability of scientists to deliver it. Even the truth of science is not entirely independent of social conditions; it will have a hard time getting discovered or taught unless practical people favor the scientific enterprise, usually on the basis of faith in its utility. There is a disturbing possibility of a vicious circle in that relationship.

Diversification of crops was the hallmark of modern farming as it emerged in Western Europe in the seventeenth to nineteenth centuries.[1] Medieval rotations of grain with fallow were transformed into complex rotations of grain with leguminous herbage (such as clover and alfalfa) and intertilled crops (such as turnips, potatoes, and sugar beet). Propagandists of the intensive cropping systems and of improved implements and breeds became strikingly numerous and effective in the eighteenth and early nineteenth centuries, as the industrial revolution intensified the demand for farm products and shifted much of the labor force away from agriculture. There was much talk of "scientific" farming, though the new methods were largely the result of unexplained trial and error. The "scientific" element consisted in the widespread effort to break free of unexamined custom, to discover methods that could be justified by rational analysis of empirical results. Genuinely scientific farming emerged only in the second half of the nineteenth century, as chemists and biologists learned how to explain what farmers were doing, and began to tell them how they might do it more effectively. When, to take the most notable instance, Liebig's mistaken guess about the source of nitrogen in plant nutrition was corrected, his dream of agriculture as applied chemistry could proceed toward realization.[2] Where the price of mineral fertilizers and the farmer's mentality permitted, the fertilizers could be used with startling effectiveness, doubling average grain yields on top of the doubling achieved by the spread of modern crop rotation.[3] That belated justification of faith in the utility of science helped to carry other disciplines through critical periods. Geneticists, for example, thought at the beginning of the twentieth century that they were experiencing a great breakthrough not only in pure science but also in practical applications. Actually they were establishing first principles in pure science. They could make only modest contributions to the breeder's art of trial and error, but their new science kept its reputation of great utility, even while the early exaggerated hopes were being disappointed.[4]

Russia experienced an anomalous reversal of the Western pattern.[5] While Russian propagandists and students of improved farming methods became sufficiently numerous to form societies, beginning with the Free Economic Society in 1765, the overwhelming majority of Russian farmers showed almost no interest, and were still showing extremely little interest in 1929, when Stalin's impatient government tried to force the modernization of farming by collectivizing the farmers. At the time, the overwhelming majority of Russian farmers were still following the grain-fallow rotations of their medieval ancestors, while Russian soil scientists had won an international reputation as the virtual founders of their discipline.[6] Russian peasants were still broadcasting mongrel folk varieties of seed, while Russian botanists had started to organize the world's greatest collection of living plants in order to breed improved varieties according to the new science of genetics.[7]

That anomaly was in large part the result of the tsarist government's intermittent efforts to modernize the country without disrupting its social structure. Favorable conditions were created for the rise of a vigorous scientific intelligentsia, while very little was done to encourage the rise of improving farmers. One can perceive this pattern already in the reforms of Peter, Catherine II, and Alexander I, all of whom gave considerable help to higher learning and very little to primary education of the peasants. As for the burden of serfdom, which was more powerful than illiteracy in stifling agricultural improvement, Peter and Catherine *intensified* it. Alexander I toyed with the thought of relieving it but did almost nothing. The model farms and agricultural schools that those rulers established had, therefore, a negligible effect on farming. Even noble landlords, who could read the publications of agricultural societies and could afford to experiment as the peasants could not, were usually disenchanted by their fitful efforts to modernize farming. Complaining of unsatisfactory labor, weak markets, and costly transportation to those markets, most of the nobles stuck to the traditional methods. In short, nineteenth-century Russia was the prototype of a species that has since become very numerous: the underdeveloped agrarian country that produces a great deal of talk about modernization and very little effective action.

Alexander II took the first plunge into basic social reform by abolishing serfdom in 1861, but this bold act aided the progress of scientific agriculture much less than the progress of agricultural science. The Emancipation reorganized rather than alleviated the great load of obligations that stifled peasant initiative. In effect, the government began collecting the dues that the privately owned peasants had formerly paid to their masters. The less onerous obligations of the state peasants were also converted into "redemption" dues. The large compensation paid to the noble masters was not, for the most part, used to modernize their estates. Only a minority of the landed gentry, and a much smaller minority of peasants, were stimulated by the Emancipation to adopt improved methods.[8] But the contemporaneous creation of the zemstvos, elected organs of local government, was eagerly exploited by the intelligentsia. The zemtvos were charged with the task of educating the peasants and extending "agronomic aid," building an agricultural extension service, to use the American expression. Near Moscow, the Petrovskaia Agricultural Academy was established to train the necessary specialists and to carry on scientific research. The Academy and the zemstvos that took their tasks seriously became centers of liberal and radical agitation.[9] It is hard to imagine how they could have taken their tasks seriously without the sustenance of liberal and radical ideology, that is, without the conviction that further reforms would bring a great release of creative energy in the countryside.

Even with this conviction they were dogged by a sense of futility. As late as 1914 an official history of the extension service referred to "the constant — until recently — vacillations and doubts among zemstvo people concerning the usefulness of this activity." [10] The most steadfast zemstvo people carried their doubts just below an earnest exterior. When the thirtieth anniversary of Emancipation was marked by a catastrophic famine, specialists stubbornly convened to discuss methods of drought control, but their underlying mood was hardly optimistic. The chairman was an exceptional enthusiast. He took his colleagues to an impoverished village in a badly eroded ravine, where he expatiated on the affluence that could be realized by planting trees here, digging a pond there, and running irrigation lines there and there. "And here," interjected a friend, "they will place a monument to Jean Jacques Rousseau!" "Why?" "Because he too was a great dreamer." The excursion broke up in laughter, probably debating the causes of peasant resistance to improvement.[11]

Far from seeking a release of rural energy by further reform, the central government recoiled against the liberalism and radicalism that it had stimulated by the limited reforms of the 1860s. In 1890 the autonomy of the zemstvos was curtailed — liberals considered it a triumph that the zemstvos were not abolished — and the Agricultural Academy was ordered to accept no more students. With the graduation of 1893 it was to be dissolved. However, because of the famine, even the most retrograde bureaucrats could not blind themselves to the need for some effort to modernize agriculture. In the buildings of the defunct academy, the Moscow Agricultural Institute was created. Every effort was made to block a resurgence of liberalism and radicalism. No Jews were to be admitted, and student fees were to be paid in advance, in order to keep out poor and dissatisfied Christians. The former faculty was not to be rehired. (One of the few exceptions was V. R. Williams, a politically reliable soil scientist who had been named curator of the empty buildings.)[12] These precautions proved as futile as the curtailment of zemstvo autonomy. Liberals and radicals continued to dominate agricultural research, education, and extension service — indeed a great upsurge of zemstvo liberalism came at the turn of the century — still convinced that they would see "the earth renewed" under "the solicitude of a really democratic government of a free country." [13]

The Revolution of 1905 jolted the tsarist government into its last effort at reform. The peasant effort to improve agriculture by forceful seizure of the nobles' lands was crushed, but the burden of peasant obligations was lightened, and Premier Stolypin undertook his own agrarian reform, a gradual reorganization of landholding designed to encourage the minority of improving farmers. They were encouraged. In the few years of reform, before the government took its backward nation into the first total war, agricultural productivity registered modest progress.[14] But the progress of agricultural science, stimulated by a sharp increase in government appropriations, continued to be much more rapid than the progress of scientific agriculture.[15] The gap between the affluent possibilities disclosed by the experiment stations and the impoverished reality of the surrounding countryside continued to grow. Dokuchaev's school pursued their work on the genetic classification of soils, which was winning Russia the reputation of founding soil science. Prianishnikov was creating a school of agricultural chemists fully abreast of the German and French pioneers. The Institute of Applied Botany was beginning to scour the world for plant material of possible use to Russian breeders, who were learning the new science of genetics to speed the creation of valuable new varieties. But the overwhelming majority of farmers were not using the improved varieties that were already available. In 1913, the peak year of prerevolutionary agriculture, virtually all of Russia's grain fields were still sown to mongrel local varieties so heavily infested with weeds as to maintain the folk belief that the seed of wheat can engender rye or wild oats.[16] Prianishnikov, working on the most effective use of mineral fertilizers, was painfully aware that they were far too expensive for Russian farmers, who had not yet learned to use leguminous herbage, the cheap green manure that feeds the soil and the livestock and feeds the soil once again, if the increased supply of animal dung is properly collected and spread.

In such circumstances it is hardly surprising that Prianishnikov had to argue with skeptics who doubted that mineral fertilizers would ever be useful on Russian soil.[17] Although this particular argument turned on technical questions that were resolved in Prianishnikov's favor, it foreshadowed the conflicts that would arise when scientists could no longer blame a sluggish government for the great gap between advanced research and backward peasant practices. Scientists would begin to blame each other, as a few were doing even before the Bolshevik Revolution. For example, Prianishnikov took for granted the practical value of his own discipline, agricultural chemistry, but doubted the utility of Dokuchaev's school of soil science. Russian farmers needed to know, Prianishnikov suggested, not how soils had originated but how they could be used to best advantage.[18] K. A. Timiriazev, the dean of Russian plant physiologists, made the same point with greater harshness:

> Who has not heard of our school of soil science headed by Professor Dokuchaev? It has consumed tens of thousands of rubles from the zemstvos and the central government, but what has it given to Russian agriculture, and peasant agriculture in particular? What has it given in answer to the question, how to raise two heads of grain where one grows now? And all the while, if these funds were spent on the simplest field tests, if we had not one experimental field per county [*uezd*] but dozens, hundreds of inexpensive experimental fields, then our peasant would know, the plant itself would tell him, what is necessary in each individual case. To see in the soil independently of the plant a self-sufficient object of study is certainly, from the farmer's point of view, an enormous mistake.[19]

The demand that theoretical research be cut back in favor of immediately practical extension work (or that a rival discipline be cut back in favor of Timiriazev's) provoked an argument rather than a savage fight, for the opponents recognized each other as scientists. They were demanding more or less support for allied but rival disciplines, not an all-or-nothing choice between science and pseudoscientific panaceas.[20]

Some people in prerevolutionary Russia succumbed to the lure of pseudoscientific panaceas. At Iur'ev University the Department of Soil Science was controlled by a nativistic obscurantist, S. K. Bogushevskii, who believed that Western methods were applicable only in the West; unique Russia required its unique agricultural science. When he refused to approve a dissertation based on standard science, the angry candidate published a long diatribe against "'True,' 'Russian,' 'Native' Science and its University Representatives." He declared that his case was symptomatic; all of Russian science was being debased by political watchdogs, who supported nativistic obscurantism because it seemed patriotic:

> *As an exception* Bogushevskiis are possible anywhere; anywhere it is possible that a man without knowledge, without a proclivity for science, without scientific scruples and the scientist's capacity for work, might, by some exceptional circumstances, obtain a chair,

> students, a certain authority. It is even possible that his naive concoction — resembling scientific work as little as a snowman resembles the artistic creation of a sculptor — might be considered worthy of some kind of "earned degree."
>
> But only in the exceptional general conditions in which we live are Bogushevskiis possible not as an exception *but as the rule.*[21]

One should not be swayed by this outraged victim of academic tyranny. It is fairly clear that his case *was* exceptional. The official journal of the Association of Soil Scientists, which published his philippic, took the victim's side, not the local tyrant's. And the further career of the victim, A. A. Iarilov, is proof that nativistic obscurantism was not stifling science in prerevolutionary Russia. He became the chief Russian historian of soil science, proudly chronicling his countrymen's part in the cosmopolitan enterprise until Stalin's regime forced him to join in a chorus of nativistic obscurantists.[22] Before that time pseudoscientific cranks were isolates. They might cause an occasional furor — as Luther Burbank and Niels Hansen did in the United States[23] — but they did not seriously influence the scientific enterprise.

A revealing case in point is that of V. R. Williams, the politically reliable soil scientist who kept his job at the Petrovskaia Agricultural Academy when it was reconstituted as the Moscow Agricultural Institute. An odd man in many respects (he owed his unlikely name to an American engineer who came to Russia to help build the first major railway), he was a combination of academic types rarely joined in a single person: the flamboyant classroom orator who captivates students and embarrasses colleagues with cosmic nonsense, and, at the same time, the adroit politician who becomes a very important person. Grandly disdainful or ignorant of the accumulating complexities in soil science, agricultural chemistry, and ecology, he tied everything from solar radiation to soil structure into one great devolutionary process that would produce a universal desert unless the world's farmers checked disaster, by planting grass. His colleagues found such claptrap beneath criticism, yet they elected him director of the Moscow Agricultural Institute when the Revolution of 1905 won them the right to elect their director. If Williams' special influence in governmental circles was the reason for their choice, they and he were soon disappointed. Williams was caught between a student body determined to use the dormitory for political meetings and a police force equally determined to break up such meetings, regardless of the government's recent promise to respect academic autonomy. He suffered a stroke, which forced him to resign the directorship (and left him with a grotesquely distorted face). But his classroom eloquence was undimmed, and in 1914 several devoted students got out a *Festschrift* celebrating his ideas.[24] At this point the learned journal of the soil scientists found it necessary to depart from quiet toleration of an eccentric but influential colleague, and carried a quietly devastating critique by a brilliant young ecologist, Vladimir Sukachev.[25] Williams remained an isolated crank, ready as ever to play a winning game of academic politics but incapable in prerevolutionary Russia of making a mark in science. In that case too, it would take the Stalinist revolution of the 1930s to force scientists, including Sukachev, to profess support for pseudoscience.[26]

Much more isolated, indeed virtually invisible, was another crank with a Stalinist future, the impoverished nobleman I. V. Michurin. Without scientific training, with nothing more than a grade school education, he turned himself into a breeder of fruit trees. When he petitioned the Ministry of Agriculture to turn his unprofitable nursery into a state experiment station, he was turned down. Two medals and an offer of a job in the Ministry did not mollify his resentment against men with diplomas, men who were so arrogant as to believe that their cabalistic learning would help the breeding of fruit trees. Michurin did not develop those dark thoughts into a revolt against the new science of genetics. He withdrew into sullen isolation, from which he would be rescued by a postrevolutionary generation of leaders eager to overleap the bewildering complexities disclosed by science.[27] The prerevolutionary regime was not run by men determined to smash all obstacles to modernization. With a few notable exceptions, they were conservatives and mindless bureaucrats, who tried to muddle through the anomalous perplexities of playing the great power game with a backward country in a century of total war. They suffered disaster, and were succeeded by the most willful radicals who have ever managed to get power and keep it.

In agriculture, the Bolsheviks began by sanctioning the universal leveling that the peasants had long dreamed of, even though its economically regressive aspects were apparent. The islands of modern farming established by a minority of improving noblemen, the "model estates" that Lenin made a gesture to preserve, were dissolved into a sea of primitive peasant communes. The peasants even destroyed some agricultural experiment stations, whose fields were cut up in the usual crazy quilt of strips allotted to peasant families, more anxious for an equal share of land than for modern efficiency.[28] From such farmers, in a time of civil war, the Bolsheviks were audacious enough to take grain without payment, by force, which saved the towns and the Red Army from starvation but pushed economic crisis to utter collapse. The peasants simply went back to subsistence farming — their ultimate weapon, which would drive Stalin to the frenzy of collectivization. At the same time, one of Russia's periodic droughts set off a famine more lethal than the famous one of 1891–92. Throughout the violent descent into chaos the Bolsheviks not only held on to political power; they maintained the conviction that they alone knew the way to the modernization of Russia. Indeed, their astonishing victory greatly strengthened their faith. If they could win out over armed opponents supported by advanced foreign powers in a time of complete economic collapse, nothing seemed beyond their power.

Such breathtaking willfulness was clearly a portent of trouble for scientific advisers who might try to remind the Bolsheviks of realistic limits. President Kalinin must have stunned and worried the audience of agricultural specialists to whom he announced — in 1922, when stories were still circulating about people eating people for lack of grain — that the prewar level of agricultural production would be doubled "in a relatively short time. . . . In two or three years we will have enormous grain surpluses." Kalinin assured his audience that he was serious:

> Inasmuch as the Soviet regime has successfully grappled with more difficult tasks [chiefly the retention of power] it will without a doubt grapple successfully with this special task. This is not the self-conceit of individuals. This is a *profound conviction* that, if hundreds, thousands, and sometimes even millions of individuals participate in a job, then exceptional results are always achieved. History forces leading individuals *to do exceptional* things. And I have no doubt, I wish, that this conviction should be shared by you too.
>
> What we need is a burning faith that our Republic, which is backward in agricultural matters, will not only catch up with the agriculture of Western Europe, *but will also surpass* it.[29]

Here was the unreasoning passion that would drive peasants into collective farms and force scientific specialists to profess belief in the impossible. It is hardly surprising that the first years of Bolshevik rule caused the faculty at the Agricultural Academy (the old title was resumed in 1917) to fear the "barbarization" of science. When they talked this way President Kalinin appeared before them to recall the beneficial effects of the barbarian invasions of Rome. "There must be barbarism," he declared, "so that, from this soil, democratic, simple science can emerge." [30]

But this kind of talk was only a portent of trouble to come; it was not characteristic of the Bolshevik approach to improved farming during their first decade of rule. With the New Economic Policy (NEP) in 1921, the Bolsheviks abandoned forced requisitions of grain and adopted a cautious, realistic approach to the chronic agricultural problem. In retrospect it is not hard to perceive during the brief period of NEP the growing anxieties that would erupt into forcible collectivization of peasants at the end of the 1920s. But it is hard to discover signs of the crisis that would simultaneously break upon agricultural scientists. Almost everything the Bolsheviks said and did during their first decade of rule pointed in a different direction; dictating truths to scientists was the farthest thing from their minds. Kalinin's taunting comment on the barbarization of science was part of a serious conflict between the new political bosses and the old scientists, but it did not touch the content of natural science. Politics and its bearing on academic autonomy were the issue, as they had been in the prerevolutionary conflicts between Russian rulers and Russian scientists.

The overwhelming majority of scientists were hostile to the new regime, more hostile than they had been to the old regime, but neither side to the conflict could do without the other. With savage joy Lenin exiled religious philosophers and sociologists like Sorokin, but he repeatedly warned his comrades that natural scientists could not be treated the same way, however deviant their politics might be. "Communism cannot be built without a fund of knowledge, technology, culture, but they are in the possession of bourgeois specialists. Among them the majority do not approve of the Soviet regime, but without them we cannot build Communism." [31] The academic autonomy that these indispensable aliens had won by the Revolution of 1905 had been curtailed in the aftermath, but it was reasserted and expanded to the point of complete academic self-government as soon as the tsar's power collapsed in 1917. When the Bolsheviks seized power, the Russian academic community was enjoying greater freedom than it had ever known — thanks to military defeat, political anarchy, and mounting crises in social and economic relations.

The Moscow Agricultural Institute, which had proudly renamed itself the Petrovskaia Academy to celebrate its emancipation from tsarist restrictions, was particularly troublesome for the Bolsheviks. Though the faculty and students were divided in political allegiance between a majority favoring the liberal tradition of the zemstvos and a minority of Socialist Revolutionaries, they presented a solid front of hostility to the Bolsheviks. Harassment forced the elected rector to resign, and bloodshed was narrowly avoided when a local Bolshevik organization took physical possession of the buildings. But the effective subjection of the faculty and student body went very haltingly until Professor Williams decided to collaborate with the new authorities as he had with the old. The specific issue that aroused cries of barbarization from his colleagues was the establishment within the Academy of a *rabfak,* or workman's faculty, a preparatory school for lower class students thrust upon institutions of higher learning without the usual academic prerequisites. In the course of the struggle over the *rabfak,* Williams became rector, the autonomous government of the Academy was subverted, and once again a new name was adopted to signify the political change. This central institution of higher education in the agricultural sciences became, and remains, the Timiriazev Agricultural Academy, honoring the eminent plant physiologist who "quite enraptured" Lenin by announcing his support for the new regime.[32]

By 1923, when Kalinin came to taunt the subdued Academy with talk of barbarism producing simple, democratic science, similar conflicts with many institutions of higher learning — the staff of Moscow University even went on strike in 1922 — had been won by the Bolsheviks, who were anxious to effect a reconciliation once they had assured themselves of political control. In the fall of 1923, a large "Congress of Scientific Workers" gave ceremonial recognition to the "union of labor and science." Dissension, the Congress resolved, was a thing of the past; scientists and the working class would now cooperate in the great cause of rebuilding the native land. Scientists who feared that the Bolsheviks might abridge their freedom not only in politics but in science itself were assured that this could not happen. Zinoviev made this clear in the main speech, and Trotsky provided the theoretical explanation in a long letter on "Darwinism and Marxism," arguing that Marxism was simply the latest stage in the development of scientific method, its application to the study of human society.[33] In a subsequent speech to chemists Trotsky was quite explicit:

> When any Marxist tried to convert Marx's theory into a universal skeleton key and flitted through other fields of knowledge, Vladimir Il'ich [Lenin] would rebuke him with the expressive little phrase, "Communist conceit." This would signify in particular: Communism does not replace chemistry. But the converse theorem is also true. The attempt to step over Marxism, on the pretext that chemistry (or natural science in general) must solve all problems, is a peculiar *chemical* conceit, which is theoretically no less erroneous and practically no more likeable, than *Communist* conceit.[34]

In short, if the scientists would leave politics to the Communists, the Communists would leave natural science to the scientists. At a conference of agronomists in 1924, President Kalinin spoke with a new benevolence, even calling the agronomists *vozhdi i obshchestvennye deiateli* ("chiefs and social leaders"), honorific terms usually reserved for Communist officials. To be sure, he voiced once again his desire for simplified science; but this time he did so gently and sensibly, pointing out that only simplified agronomy could win peasants to modern farming.[35] To a conference of rural physicians Kalinin expressed a cheerful indifference even to anti-Communist political feelings, as long as the specialist who expressed them worked conscientiously at his profession.[36]

It is often said that Soviet leaders were eager from the start to impose peculiarly Marxist ideas on natural science, and for this reason favored cranks like Williams and Michurin. That is simply not true. It is true that Williams was their darling because of his help in subduing the fractious Agricultural Academy. Even after he was eased out of the rectorship in 1924, he remained a member of the Agricultural Committee within the State Planning Commission, where he kept insisting that his simplistic scheme of crop rotation must be applied all over. But the overwhelming majority of soil scientists and crop specialists were against him, and the Bolshevik government deferred to *them* (until the 1930s).[37] It is also true that Michurin's nursery won recognition from the Soviet government as a research center, which the tsarist government had declined to grant. But it was genuinely competent specialists who decided, after considerable argument, that the practical breeder deserved support even though his "scientific" views could not be taken seriously (until the 1930s).[38]

In matters of natural science, Lenin set a pattern of deference to the authority of autonomous specialists, and his successors unquestioningly adhered to this pattern until 1929.[39] Scientists were at first apprehensive that the Bolshevik regime might be a creature as well as a creation of "the people's darkness and ignorance."[40] At their first meeting with the Commissar of Agriculture, worried agricultural specialists were relieved to learn that Bolshevik endorsement of peasant revolution did not include approval of the liquidation of experiment stations by peasant rebels. On the contrary, the government was wholeheartedly dedicated to scientific farming. Even the manners of the new leaders were a reassuring surprise, contrasting favorably with the haughty bearing of tsarist Ministers. While soliciting the advice of specialists, the Commissar of Agriculture sat at a common table with them, sharing their weak, barely sweetened tea in a tin cup, taking no more food than the standard ration (an ounce of bread). Tulaikov, the country's leading specialist on dryland agriculture, spelled out the significance of this amiable meeting.

> Thus, at a moment that was very grave for the directors of our life [the Civil War had begun], a close connection was formed between agricultural specialists and the new regime, which saw in them, first, workers dedicated to their cherished cause, honorably, without any duplicity, offering their knowledge and experience to the new upbuilding; and second, specialists who loved their work, who were seriously prepared to defend it against any underserved attacks, wherever they might come from.[41]

Tulaikov was one of the first of several leading specialists to be sent to the United States, whose system of agricultural research and extension work was fixed in many Russian minds, Lenin's included, as a model.[42] Tulaikov's popular report of what he learned was given wide publicity, with a prefatory endorsement by Lenin's wife, Krupskaia.[43] In 1922 the government began to plan a Soviet version of the U.S. Department of Agriculture's (USDA) federation of research institutes and extension centers. The man in charge of the planning, N. I. Vavilov, was nominated for the post by his fellow specialists.

But Vavilov, and the manner of his nomination show that the prerevolutionary scientific establishment did not continue, under the new political masters, essentially unchanged, with the same old leaders, methods, and attitudes. The new masters were passionately determined to establish scientific farming as soon as possible, and many specialists were ready to respond with utopian strivings of their own.

Vavilov, who was only thirty years old at the time of the Revolution, became chief of agricultural research because he appealed to millenarian yearnings among agricultural specialists as well as Bolshevik politicians. When Lenin was preparing to seize power in Petrograd, Vavilov was lecturing to his first class at the Saratov Agricultural Institute, enthralling his students with a world-sweeping picture of the task that lay before them: the fusion of all disciplines bearing on agriculture, the collection of plants and experience from all countries, *"the planned and rational utilization of the plant resources of the terrestial globe."* He declared that the new science of genetics showed men how

> . . ."to sculp organic forms at will," and constant forms at that. In the near future man will be able, by crossing, to synthesize such forms as are entirely unknown in nature. Biological synthesis is becoming as much a reality as chemical.[44]

That simile — it is still, a half century later, more of a hope than a reality — was not a teacher's momentary exaggeration to make a point. At a conference of breeders in June 1920, Vavilov outlined his "law of homologous series," the biological equivalent, he said, of the periodical table of chemical elements.[45] Not only was Vavilov bold enough to make this claim; the conference was so exalted as to accept it. "He spoke with inspiration," a witness recalls; "everybody listened to him with bated breath; we felt that something great, something new was opening before us in science." [46] In the ovation that followed, the exclamation of one delegate — "Biologists hail their Mendeleev!" [47] — was caught up, incorporated in a resolution, and wired to the Commissars of Education and Agriculture, calling on them to support Vavilov's work on the broadest scale.[48] He was called to Petrograd to become Director of the Institute of Applied Botany (subsequently VIR, the All-Union Institute of Plant Industry), and chief organizer of the other institutes that would ultimately be federated in the Lenin Academy of Agricultural Sciences.

In sober fact, as Russia's leading geneticist, Iu. A. Filipchenko, pointed out, Vavilov's law of homologous series was not the biologist's equivalent of the chemist's periodic table.[49] The embryologist, M. M. Zavadovskii, was closer to the truth when he compared biology with the condition of chemistry *before* basic theoretical principles had drawn great masses of factual data into predictive schemes.[50] Biological science is still, almost fifty years after Vavilov's memorable speech, far from providing the breeder with a neat chart of the basic elements of all plants and animals, whether morphological, which was Vavilov's old-fashioned approach, or biochemical, which is the focus of contemporary efforts in a similar direction.[51] And in sober fact, Russia of the 1920s was in no position to create an equivalent of the USDA's famous network of research, extension, and educational institutions. The scientific enterprise could be created. There was a rich prerevolutionary heritage to draw on, and Vavilov's extraordinary talent for organizing the scientific intelligentsia, and for touching the complementary romanticism of Bolshevik leaders, could overcome the shortage of funds and trained personnel. With the wholehearted support of Lenin's government, and even more of Stalin's, Vavilov could and did create centers of advanced research that won great respect in the international community of scientists. The exaggerated importance that Vavilov ascribed to his "law of homologous series" was no hindrance. Self-conceit and academic tyranny were entirely foreign to his nature; indeed, he had a genius for sensing the importance of other people's work and organizing support for its development. But practical benefit to agriculture was supposed to be the chief purpose of the Lenin Academy of Agricultural Sciences,[52] and Russian agriculture was less ready to receive that benefit than it had been before the Revolution.

There was a painful element of truth in Williams' nativistic appraisal of the Revolution's effect on the countryside:

> Only that has remained which proved to be adapted to the real conditions of our actuality. "Corners of Europe" have perished. They were planted either by the stupid willfulness of parasites [noble landowners] or by the sentimental, vain attempts of those same parasites to contribute their portion of benefit to the cause of enlightening the dark masses of Russian farmers, without taking into consideration the actual requirements of their economy.[53]

Williams' satisfaction with the primitive, leading to his cheap panacea of grass planting, may be ignored, as it was by the original chiefs of Soviet agriculture. But the fact cannot be ignored: Russian agriculture was more primitive after the agrarian revolution than before. The gap between the backward realities of Russian farming and the advanced recommendations of science was greater than ever; and the government's wholehearted support for Vavilov's grand project was accelerating the development of science, widening the gap even more. The most dangerous element in the situation was the government's utopian motive: a growing conviction that the world's most advanced scientific enterprise was needed because collectivization would create the world's most advanced agricultural system.

Scientists perceived the gap between their work and the peasants', though they did not foresee the crisis to which it was leading. Tulaikov, for example, noted that the primitive condition of the average peasant farm — many weeds, little fertilizer, no seed drill or proper rotation — would rob improved seed of its advantage, even if the peasant could be persuaded that it was worth buying. Tulaikov's academic prose defined the problem ponderously but clearly:

> Improved varieties of wheat must be introduced in practical use together with improvement of the technique of growing it. The present task of agronomists and plant breeders . . . consists precisely in the necessity of finding and achieving this harmonious linking of biological adaptation of the cultivated plant to economic tasks and possibilities.[54]

It was far easier to define the problem than to solve it, unless the scientists were able to change the peasants' methods of cultivation — and they were not — or willing to bring their work down to the primitive level of the peasants. They might, for example, have limited themselves to identifying the varieties of grain that would grow best in grain-fallow rotations on weedy fields without fertilizer. But that was just what the peasants had already achieved with their mongrel local varieties, without benefit of science.

It is hardly surprising that Doiarenko, the leading professor of agronomy at the Timiriazev Academy, was worried lest the intensified support for agricultural science generate a disappointed reaction, as it had repeatedly in the past. He feared that government leaders might begin to ask why they should support experiment stations running far ahead of what peasants could or would accomplish. Of course he saw no danger in the government's intense commitment to improving peasant practices; that intense commitment seemed to assure the ultimate utility of the scientists' work. Doiarenko did not presume to lecture the government on the best way to handle the peasants; at least he did not do so in public. He was only concerned that government leaders recognize the proper function of science: to run ahead of current needs and possibilities, to anticipate the future.[55] During the 1920s, government leaders did recognize that. They took it for granted that their investment in agricultural science would not pay off immediately, only in the future, when socialist farmers would take eager advantage of all that modern science had to offer. While preparing for the breakthrough into that future, they were glad to hear Vavilov tell about the affluence that would flow when biologists, working with a periodic table of biological elements, would "sculp organic forms at will."[56] As long as realization of these dreams was deferred, scientific and Communist romanticism did not conflict; they complemented and reinforced each other.

Chastened only temporarily by the shocks of the Civil War and economic collapse, Bolshevik willfulness revived at the end of the 1920s. As it became clear that economic levers would take a very long time to raise peasant agriculture to the level necessary for rapid industrialization, and might also make the peasantry an independent power in Soviet politics, impatient hands reached for the political levers that would force the process and achieve the complete subjection of the countryside.[57] The Bolsheviks revived forceful campaigns to collect grain and moved toward forceful collectivization of the peasants. At the same time, true to their utopian expectations, they intensified their investment in agricultural science. They really expected that old peasants in new collectives would be capable of applying the most up-to-date methods science could recommend. It was at this time, 1928–29, that the Lenin Academy of Agricultural Sciences was quickly transformed from paper projects into a real federation of research centers, part inherited from the old regime, part newly created, and all swiftly expanding their staffs and projects.[58]

While the archives are locked, it is impossible to know how much of that runaway utopianism came from scientists and how much was forced upon them against their better judgment. Tulaikov may or may not have been expressing his own calculations on the tenth anniversary of the Bolshevik Revolution, when he made this startling forecast for the next decade:

> Under the social scheme that we have outlived we extracted from agriculture perhaps 25 to 30 percent of what it could yield with a rationally organized utilization of the forces of nature. With an ardent desire to achieve more, by the combined powerful exertions of all agricultural workers, by a close union of science and practice, we will strive to achieve, at the next, the twentieth jubilee of the new life in our country, an inversion of the figures just cited. . . .[59]

In other words, within ten years there would be more than a doubling of agricultural efficiency, a jump from 25 to 30 percent of the maximum to 70 to 75 percent. Tulaikov published this prediction about the same time that Iakov A. Iakovlev, the chief Party propagandist for agrarian problems, forecast a doubling of grain yields within a decade and criticized unnamed specialists who had predicted an increase of only 6 to 11 percent.[60] The conservatives were economists. Prodded by impatient Bolsheviks, they raised their most optimistic forecast to a 19.4 percent increase in grain yields within five years, whereupon Iakovlev denounced them once again for caution unbecoming to revolutionaries. Openly and proudly proclaiming his faith in revolutionary fantasy (*fantastika*), he insisted on a doubling of grain yields within a decade.[61] It may be significant that agricultural economists rather than other specialists were the chief victims of the Bolsheviks' repeated outbursts against pessimists and skeptics. Perhaps specialists such as Tulaikov, whose attention was focused on plants, soil and water, were almost as insensitive as Bolshevik propagandists to the social and economic determinants of yields. But the archives are closed, and it is futile to argue whether agricultural scientists were initiators or victims of the utopian fever that seized Bolshevik leaders at the end of the 1920s. Most likely they were some intricate combination of both.

In any case, the public record makes it clear that many leading scientists were active agents of the utopian fever. Even N. K. Kol'tsov, the dean of experimental zoologists, whose intellectual independence repeatedly angered the Bolsheviks, published enthusiastic articles on the great opportunities presented by collectivization. He expected centrally directed farms to carry out a nationwide experiment in the improvement of livestock, enriching science and agriculture at the same time. His student, the geneticist A. S. Serebrovskii, undertook the actual planning of the grand experiment, with the warm support of Iakovlev, who became Commissar of Agriculture when the frenzy of collectivization reached its height.[62] Serebrovskii, like Tulaikov, was moving toward Party membership,[63] but even Vavilov, who was not, caught the fever of the times. He appeared before the Sixteenth Party Conference in 1929, which anathematized the Party's timid right wing, to endorse the agrarian offensive and promise the active aid of science: "Enormous energy is needed to get our enormous country moving, the whole peasant mass of it, to rouse to willful action all the creative forces of our country." [64] Whatever reservations he may have kept to himself, it is hard to doubt that there was an important measure of sincerity in such declarations of faith, for he made them not only in public speeches to Bolshevik audiences but even in the privacy of an American hotel room, pleading with the émigré geneticist Dobzhansky to go home and help the great cause.[65] The ultimate proof of Vavilov's sincerity is his own return from his repeated foreign trips, even in the early 1930s, after the first signs of a serious conflict between Bolshevism and agricultural science had appeared. In the collectivized countryside science would at last realize its great potential — he said it repeatedly, and to some degree he believed it.

In January 1929, at the All-Union Congress of Genetics and Breeding, Vavilov and hundreds of colleagues unwittingly celebrated the climax of the heady marriage between Bolshevism and agricultural science.[66] They thought they were preparing for a long and fruitful union. Draped along the front of the hall was a banner, *Shire v massy dostizheniia nauki!* ("Spread the Achievements of Science among the Masses!") Party chiefs were on hand to give their blessing. S. M. Kirov, who was one of Stalin's right-hand men, and N. P. Gorbunov, who was still, as he had been under Lenin, the highest official for science and technology,[67] expressed their complete confidence in the ability of the scientists to help achieve the basic agricultural goal of the Five Year Plan — a 35 percent increase in average grain yields per hectare. Only a minor trade-union official gave a hint of trouble, vaguely suggesting that the banner should call for a more critical approach to science.[68] Characteristically, it was the scientists themselves who called attention, in their resolutions, to the gap between agricultural science and peasant practices and made recommendations on ways to overcome it. They were even confident enough to complain about inadequate funds and insufficient opportunities for foreign travel, and in the next few months the government further increased its appropriations to them.[69] But before the year ended, the honeymoon was over.

With the crisis of forced industrialization and the agrarian offensive, Bolshevik hostility to "bourgeois" specialists was reviving. The minor trade-union official at the January conference had vaguely suggested it; by the end of 1929 it had turned into a furious drive to eliminate "bourgeois" specialists altogether. The tolerant compromise that Lenin had arranged was cancelled. Henceforth all specialists were to be, like Vavilov, Tulaikov, and Serebrovskii, active participants in the renewed revolution. Any critical comment, any silence that could be interpreted as criticism or mental reservation, became grounds for dismissal, or jailing, or even shooting.[70] For example, Doiarenko was dismissed and imprisoned because he would not join in the paean to the agrarian offensive. A colleague who had sung at the recent funeral of Doiarenko's wife was threatened with dismissal for this manifestation of religion. (His explanation, published in the school newspaper, may not have helped him. He said he loved music, not religion, and could not resist underlining the girls' beautiful voices with his bass counterpoint. If he really loved music, the editor retorted, he would have joined the school chorus.)[71] Out of the Academy's faculty of 168, 20 were dismissed and 5, mostly economists, went to jail or execution as "wreckers."* [72] All academic institutions were tormented by similar violence in 1929. Indeed, it became an intermittent plague for twenty-three years, until Stalin was laid to transient rest in Lenin's tomb.

The long crisis that began in 1929 was not only a cancellation of Lenin's compromise with "bourgeois" specialists. At the very end of 1929 Stalin went further. He erased the fundamental distinction on which that compromise rested, the distinction between the political views of specialists and their professional work. Stalin told a conference of agricultural economists that their studies were being proved useless by practical Party workers in the countryside, who were pushing collectivization much faster than any economist had believed possible. Henceforth such practical achievement was to be the test of scientific truth.[73] The implications for agricultural economics were very clear; it was immediately transformed from autonomous scholarly inquiry into sycophantic commentary on the snap judgments of political bosses. But the implications for the other sciences bearing on agriculture were clear only in principle. They too were to be subject henceforth to the test of Stalinist "practice." In administering the test, Party officials were to emulate the spirit that Stalin had expressed in a telegram to Lenin back in 1919, during the Civil War. (It was dug out of the archives and published in *Pravda* a week before Stalin's speech to the agricultural economists.) Going against the advice of naval specialists, Stalin had ordered a mutinous fortress taken by amphibious assault, and he had won.

> The naval specialists declare that taking [the fortress] by sea subverts naval science. All I can do is bemoan so-called science.

* We are stuck with this clumsy nontranslation of *vrediteli*, a deftly venomous epithet, whose basic meaning is "pests," creatures like Japanese beetles or Hessian flies.

> The swift taking [of the fortress] is explained by the roughest interference on my part . . . .
>
> I consider it my duty to declare that in the future too I will act in this way, in spite of all my reverence for science.
>
> Stalin[74]

He did not spell out the form that "rough interference" in science should take, except for agricultural economics. The propagandists who tried to apply his new line to the other agricultural sciences were accordingly vague. They did not explicitly reject the slogan that had been so confidently displayed in January — Spread the Achievements of Science among the Masses! Very likely they did not even realize that they were subverting that old slogan with their new ones: Science must be infused with the achievements of practical leaders. Hope of benefit from science was to be deferred no longer, it was to be satisfied immediately. Agriculture was being reorganized on socialist principles, Bolsheviks were taking direct control of farming. Let no scientist give them advice they could not use. Agricultural science could no longer be described as running ahead of primitive peasants. It was lagging behind the practical chiefs of successful collective farms. Its task was to catch up, to make itself useful by transforming itself.[75] The propagandists who mouthed these slogans were not sure what they might mean, but a number of cranks and promoters were eager to rush forward with specifications.

3

Harmless Cranks

Many people still believe that Ivan Vladimirovich Michurin bred a lot of commercially valuable varieties of fruit but was unaccountably neglected until discovered by Lenin himself; that Trofim Denisovich Lysenko made a brilliant discovery in plant physiology (vernalization) while developing an economically valuable method of seed treatment (also called vernalization); that the doctrines of Michurin and Lysenko grew out of or were grafted onto some mighty trunk of Communist theory (faith in environment as opposed to heredity is the usual choice). Those are legends, as the present chapter will show. But there is another and more important reason for exhuming the original Michurin, Lysenko, and forgotten "peasant scientists" of the period before December 1929, when Stalin elevated their kind of "practice" over scientific "theory." Uncorrupted by power, these men reveal in original purity the crank's understanding of a practical approach to agricultural problems.

The word "crank" is a pejorative, but it cannot be avoided, for cranks are quite different from scientists. Fantasy is not the distinguishing feature. Cranks may have feeble powers of imagination, while genuine scientists may indulge in extravagant flights of fancy. The essential difference is the crank's individualistic self-assurance versus the scientist's collectivistic self-doubt. The scientist looks for tests that will convince his fellows; he submits his fantasies to the disciplined examination of a professional community. The crank does not and usually cannot. Of course, the distinction is hard to make before a discipline has emerged from formative chaos, but this will not be a problem here, for biology and the agricultural sciences had emerged by the 1920s. In the Soviet Union as elsewhere those disciplines were in the charge of professional communities of specialists, which enjoyed strong government support, indeed, overeager support. Cranks held marginal places, partly on the sufferance of the professionals, partly by techniques of public relations — "going out on the broad road of public opinion (*obshchestvennost'*)," to quote one of the many journalistic admirers of Michurin.[1] They learned no truths of biology on that road, but they did discover some basic qualities of the Stalinist mentality. To be more precise, they anticipated the new Stalinist mentality, which astounded everyone at the end of the 1920s by its abrupt rejection of the realistic practicality that had previously been Stalin's hallmark. Suddenly Stalin announced that the essence of practicality was not calculating realism but extreme willfulness. The cranks of agrobiology could have claimed that they had discovered this truth before Stalin, but even their boldness had limits.

MICHURIN

The life of Ivan Vladimirovich Michurin began as a bleak variant of *The Cherry Orchard,* as if Gorky or Bunin had written the play instead of Chekhov. Rather than sell the orchard, the impoverished noble family went to work raising fruit for sale and suffered a painfully protracted failure. The urban markets of postreform Russia grew too slowly to absorb both the luxurious southern fruit brought in by the new railroads and the scraggly varieties that Riazan province (*guberniia*) had long been sending in by wagon. Or perhaps temperament and luck were more important causes of the Michurins' failure than sickly markets and harsh climate. Though Riazan is as far north as Saskatchewan (and more distant from the sea), some noblemen maintained commercial orchards there until they were wiped out by the Revolution, and some peasants until the collectivization drive. Michurin's father Vladimir, growing ever more indrawn and morose, had already failed in the 1870s. Tuberculosis killed his wife, his half-mad mother tyrannized the family, and six of his seven children died. At his wife's funeral he came out with a dance song instead of the dirge and was taken away for the first of several stays in the "yellow house," the Russian bedlam. Ivan, the one child who survived (though he did not escape the nervous disorder that ran in his family), aspired to an aristocratic *lycée* in

St. Petersburg but was not even allowed to finish his first year in the Riazan *gymnasium*. He was sent to work in a depot town in nearby Tambov *guberniia*, first as a railroad clerk and then, when he showed mechanical ability, as a signal repairman. When he married the daughter of a low-class *meshchanin*, the drama of social decline would seem to have ended.[2] But Ivan Michurin would not reconcile himself to such an end. He determined to change the trees that had failed his father, to wrest from plants a victory that geography, economics, and temperament seemed to prohibit.[3]

This willful approach to fruit growing would be the unchanging essence of Michurinism, preserved through all its metamorphoses. In the first stage, the long period from the late 1870s to 1929, it was little more than a one-man movement. At first it was not even that, but the afterhours hobby of a railroad employee who also kept a watch-repair shop in Kozlov, Tambov *guberniia*. In 1888, apparently feeling his life slipping by as he approached thirty-five, Michurin plunged for all or nothing: bought thirty-four acres with borrowed money, set up a nursery, quit his job and shop. He would develop and sell to the orchardmen of central Russia varieties of fruit that could compete with the Crimean and West European varieties.

Through years of drudging trial that yielded only error, Michurin came to reject the method, then popular, of acclimating southern varieties by grafting them on northern stock.[4] He turned to hybridization, a highly uncertain method, since most cultivated varieties of fruit are very complex and often unstable hybrids to begin with. (It is a gardener's rule of thumb that fruit trees must be propagated vegetatively; if raised from seed, they will be wildings.) Michurin lacked the acreage and personnel necessary for large-scale breeding on the expectation of a tiny percentage of improved varieties to be selected from a great mass of wildings. Forced to work with comparatively few specimens at a time, he became convinced that many seedlings which appeared to be wildings should not be thrown away. Signs barely perceptible to his intuitive eye told him which ones might be improved by training. (Even so, he saved only one in a thousand.) A favorite method of training was to graft the promising seedling to the variety he wished it to resemble, or vice versa, on the conviction that the superior variety would act as a "mentor" to the young hybrid. To predispose two different varieties or species for mating he practiced "vegetative blending" (*sblizhenie*), grafting one on the other as a preparation for cross-pollination. In the actual work of cross-pollinating he scorned excessive rigor; indeed, he took to mixing pollen when two species resisted hybridization.[5] When he did this, he had no way of knowing whether he was really crossing the two species or simply fertilizing the maternal plant with its own pollen. The important thing was not to establish accurate genealogies but to get commercially valuable combinations of such qualities as winter hardiness, good flavor, and durability in storage. Such combinations, once achieved, could be propagated vegetatively without concern for hereditary stability.

At a time when the new science of genetics was groping toward its basic conception of heredity as mathematically predictable combinations and separations of particulate qualities, Michurin was growing ever more committed to the ancient folklore of gardening.[6] Blending versus particulate heredity, the inheritance or noninheritance of acquired characters, little distinction or much between genotype and phenotype — these issues would claim most attention when the two schools would come into full conflict. But underlying them all would be the gardener's conviction of the *immeasurable* diversity and plasticity of living things, pitted against the scientist's proof of their *measurable* diversity and plasticity. In response to scientific criticisms of his methods, Michurin would repeatedly argue that heredity "does not yield and in essence cannot conform to any patterns worked out by theoretical science and determined in advance." [7] A science of heredity was not possible because there was no regularity in hereditary phenomena:

> If we cross two plants and get hybrids with a combination of certain characters, then, no matter how many times we repeat the cross between this pair of plants, we will never get the same pattern of hybrids. Even seeds from one and the same fruit that is obtained by crossing give seedlings of absolutely different varieties. It is evident that nature, in its creation of new forms of living organisms, gives infinite diversity and never permits repetition.[8]

It must not be imagined that Michurin thought or wrote extensively about theoretical science. For the most part he ignored it, convinced that the mastery of plants could be only an art, developed by years of practice and essentially incapable of communication in abstract words or formulas. He had the instinctive, unlettered vitalism of many gardeners; at times, indeed, he was anthropomorphic, ascribing to "every living organism an intelligent [*razumnuiu*] power of adaptability in the struggle for existence." [9] He also shared the latent disdain that many practical breeders have for academic theorists, "the caste priests of jabberology," as he called them in one outburst.[10] Usually he ignored them as they ignored him. His business was to produce improved varieties of marketable fruit, and he felt that he was doing this very well without benefit of diplomas. By 1914 he was claiming "several hundred new varieties suitable for cultivation in our orchards." [11]

Michurin's occasional criticisms of genetics were chiefly the result of his collisions with academic specialists, which were in turn the result of his efforts to become the head of a government experiment station. He began these efforts in November 1905, just after strikes and demonstrations had wrung a pledge of constitutional government and civil liberties from the Tsar. The long communication that he sent to the Ministry of Agriculture at that revolutionary moment has not been published, but its contents and tone are fairly clear from later references to it. "As a true Russian man," considering it his "sacred duty to offer his humble labor for the good of the fatherland," he probably decried the revolutionary disorders and described his type of work as a method of preventing renewed upheavals by improving Russian agriculture.[12] Very likely he deplored the decline of the large noble estates, the best market for improved fruit stock. With some such preamble, he asked the Ministry of Agriculture to make his nursery a state institution and provide him with apprentices so that he might be free of dependence on rebellious, expensive day laborers, and might pass on his methods to a new generation of plant breeders.[13]

The Ministry of Agriculture replied to Michurin two and a half years later, without apology for the delay, blandly commenting that they had inquired about his work and convinced themselves of its merit. Nevertheless, the government maintained experiment stations for fruit only in the southern extremities of the country. In rare cases, subsidies were given to private experimenters, and the Ministry would give Michurin one, if he would name a sum and agree to carry out a few projects at their direction.[14] Michurin replied quickly and at great length. More than a small subsidy was required, since the great rise in prices resulting from "the workingman's false interpretation of various liberties" and the decline of the great estates had ruined his business. He had not been in the black since 1905 and was ready to sell out, waiting only for a good offer. If small holders were to take the place of noble estates in developing fruit culture (this was the time of Stolypin's reforms), the government would have to distribute fruit stock without charge and propagandize proper methods of cultivation through a journal free of "theoretical chatter." Careerists with diplomas should not be allowed to take posts for which dedicated men of great practical experience were most suited. And Michurin closed with a renewed offer of his services to the government.[15]

In the years that followed — the eve of total war and revolution — the Ministry of Agriculture financed a considerable expansion of research.[16] Michurin, anxious to be placed in charge of a station breeding improved varieties of fruit for central Russia, was involved in protracted negotiation. The fragments that have been published show him writing about to men of influence, asking for advice and help, insistently calling attention to the high praise he received from the Plant Explorer of the USDA.[17] The scientific consultants of the Russian ministry are not even named in the published record, but one gathers that they praised Michurin as a practical plant breeder while doubting the wisdom of putting an experiment station under his control. Turning from earnest pleading to proud indignation, Michurin refused even to name a figure for "a one-shot subsidy" and turned down a minor job that was offered him. All that he had to show for his trouble were two decorations, apparently given as consolation prizes. In 1914, verging on sixty, when he was asked for an autobiography by a journal of horticulture, he replied with a bitter lament. He contrasted the importance of his work with the meagerness of his rewards, justified his increasing seclusion, scolded horticultural journals that paid little or nothing for articles and scientists who sent him questions in language that he could not understand even if he could decipher their handwriting. A morose, self-pitying old man emerges clearly in this self-portrait, willing to twist things in order to make his angry points.[18]

Some of Michurin's twisting was merely pathetic. He left the impression that he had graduated from the secondary school where he had actually studied for no more than a month or two. He boasted that he "bore all possible hardships in silence, never asked for help from the government."[19] Even the tale he wove about himself and the USDA is pitiful rather than disturbing. The facts are that he was visited once in 1911 and again in 1913 by Plant Explorer Frank Meyer, who tried to arrange regular purchases of plant stock from Michurin's nursery. Michurin demanded a written request from

Washington, a fixed annual fee, payment for his reports, and an agent to do the packing and shipping. When Meyer's chief — David Fairchild, the well-known organizer of American plant introduction — decided that all this was too expensive, Michurin reduced his demands, but the little project fell through anyhow. Those are the facts.[20] The story spread by Michurin grew with the years until he had agents of the USDA visiting him regularly for eighteen years before 1914, repeatedly begging him to come to the U.S., offering a special ship to carry his plant collection, the directorship of a plant-breeding station (one account places it in Quebec), one hundred assistants, and a salary of $32,000 a year.[21] (In 1913 the top salary allowed by the USDA was $4,000 a year.[22]) All that was a poor man's castle in the air, probably built on the remark that Frank Meyer made to a friend of Michurin's, who promptly published it in a journal of horticulture: "'If Michurin were in America, they would make him wealthy.'"[23] Maybe they would, considering the boom in the advertising industry. Luther Burbank was doing very well for himself, despite the hostility of genuine scientists. Frank Meyer, an excellent botanist, thought that Michurin had "slipshod methods, à la Burbank," but had achieved more than Burbank in the production of interesting hybrid stock, of possible use in further breeding.[24]

In one respect Michurin's boasting had dangerous implications. He claimed that he had bred "several hundred new varieties suitable for cultivation in our orchards."[25] In fact, only experimenters liked his creations. He could not make a commercial success of his nursery, and as late as 1931 a Soviet authority would find only one of his varieties worthy of certification for use in commercial orchards.[26] From first to last Michurin simply refused to believe such evidence. He blamed the commercial failure of his varieties on the decline of noble estates, on the indifference or hostility of diplomaed specialists and illiterate peasants; ultimately, he tacitly concurred in a charge of "wrecking" against such specialists as found only one of his varieties worthy of certification.[27] And he cannot be charged with conscious twisting or even with illogic; he was convinced beyond any reasonable doubt that his practical intuition was the ultimate test of truth. If he liked the performance of a hybrid in his nursery, he added it to his list, a crankish habit and harmless, until the Commissar of Agriculture — and the police — took the side of such practical judgment.

On many important points the biographies of Michurin are vague, especially on the Soviet government's "discovery" of him. However poor, he was a nobleman with something like thirty acres in a region of extreme land shortage, "the center of poverty," as Tambov was called.[28] All but his fenced-in household plot had fallen into a wild, overgrown condition by the time of the Bolshevik Revolution; indeed, just before the Revolution he rented out a portion of his land.[29] The local Soviet government seems to have ordered that his surplus land be divided among the surrounding peasant families in proportion to the number of "eaters" (*edoki*) that each contained. Even after the Bolsheviks took control of the local Soviet from the Social Revolutionaries, it was not clear that Michurin's nursery would qualify as one of the model farms that the Bolsheviks were trying, vainly for the most part, to preserve against peasant division. Michurin seems to have had the support of two local agronomists, who won recognition for his nursery as a state institution with Michurin as the salaried chief.[30] But contention with the local peasants continued, especially when a young horticulturist named Gorshkov was converted to Michurin's cause and began agitating for an increase in funds and *expansion* of the land allotted to the nursery. Gorshkov's demand for the lands of the Trinity Monastery in Kozlov excited the most acrimony.[31] The quarrel moved from Tambov to Moscow, and the prerevolutionary history of conflict between Michurin and academic specialists seems to have been repeated, with the difference, this time, that Michurin won a modest amount of government aid.

Learned specialists (perhaps the same ones as before the Revolution) studied his record and apparently could not agree on how much aid he deserved, if any. Some dismissed him as "a crank [*chudak*] who dreams of growing pears on pussy willow." [32] The chaotic, weedy state of his nursery told against him; so did his lack of scientific education and his surly reticence, which barely concealed a basic hostility to any principles save those he had worked out for himself. But fairminded specialists could overlook or explain away such things. Vavilov, for example, who visited Michurin in 1920, was impressed by the old man's notes and added a decisive voice to those endorsing Michurin as a practical breeder who deserved modest support.[33] The most serious point against him was the failure of local peasants to use his varieties, but here too a rebuttal was possible. The peasants, it was said, thought that only a magician like Michurin could succeed with new varieties.[34] In any case, there could be no doubt that Michurin produced new plant stock of value for further experimentation, which could in turn produce commercially useful varieties for the orchards of the future. That characteristic concern for the long run was probably the decisive argument, which won Michurin's nursery very modest support as a breeding station with a staff of seven.[35]

There the matter would have ended but for the persisting dissatisfaction of Gorshkov and Michurin. In the long run, they hoped to transform the nursery into the command post of a network of stations breeding regional varieties of fruit and transmitting Michurin's precious methods to another generation of breeders.[36] Immediately they wanted funds to enlarge the tiny staff and acreage in order to produce commercial stock in quantity for the orchards of central Russia. Somehow they got Kalinin, the President of the Republic to visit the nursery in September of 1922.[37] Within a month, the Commissariat of Agriculture showed increased favor. It decreed celebration of the forty-fifth anniversary of Michurin's "scientific and practical work," publication of his writings under the editorship of Vavilov, and a slight increase in financial benefits.[38] In November N. P. Gorbunov, Lenin's administrative assistant for matters of science and technology, wired the local government in Tambov, demanding a report on Michurin's work, which seems to have touched off another round of investigations and conflicting reports.[39] Local officials increased their allotment of money and land to Michurin, but he and Gorshkov were still dissatisfied. They were demanding for their one station, the Commissariat of Agriculture wrote to Gorbunov, more than had been allotted for all experiment stations in the entire Central Black Earth Region.[40]

It was impossible for Michurin's nursery to get preferred treatment without appeals over the heads of the learned specialists in the Commissariat of Agriculture, appeals beyond rational calculation, to the sentiment that the poet Mayakovsky expressed with comic extravagance in his

> New
> sermon
> on the mount! . . .
>
> I do not roar to you about Christ's paradise,
> where lenten fasters lick at sugarless tea.

I roar about genuine
earthly heavens . . . .

There the sun pulls off such stunts
that every step is drowned in seas of flowers.
Here for ages gardeners drudge
with glass-topped frames and heaps of dung,
but in my land
on parsley roots
six times a year pineapples grow . . . .

My heaven is for all
except the poor in spirit . . . .

Throw off nature's insolent yoke![41]

Until the crisis of collectivization, that mood would have little influence on the specialists in the Commissariat of Agriculture or on Stalin, who showed no interest in such things, or even on Kalinin and Sosnovskii, who occasionally revealed their genuine sympathy with crankish romanticism (without the saving comic sense of Mayakovsky).[42] As for Lenin, Gorbunov's telegram is evidence, when viewed in historical context, of nothing more than a flicker of interest in a marginal crank. Michurin was seriously "discovered" by the Soviet government, that is, he won a specially favored claim on the Commissariat of Agriculture, late in 1923 when he enjoyed his first major triumph in public relations.

The First All-Russian Agricultural Exhibition in the fall of 1923 set the stage. For days, the central papers were full of news about outstanding scientists and practical innovators, but Michurin got no prominent mention. A meeting of agricultural experimenters, Vavilov presiding, did wire its respects to him for his many years of self-sacrificing work.[43] But this fact was barely mentioned in the central press, perhaps because of the trouble at Michurin's exhibit. Learned specialists denied the authenticity of some hybrids on display (especially an alleged hybrid of melon and squash) and denied that "vegetative blending" (*sblizhenie*) and "mentors" facilitated hybridization between species. They argued that grafting, the essential feature of both techniques, does not alter the germ plasm. Gorshkov, who officiated at the exhibit, came off poorly in the disputes. Michurin sent him arguments, including a leaflet entitled "Mendel's Law Is Not Applicable in Fruit Breeding." He knew how Gorshkov must feel at the prospect of speaking "before such a congregation [*sonm*] of professors and various *spetsy* [slang for specialists]. But it will do you good, if only by showing you the forces of the enemy camp, and their weak side in the sense of ignorance of practical matters."[44] Michurin resumed his own angry complaining. The nursery was in a financial crisis; if Gorshkov could not get extra help from the Commissariat of Agriculture — and it looked as if he could not — Michurin felt he might as well throw it all up and take the lucrative offer that the USDA had long ago made to him.[45] As the exhibition ended, Gorshkov won the support of *Izvestiia,* which broke the story on October 14, 1923, under the headline, "Kozlov or Washington?" Where would the potential of Michurin's great work be realized, in Soviet Russia or in the United States?[46] Michurin was so thrilled that he presented Gorshkov with a watch engraved by his own hand (he had learned the skill in his days as a watch repairman): "From I. Michurin to I. Gorshkov. For Victory Over The Old Fogies of Horticulture [*rutinerami v dele sadovodstva*]. October 14, 1923."[47]

Within a month, the Council of People's Commissars decreed that Michurin's nursery was an institution of nationwide significance, and its area, staff, and financial support were considerably enlarged.[48] What was more important for the long run was the adoption of Michurin by Soviet journalists. Their celebration of the wizard of Kozlov made Michurin, like Edison or Burbank, a twentieth-century folk hero, especially honored — by "the common man" in America, by the "toilers" in Soviet Russia — for performing scientific marvels without benefit of diplomas. His seventieth birthday in 1925 became a sort of Soviet arbor day; honoring the "Father of Apples," the Soviet press honored all those who were restoring horticulture to its prewar level.[49] Michurin was given space in *Pravda* to declare that his varieties — "more than one hundred" was the modest estimate this time — increased the orchardman's profit *tenfold.*[50] The volume of orders grew sharply, and complaints of poor performance grew with them. Michurin testily replied that his varieties were suitable only for regions like Tambov. The important thing, he stressed, was to apply his *methods of breeding* in all the regions of the Soviet Union.[51] He was still dissatisfied. His staff had risen to forty-five and the former Trinity Monastery had been entirely assigned to his nursery, but those favors only guaranteed that his work would flourish for a few more years in Tambov. A few disciples were doing his kind of work in other places, but they met with no official support. Unless a nationwide network of nurseries and training centers were put under his direction, his school of plant breeding would die with him.

What exactly was his school of plant breeding? How did it differ from others? What significance did it have for scientific principles of breeding or for the theoretical science of genetics? Such questions could not be answered precisely then or since, in part because Michurin's harmonious public relations with learned specialists required him to keep certain views to himself. In a letter to Gorshkov, Michurin referred to "professers and other *spetsy*" as "forces of the enemy camp." In his notebook in 1922 he attacked Mendelian genetics and hoped that "a dispassionately thoughtful observer" would recognize "the present conclusions as a foundation that we are bequeathing to natural scientists of future centuries and millennia." [52] But in a report to Vavilov in 1922, Michurin attempted to disarm scientific criticism by disclaiming any pretensions in science. He was a practical breeder pure and simple,

> And it may very well be that I have in several instances fallen into errors of incorrect understanding of diverse phenomena in the life of plants and the application to them, let us say, of Mendel's laws and other doctrines of recent times; but such mistakes, unavoidable in any work, cannot have great significance, for later on they will probably be corrected by other workers.[53]

The public and the private Michurin were different men in the 1920s.

Vavilov made that apologetic report the lead piece in the first edition of Michurin's "works," which was actually a ninety-page pamphlet, less than half of it by Michurin. Forty-nine pages were commentary, mostly by V. V. Pashkevich, a learned fruit specialist, who undertook to explain Michurin's relationship to the science of genetics. Pashkevich noted, as many subsequent commentators would, that Michurin was not consistent. Sometimes he seemed to accept genetics, sometimes to reject it, sometimes *in toto,* sometimes merely as applied to fruit trees. But personal conversation with Michurin had convinced Pashkevich that there was no *essential* conflict between the great breeder's methods and the science of genetics. Michurin was simply trying to work out the hereditary patterns of fruit trees on his own, without reference to Mendelian principles, thus providing an entirely independent test that could only strengthen science.[54] In fact, Michurin was annoyed by Pashkevich's effort to establish harmony between his work and genetics. He said nothing in public, but in a private letter he complained about the meager selection of his items in the booklet, and wrote that Pashkevich had got everything "all mixed up." [55] Neither in private nor in public did Michurin try to clarify what Pashkevich had allegedly confused. He simply put pressure on the reluctant publisher of agricultural literature to bring out a larger collection of his articles without critical commentary.[56]

Similarly Michurin had nothing to say when Vavilov published an article that gently but firmly dismissed Michurin's imaginary hybrids of melon and squash.[57] Silence was also his response to an article by P. N. Shteinberg, a fruit specialist who had befriended Michurin before the Revolution and in 1926 tried to protect him from sensationalism. Shteinberg felt sure that Michurin was dismayed by journalistic exaggerations of his accomplishments. Surely he wished to see other Russian breeders get proper recognition, and Shteinberg rehearsed their names and achievements. Surly Michurin wanted farmers to know that most of his varieties could not be recommended for commercial use because they had not as yet been properly tested.[58] Silence from Michurin, and silence again when E. A. Aleshin, another fruit specialist, published an article called "Michurin and Science." He opened with a caution against misunderstanding: of course Michurin was not a scientist; but his data, if tested and analyzed by scientists, might have significance for genetics. The criticism of Michurin by Vavilov and others was therefore, in some respects, hasty. As American scientists twenty years previously had looked forward to an investigation of Burbank's work that Shull, a well-known geneticist, undertook for the Carnegie Foundation, Aleshin hoped for great results from the scientific checking of Michurin's methods and data.[59]

In both cases the results were never published. Shull found Burbank's data worthless; Aleshin, who joined Michurin's staff, was driven away with bitter recriminations after two years.[60] But that was already in the 1930s. In the 1920s, Michurin maintained steady public silence in the face of learned criticism and condescension. His silence could be regarded as deference to science. Other evidence could also be construed that way. Within Michurin's nursery his most questionable method, the use of "mentors" or grafts to alter heredity, was being subjected to rigorous tests. One of his chief disciples, P. N. Iakovlev, was sent off to study for a doctorate under Vavilov, with the understanding that it would be his dissertation task "to put a scientific foundation under Michurin's work." [61] In the meantime

Michurin's disciples would point to an article by a German geneticist as the best theoretical explanation of Michurin's achievements. (The article actually made no mention of Michurin; it was simply a Mendelian analysis of interspecific hybridization.)[62] In 1927 Gorshkov, the chief disciple, publicly endorsed the official image of his master as a modest breeder without scientific pretensions. "In his work [Michurin] has pursued and pursues only a practical goal. He does not set up rigorous experiments for the study of heredity, and therefore it is possible that his judgments on heredity may be untrue. . . ."[63]

In spite of such evidence, one can easily destroy the pretty picture of a harmonious division of labor between the practical breeder and the theoretical scientists. It is not even necessary to look forward to the bitter clashes that erupted in the 1930s. In 1927, when two scientists in Michurin's nursery were just beginning a rigorous test of "mentors," they were harshly rebuked in a private memorandum that Michurin sent to Gorshkov. The language was extremely clumsy, but the master's great anger was clear enough. There was even a portent of criminal accusations.

> The slipshod performance of the tasks they have undertaken to fulfill can have harmful repercussions on the development of the general work of the Nursery itself, by undermining the faith in it of many excursions of student youth, and will also bring inevitable harm to the trainees [*praktikantov*] when they see failures in the application of my methods, failures that in essence derive exclusively, I assume, at the least from careless execution of the work, indeed, so careless as to give rise to the suspicion that they are pursing the provocational goal of subverting the true value of my methods.[64]

The most astonishing feature of this outburst is not its paranoid crankishness — one grows used to that in reading Michurin — but its basic innocence. Michurin was not only violating the cardinal rule of team research (that the chief assigns tasks but not results); he was so completely unaware of the rule that he did not even think of paying lip service to it. He simply assumed that honest scientists would not question what he had proved by fifty years of practical experience. In short, Michurin was not genuinely cooperating with the scientific community. He was too much the semiliterate individualist to do so. Scientists must accept his practical wisdom. Good for them if they did. So much the worse for them if they did not.

At bottom, Michurin and his disciples were indifferent to the first rule of discourse in a scientific community, the requirement of consistency and rationality. Gorshkov, for example, after summarizing Michurin's methods of breeding, made the usual glib assertion that they had "received a scientific foundation in Germany." [65] Though he was referring to a Mendelian analysis of interspecific hybridization, he declared that Michurin had proved the inheritance of acquired characters (by watering a pear tree with sweetened water he had made the pears sweeter unto the second generation). On the next page Gorshkov returned to the Mendelian framework. He gave a very crude version of the argument that environmental influences affect heredity by modifying the penetrance or the dominance of various genes.[66] Sophisticated versions of this argument are still being used by Soviet biologists who wish to prove that there is no conflict between the science of genetics and the beliefs of Michurin, most notably his belief that grafts can alter heredity in desired ways.[67] They are wasting their time on a hopeless task.

Michurin believed that heredity "does not yield and in essence cannot conform to any patterns worked out by theoretical science and determined in advance." He believed in practical intuition, a semimystical faith, as Gorshkov expressed it in 1925.

> I. V. Michurin has become so intimate with plants and has studied their life so much that one glance is enough, and he can already predict the qualities of a plant and its future fruit. I. V. Michurin cannot even express in words the characters [by which he judges]. He feels them instinctively. By virtue of all this he can subject plants to his will.[68]

Obviously such a man had nothing to tell scientists and nothing to learn from them. He could produce plant stock of interest to breeders and occasionally of use to farmers. Advertising could win him an undue share of fame and funds, but he could have no influence on theoretical science, not even on its applications in agriculture, as long as genuine scientists were in charge of those enterprises. They could not talk with each other; they could at best exchange ceremonial gestures.

In such a context, the treatment of Michurin at the First All-Union Congress of Genetics and Breeding becomes understandable. That climactic meeting in Jauary 1929 honored D. L. Rudzinskii as "the founder of Russian plant breeding." [69] (He had founded the first breeding station in central Russia, at the very time when Michurin was being rebuffed in his efforts to establish one.) The Congress heard Gorbunov reminisce, without mentioning Michurin, about Lenin's part in the development of Russian agricultural science, though Gorbunov, as Lenin's administrative assistant, had sent the telegram of inquiry that is supposed to prove the "discovery" of Michurin by Lenin himself.[70] Of the scores of papers read in the special sections of the Congress, only two, by Michurin's disciples, paid attention to his methods of breeding. Lysenko, who presented a paper, had nothing to say about Michurin.[71] But the old man was not overlooked. In plenary session, the Congress sent two telegrams of warm salutation: one to the Central Committee of the Communist Party, the other to Michurin.[72] Clearly both telegrams were an expression of respect for Soviet public opinion (*obshchestvennost'*), rather than a part of the Congress' professional work. Public relations men had made Michurin a popular hero of plant breeding, but they could not make him a member of the professional community of scientific breeders and geneticists. Only Stalin's mighty hand could accomplish that feat.

LYSENKO AND OTHER PEASANT SCIENTISTS

Michurin was the famous agrobiological crank of the 1920s, by no means the only one. The newspaper *Bednota* (*Poor Peasants*) organized an army of "peasant scientists" in "hut labs," claiming over 23,000 participants by 1929.[73] That was not entirely a crankish project. In part it was a backward country's makeshift for an agricultural extension service. There were not nearly enough trained people or funds to place an agronomist within reach of every village. The minority of peasants who wished to put questions to experts wrote to the newspaper, which gave general replies and urged its correspondents to discover specific answers for themselves by organizing experiments in groups of "lab correspondents," in their own "hut labs." The phrases have the tinny ring of journalistic promotion, but it would be unfair to overlook the fact that the editors (chiefly Ia. A. Iakovlev) were trying to promote a break from hidebound routine. Weed control, proper collection and spreading of manure, introduction of clover, purchase of certified improved varieties of seed, sprouting potatoes before planting—the newspaper carried reports from peasants who were trying out such tried and true innovations. But the plain truth is that this was a minor aspect of the hut lab movement. The main stress was on the bizarre and flashy. Most hut labs concentrated on efforts to stimulate plant growth by soaking seed in salt solutions, in plain water, or in the juice of dung. Most of the peasant scientists were not struggling to introduce modern farming methods; they were trying to prove that their favored seed stimulant was the best.[74]

Hut labs were not the only places where seed stimulants were being tried in the 1920s. Doctor P. I. Shpil'man — or Spielmann, a German who wandered into the Soviet Union in 1921 — won the support of a state farm (run by the OGPU) and ultimately of *Pravda* itself for his work on "biorization" or "biontization." [75] He coined these terms to evoke a complex process set off by his special solution for stimulating seed. A different and even more stirring approach to seed stimulation was promoted by the native philosopher A. L. Chizhevskii. He had published a little book demonstrating the influence of sunspots on human history, and Soviet Marxists had of course rebuked him for such a mechanistic analysis of social evolution.[76] So Chizhevskii changed his field; he studied the influence of radiation on the growth of plants and subhuman animals. With boosts from journalists he won the support of the Commissariat of Agriculture, which provided him with a laboratory to investigate the improvement of yields by "ionizing" plant seed and animal semen.[77]

In the mid-1930s *Pravda* would turn its special kind of death ray on Chizhevskii,[78] but in the 1920s the Central Committee and its newspaper did not claim ultimate authority in the field of natural science. Scientists were free to publish criticism of the seed stimulators, though they had the backing of Ia. A. Iakovlev, the editor of *Bednota,* who was becoming one of the major Party chiefs of agriculture in the late 1920s. The result was a mild debate over the kind of issue that would produce stormy fights in the next three decades. N. A. Maksimov, for example, the leading plant physiologist in the Soviet Union, carefully explained why growth stimulants were a poor choice for research efforts, though a popular one. They seemed to be a cheap and easy way to increase yields. The usual methods — weed control, diversification of crops, fertilization, purchase of certified improved seed — required initiative, knowhow, work, and money. A magic dip would obviously be easier and cheaper. Unfortunately, Maksimov warned, no such dip existed or was about to be discovered. The treatment of seed before sowing was an important element of modern farming — sorting, cleaning, testing for germination, drying and heating, dipping in chemical solutions to destroy pests — but no researcher was anywhere near the discovery of an economically useful growth stimulant. That was music of the future; and anyhow, Maksimov argued, scientific research was not a job for peasants.[79]

The response of N. I. Feiginson, chief of the hut lab movement and future lieutenant in Lysenko's army, revealed the logic that would sustain the Lysenkoite faith through every disappointment. Feiginson reasoned this way: The scientific skeptics might be wrong; the peasant recipes for seed stimulation might be useful. Therefore, they ought to be applied immediately on the widest possible scale, lest their possible benefit be lost while scientists drag out time-consuming tests on little experimental plots.[80] Such reasoning invites contemptuous laughter, but the urge should be resisted. The consequences for Soviet agriculture and science were far from laughable, and we have no right to be smug. We have stumbled on the logic of desperation. In the 1920s it was mild and somewhat amusing — like a poor man's use of his only dollar to buy a lottery ticket. In the 1930s it would become severe, even terrifying — like a cancer patient's insistence on quack nostrums. In any form or degree the occurrence of this logic suggests the existence of a problem that laughter cannot dismiss.

A dendrologist who admired Michurin approached a rational description of the kind of problem that provokes the logic of desperation. German tree specialists had framed the maxim, *Erst studieren, und dann probieren.* And it seems to be obvious wisdom: orchards and forests should be planted only with varieties that have been thoroughly tested for the given soil and climate. That is sound reasoning, the Russian dendrologist commented, for a comfortable country with a good climate; but should not a backward country with a very harsh climate try first and study afterward?[81] He did not spell out the implications, but they are fairly obvious. Time and need are important factors in decision making; the greater the urgency, the greater the chance one must take. G. K. Meister, one of the chief Soviet plant breeders and a genuine scientist, used a similar argument while defending his crosses of wheat with rye. He was aware that Swedish breeders had found such distant crosses too chancy and had abandoned them in favor of the sure, slow gains that could be achieved by crossing closely related varieties and species of wheat. But Russia had a harsher climate and far lower yields than Sweden; therefore, Meister reasoned, it was worth taking an outside chance with rye-wheat crosses, on the hope of obtaining quickly a stable hybrid that would combine the extreme hardiness of rye with the high yield and fine quality of wheat.[82] (The hope was disappointed, as most long shots are, but that is beside the present point.)

That clear, self-conscious form of the logic of desperation is adduced merely to point up what was obscure in the minds of Feiginson and the peasant scientists. Meister recognized that he was taking an outside chance. Feiginson would not or could not think of experimentation as taking chances. To rebut the argument that only trained scientists should experiment, he quoted a peasant correspondent:

> If just the experiment stations alone carry out experiments, what good will it be to the peasants, if, only five or six years later, the experiments move over to the peasants' fields? That is a very long time to wait. Let everyone all together carry out experiments, everyone as he can and as best and from whatever experiments we get the best harvests.[83]

The messy syntax suited the fuzziness of thought. Neither Feiginson nor his peasant correspondent saw agricultural experiment as a statistical comparison of results with expenditure of funds and labor. In their minds experiment was a means of getting something for nothing and right away. "Here in the USSR," wrote another leader of the hut lab movement, "agriculture is at such a low level of development that there are still many possibilities of increasing gross agricultural production by 40 to 45 percent almost without financial expenditure." [84] Naturally he put no price on the labor given to experiments with seed stimulants. Neither did he consider the possible cost of spoiled seed and wasted acreage, nor — what is most important here — the psychological opportunity cost, that is, the organized effort that might have been given to stimulating peasants away from panaceas, toward proved methods of raising yields. In the literal sense of the word, the peasant scientists and their leaders were reckless men; they spent their energy on impulse, they rejected calculating expenditure on principle.

It is misleading to draw a sharp, total contrast between these cranks and the notoriously conservative majority of peasants. Of course, the peasants who were drawn into the hut lab movement were a tiny minority of exceptional individuals, even more exceptional than the slowly growing minority who were quietly adopting modern methods that seemed tried and true. The man who merely bought a modern plow and seed drill and planted winter wheat in neat level rows could arouse hoots of derision from neighbors who wastefully broadcast seed of winter rye on ground roughly broken with the archaic *sokha.* In one case — it happened to be recorded by anthropologists investigating the strange life of peasants — a midwinter thaw followed by a hard freeze did not kill all of the winter rye planted in the primitive way, but the innovator's uniformly planted field of wheat was uniformly destroyed beneath the ice. Village laughter became so abusive that he moved away to an industrial job.[85] Such contretemps may have been another reason why the peasants who were drawn into the hut lab movement paid most attention to seed stimulants, which promised much for little and right away. The hostile pressure of village opinion is evident in the advice that the leaders gave to peasant scientists: form groups under the protective banner of the *komsomol* or civil defense organizations, the better to resist ridicule and hooligan jokes.[86] But even as the obvious differences between peasant scientists and ordinary peasants are noted, an essential similarity emerges. The minority's preoccupation with panaceas was simply the other side of the majority's clinging to ancient routine. Either way, the peasant, chastised by centuries of hard luck, was avoiding the anxiety and danger involved in consciously taking a chance. The exceptional peasant "experimenter" was no less devoted to sure things than the ordinary hidebound peasant. If he thought his "experiment" might fail, he would not try it himself, much less urge it on others. His greater boldness was really a greater degree of the practical man's customary dogmatism, his refusal to subject obvious truths to useless questioning and academic testing.

Trofim Denisovich Lysenko is the prime example of this peasant mentality. To be sure, he had the benefit of education, but the peasant style of thought survived the years he spent at the Kiev Agricultural Institute. What he did learn very well — unless it was mainly the gift of his genes — was the art of self-advertisement. In 1927, when he was only twenty-nine years old, working at an obscure experiment station in Azerbaidjan, he managed to get a boost from *Pravda* itself. A feature article said he had "solved the problem of fertilizing the fields without fertilizers and minerals. . . ."[87] He had proved that a winter crop of peas can be grown in Azerbaidjan, "turning the barren fields of the Transcaucasus green in winter, so that cattle will not perish from poor feeding, and the peasant Turk will live through the winter without trembling for tomorrow." Such miracles will seem trite to anyone who is familiar with the Soviet press, but this miracle worker was quite original.

> Skinny, with prominent cheekbones and closecropped hair [later replaced by a lank forelock], . . . this Lysenko gives one the sensation of a toothache. God grant him health, he is a man of doleful appearance. Both stingy with a word and unremarkable in features, except that you remember his morose eye crawling along the earth with such a look as if he were at least getting ready to kill someone. He smiled only once, this barefoot scientist. . . .[88]

The winter crop of peas evidently did not prove itself in succeeding years. It became the first of a long series of sensational triumphs, celebrated for a while in the newspapers, ultimately forgotten by the public, and carefully ignored by the Lysenkoites. But the young man's masterful way with journalists, his skill at using newspapers to make scientific discoveries of great practical importance, this was not ephemeral. It would be a constant feature of Lysenko's entire career, from the *Pravda* article of 1927 until the end of 1964, when *Pravda* and all the other newspapers would finally turn against him.

The reporter of 1927 confessed that he stared at Lysenko's notebook with ignorant awe. He did not understand "the scientific laws" by which the barefoot scientist had quickly solved his problem, without trial and error. But he passed on Lysenko's popular explanation: "Every plant needs a determinate amount of heat. If all are measured in calories, then the problem of [a fodder crop for] winter fields can be solved on a little old scrap of paper!" [89] What Lysenko meant by this "determinate quantity of heat" was spelled out in his first major article, published in 1928. It was spelled out in degree-days rather than calories, which raised the article from the illiterate to the semiliterate level. Trying to correlate the time and the heat that a given variety of plant requires to go through its phases of development from sprouting seed to reproduction of new seed, Lysenko tried to correlate data on growth, calendar days, and degree-days. He made a primitive error in statistical reasoning, and he paid almost no attention to the lessons learned by previous investigators of this problem.[90] He was courteously but firmly set straight by N. A. Maksimov, a world leader in the study of thermal factors in plant development, who could find only one or two minor merits in Lysenko's clumsy article.[91]

Lysenko then revealed another of his chief and lasting characteristics: a total, angry refusal to give any thoughtful consideration to criticism. He did take steps to make himself less vulnerable, in the first place by dropping the effort to reason statistically. Brushing aside all the other factors that affect the earing of winter grain, he laid it down that chilling of the germinating seed for a minimum number of days is the sole determinant.[92] For the rest of his career Lysenko would limit himself to the crudest kind of theory "proved" by arbitrarily selected examples. He also returned to practical triumphs in agriculture, when a new version of his scientific work, presented at the large Conference of January 1929, met with indifference and a cool dismissal by Maksimov: "The results obtained by Comrade Lysenko do not represent anything new in principle, are not a scientific discovery in the precise sense of the word."[93] Lysenko got his Ukrainian peasant father to prove that Professor Maksimov was wrong. The Ukraine was then suffering, for the second season in a row, a calamitous loss of winter wheat. Surreptitiously, lest the neighbors laugh at him, the old man soaked forty-eight kilos of winter wheat in water and buried the sack of moist seed in a snowbank, to be kept in natural refrigeration until the spring, when it was sowed alongside a field of spring wheat. Usually winter wheat will not ear when sown in the spring, but the grain subjected to "vernalization" (*iarovizatsiia*), as Lysenko named the process of soaking and chilling seed, did ear. Indeed, it yielded a better harvest than the neighboring spring grain. At least, that is what the Lysenkos claimed, and a commission sent to investigate by the Ukrainian Commissariat of Agriculture agreed. The Commissariat became enthusiastic. They ordered large-scale production tests of vernalization. And long before the results of the tests were known, indeed, *before the tests had begun*, the Ukrainian Commissariat handed out the sensational news to the papers: A solution to the problem of winter killing had been found.[94] This time Lysenko's venture into farming by intuition and press releases was an overwhelming success.

There were objective reasons, as a Soviet writer would say, for this triumph. In the winter of 1927–28 about five million hectares of winter wheat had perished, chiefly in the Ukraine. In the summer of 1928 a large gathering of experts had argued at great length over the possible causes and cures but had come to no simple feasible solution. They had even toyed briefly with the notion of treating winter wheat so that it could be planted in the spring, but they had dropped the notion as impractical.[95] The scientists at the meeting had also tried to overcome the belief of many practical delegates that certified, pure breeds of wheat were more susceptible to winter killing than the mongrel peasant varieties.[96] (On the eve of the conference, Iakovlev, the chief of agricultural journalism and propaganda, had given public support to this insidious emotion. It was too vague to be called a belief. Iakovlev did not specify which improved varieties he was comparing to which peasant types in which localities for which time periods.[97]) About the only thing everybody seemed to agree on in 1928, Iakovlev included, was the complexity of the problem of winter killing and the extreme difficulty of solving it. No one expressed disapproval of Vavilov's attitude: "One must not shut one's eyes to the great difficulties, to the enormous amount of work that alone can lead to a basic solution of the most important tasks posed by life, posed by the disaster of the current year."[98] As if to prove him right, the disaster struck again the very next winter and was even worse: about seven million hectares of winter wheat perished in the winter of 1928–29.[99] That was just the time when forced industrialization was beginning, and the Bolshevik chiefs were working themselves into a frenzy over the difficulties of collecting enough grain at a low enough price to feed the urban masses and to accumulate funds for heavy investment. Small wonder that Ukrainian agricultural officials leaped to agree, uncritically and enthusiastically, when Lysenko claimed, in the summer of 1929, that he had discovered an easy way to insure against losses from winter killing.

Unlike the "theoretical" scientists puttering about their tiny experiment plots, inhumanly patient in the face of "practical" disasters, Lysenko had a plan of action. There were other plans of action — "superearly sowing" and "sowing close to winter" — which will be examined later, when vernalization is analysed in detail. No one could claim the other plans as his invention; they were known to be old peasant recipes. Vernalization seemed to be Lysenko's discovery, and one season's success on half a hectare made him a big man. In October 1929 he was transferred from his obscure station in Azerbaidjan to the All-Union Institute of Plant Breeding in Odessa, the most important center of agricultural research in the Ukraine.

Iakovlev's agricultural newspaper began a ballyhoo for vernalization, which became especially serious in November, after Iakovlev became All-Union Commissar of Agriculture.[100] When the newspaper asked Maksimov and other leading specialists in plant physiology and wheat production to give their views on vernalization, the experts necessarily expressed great admiration and respect. Of course they qualified their statements. They admired Lysenko's ardent desire to help agriculture and regretted that they had not yet learned his skill in winning the support of practical agriculturists. They cautioned against haste and extravagant expectations. After all, Father Lysenko's one-shot trial on half a hectare, however promising, could hardly be considered adequate proof that vernalization was really a practical solution to the problem of winter killing. Indeed, the eminent specialists could think of several reasons why vernalization would *not* prove to be a practical solution. But they framed these cautions and criticisms with inconsistent expressions of respect for the practicality of Lysenko and the agricultural bosses who had ordered large-scale production trials of vernalization.[101]

Experience would prove that the skepticism was well justified, but the skeptics would not be allowed to say "I told you so." In the last months of 1929 they had already been forced into a box that would confine them for thirty-five years. Unable themselves to prescribe quick and easy solutions to terribly urgent problems, they could only throw cold water on the ardent men who did. What else but hostility could such do-nothing scientists expect from harassed agricultural officials? Harassed is too pallid a description. The end of 1929 was the time when agricultural specialists suspected of cautious realism were condemned as "wreckers."[102] In December, Stalin appeared before a conference of the cowed survivors to give his famous speech elevating "practice" above "theory." In context it was clear that he was elevating the intuitive judgment of political bosses above the reasoning of scientists and technical specialists. They must not presume to place any limits on the bosses' will to transform agriculture. Limitless possibilities were thereby opened to the bosses and to the cranks of agrobiology, no longer harmless.

4

Raising Stalin's Hand

THE FIRST CLASHES

If the Stalinist system was literally one-man rule, he was a mighty fickle man. Consider the campaign for scientific agriculture. The government gave much more support to genuine scientists than to cranks until 1935, when it abruptly subordinated the scientists to the cranks. For the next thirteen years scientists in the afflicted fields dragged out a wretched existence, until the government suddenly denied them any right of existence. Less than four years later, in 1952, the government changed once again: scientists were allowed to revive public opposition to pseudo-science. With Stalin and the terror gone in 1953, the opposition grew much bolder, but the government spent another eleven years in hesitation and one-sided compromise until it finally withdrew its support from the cranks, and restored the campaign for scientific agriculture to the control of a normally autonomous community of genuine scientists in 1964. The wavering progress from science to quackery and back again to science can hardly be attributed solely to Stalin's personal opinions. We would have to imagine him first an enthusiast of genuine science; then a sudden convert to quackery, but half-hearted in his conversion; coming close to total suppression of the scientific opposition by 1935, but holding back for thirteen years; taking the plunge in 1948, but faltering in 1952. Finally, we would have to imagine that his fickle mind survived its body and kept Soviet leaders wavering until 1964, when the ghost either gave up its control of the Soviet government or came back to support genuine agricultural science once and for all.

Such fantasies are unavoidable if Soviet history is explained by simple reference to Stalin's omnipotent free will. Much better to begin with the equally simple but realistic assumption that politics does not stop when absolute power is brought to a focus in a single person. The authority focused in Stalin was refracted to and from him through multitudinous prisms of lesser authority, all of them submerged in refracting social strata, as animal eyes are submerged in various layers of air and water. But this figure of speech is still too mechanical and simplistic. Let it be said plainly in the language appropriate to the subject: Stalin presided over an evolving political system. The object of scholarship is to discover how the system worked and, especially, how it evolved. The first step is to recognize that one of its famous peculiarities, the passion for secrecy, makes scholarship difficult but not impossible. The politics of the system, the maneuvers and deals of the people involved in it, were assiduously screened from public view. To a considerable extent they remain hidden, for the archives are closed to most scholars. But the policies — the shifting results of the maneuvers and deals — were abundantly documented in the public record, with endless explanation and justification. From those it is possible to figure out what considerations were moving the politicians as a group, now this way, now that, occasionally in different directions at the same time. Through the screens that hide the politics we can hear voices working out policies.

Forced collectivization put technical and scientific specialists in a strange position. The demand for their services was greatly increased, but the result was a deterioration rather than an improvement in their situation. Too much was demanded of them. The reader may recall Ia. A. Iakovlev's avowedly fantastic proclamation in 1928, shortly before he became Commissar of Agriculture, that yields could be doubled in ten years.[1] Translated into the official goal of a 35 percent increase during the first Five Year Plan, it proved to be fantastic in the usual sense of the word, absurdly beyond the capacity or the will of traditionally inefficient farmers thrown into violent confusion by Stalin's "revolution from above." Most of the peasants who had shown marked ability at modernization were probably branded kulaks and driven away from the new farms. Those who remained tended to shun work on the collective fields and concentrate on their private gardens. By a great expansion of acreage under collective cultivation and the imposition of very low prices on the produce of that acreage, the government achieved its main goal: enough grain was delivered to the state each year at low enough cost to feed the rapidly swelling cities and to accumulate funds for a swift growth of heavy industry. In this sense, the Bolsheviks had reason to begin, as early as 1932, the periodic boasts that they had solved Russia's grain problem.[2] In many other senses they had reason to fear that they had not solved it. They could not, for example, overlook the fact that grain yields per hectare, far from increasing by 35 percent, had declined. The average for the period from 1930 to 1934 was approximately 14 percent lower than the average for 1925 to 1929.[3] The silent "We told you so" obviously forming in the closed mouths of realistic specialists provoked Bolshevik officials to angry tirades on the disloyalty of the intelligentsia. Criminal accusations, arrests, and intermittent shooting, which marked the plunge for collective farming, became one of the chronic diseases of the new system.[4]

Even an enthusiast like Williams, who had endorsed Bolshevism before such endorsement became mandatory (and therefore suspect), had a bad moment in the early 1930s. The class enemy, *Pravda* muttered, was using Williams' scheme of crop rotation (*travopol'e*) to justify cuts in grain acreage.[5] Williams responded with the proper reassurances: grass should be planted in addition to the planned grain acreage, never in place of it. With such qualifications he still insisted that *travopol'e* (rotation based on an extended course of grass) was a panacea.[6] From 1928 to 1933 the government preferred to follow Tulaikov's advice. He had no panacea; he proposed radical measures that promised quick large returns on a small investment. Following his advice, the government did get a lot of grain quickly, while Williams was maundering about ultimate disaster and perfect, simple, cheap ways to prevent it. But the government reversed itself in the mid-1930s. It began listening raptly to Williams — an enchantment that would last more than twenty-five years — while flinging Tulaikov violently aside.

In 1928, when agricultural officials were straining every nerve to increase grain collections from peasants who would not sell unless the price was right, Tulaikov proposed a Soviet version of the giant grain farms he had observed in the semiarid plains of North America. A special commission, established directly by the Politburo with Kalinin as chairman, endorsed the proposal.[7] Stalin himself was sufficiently impressed to quote Tulaikov at great length, when he explained the plan to the Party's Central Committee.[8] Huge state farms were established on marginal land to work as "grain factories."

Following simple grain-fallow rotations, ignoring Williams' plea for grass and livestock and crop diversification, they were supposed to yield quick large returns on a relatively small investment of farm machinery, dedicated labor, and skilled management. In effect, this was the way American businessmen had exploited their semiarid plains toward the end of the nineteenth century, as soon as railroads and ships tied them to the world grain market. The initial results in the Soviet Union were interpreted as indications of similar success, though the agency in charge, the Grain Trust, ended each year in the red.[9] Stalin had decreed that profit (*rentabel'nost'*) was not to be calculated for individual enterprises but only on a national scale.[10] According to this rule the state grain farms deserved praise for delivering large amounts of desperately needed grain regardless of cost. Nevertheless, the authorities kept an eye on the cost of production (*sebestoimost'*). In 1929–30 it showed a satisfying decline from 10.6 to 8.4 rubles per centner of grain, but the next year it jumped to 17.3 and in 1932 reached 22.9.[11] Tulaikov, seeing what was happening even before the figures were in, laid out the causes with bold candor.[12] He was unwittingly describing a vicious circle that would restrict yields for decades to come, on collective as well as state farms: poor management (in those early years it was frequently chaotic), slovenly labor, too few supplies (of everything from improved seed to consumer goods), and too many weeds. (Williams said the Grain Trust should be called the Weed Trust.) Each cause of low yields reinforced the others. Correction required a considerable investment over a long period, and always echoing in official minds were the silenced arguments that much of the land was too dry anyway[13] and the continued insistence by Williams' sect that salvation lay in grass.

At the large Conference on Drought Control in October 1931, Williams' disciples argued once again for *travopol'e* as the way out of all agricultural difficulties.[14] They were once again repudiated by the majority of specialists, though they won the right to try their scheme on twenty state farms and twenty Machine and Tractor Stations (MTS).[15] The major officials who spoke at this Conference — Iakovlev, Kalinin, and Molotov — held to their policy of allowing specialists to decide such matters. Molotov, in fact, expatiated on the practical benefit of free debate among specialists. Freedom of criticism meant that mistakes would be quickly recognized and their costly effects minimized.[16] In context this may have been a gentle hint to the majority that they should give a chance to Williams' schemes, but three years after the Conference, Tulaikov was citing Molotov's speech on his own behalf. By that time agricultural officials had grown dissatisfied with Tulaikov's recommendations, and he reminded them of Molotov's pledge of support "to any grand projects, if there is a grain of truth in them, and if they can bring benefit to the toilers."[17] Tulaikov seemed unaware that wild romanticism among politicians tends to subvert free speech among specialists.

The occasion for Tulaikov's first chastisement was his suggestion that "simplified agronomy" might be the way to improve badly managed farms with desperately low yields. For example, shallow plowing might be practiced in order to get a large acreage ready for seeding at the proper time with a minimum of machines, fuel, and reliable personnel. In 1933 his suggestion was savagely attacked as en encouragement to loafers and "wreckers."[18] Kalinin had recommended simplified agronomy for the backward Russian peasants of the 1920s; but second-best techniques had no place in the socialist farms of the 1930s, the world's most advanced agricultural system.[19] Tulaikov's assistant, the economist N. G. Samarin, was arrested as a "wrecker"; Tulaikov got off with a public apology; and the trouble seemed to be at an end.[20] But within the Commissariat of Agriculture the long-standing debate over crop rotation and land use sank toward irrational turmoil, especially after Stalin entered it. In 1934 he criticized the plan for regional specialization of agriculture and announced that every district (*oblast'*) must help relieve the transportation crisis by producing as nearly complete an assortment of agricultural products as possible.[21] Of course, Stalin did not put his case that simply, which would have been an implicit admission that the Soviet economy was not ready for the most advanced kind of specialized agriculture — the wheat belts, corn belts, and tobacco roads achieved by the United States. They had long been the model for Soviet specialists and Bolshevik officials; now they were denounced as a capitalist snare. Local self-sufficiency henceforth was the mark of socialist progress in agriculture. Williams, who had long made a cult of the primitive, became the Bolshevik officials' favorite specialist on matters of crop rotation and land use. Tulaikov became a favorite scapegoat.[22]

By no means all the fussing about agricultural specialists was so clearly connected with basic agricultural issues. A good deal of it seemed to be merely sound and fury, signifying nothing more than official worry that the specialists might still have skeptical or hostile attitudes behind the masks of Bolshevism all were now required to wear. The specialists were warned that declarations of Bolshevik sympathy were not enough; they must prove their dedication by bringing Marxism-Leninism into their specialties. Ironically, this meant trouble for the minority of scientists who had shown an interest in Marxism in the 1920s, before it became mandatory. They had been mainly concerned to discover Marxist implications in the outlook and theories shared by all scientists. Now Marxist scientists were supposed to prove their loyalty by establishing unique positions in science — by showing how different Marxist science was from that of bourgeois specialists.[23] They could not do so; but they could and did add their voices to the penitent chorus of the scientific community echoing the call for a revolution in science, for an end to the ivory tower, for science that would at last be the useful and faithful servant of the toiling masses.

A fine specimen of this litany came from the committee that organized the Congress for Planning Genetics and Breeding in 1932. The committee called for

> the ideational arming of the proletariat in its struggle for the strengthening of socialism against wrecking, against idealist and mechanistic distortions of Marxism. Along with this, faced by the tasks standing before us, we must raise more decisively the question of the reconstruction of our science itself, the rethinking of its methods of work, the introduction of the principle of the classness and partyness [*klassovost' i partiinost'*] of science on the basis of Marxist-Leninist methodology, the rethinking of trends and interrelationships with other sciences, which have been brought to the fore as a result of the tendencies in the development of bourgeois science toward a condition of fragmentation [*razdroblennost'*], of mutual isolation and misunderstanding, toward a condition of decay, impotence, and nit-picking.[24]

At first sight this stupefying effusion of slogans seems to be pure gibberish. The gratitude that the committee expressed for recent government decrees to geneticists and breeders is comparatively grammatical, but similarly without content. Those decrees, the committee declared, "open possibilities, never before seen in the world, for the penetration of our science into production, into everyday work on the construction of socialism, and by that very fact [they open possibilities] for the flowering of Soviet genetics as well." [25]

Such declarations of the scientist's subservience to political authority seemed to have no bearing on the content of science. The appeal just quoted brought together a Congress that fashioned a genuinely scientific program of coordinated research in genetics and breeding. Indeed, scientists at Britain's Imperial Bureau of Plant Genetics were so impressed that they published a translation of the program with an admiring foreword.[26] They simply omitted the esoteric passages declaring submission to the "classness" and "partyness" of science. Hindsight, and a stomach for exhuming sordid little quarrels in obscure places, disclose the meaning of such chatter. To denounce the principle of scientists' autonomy and fire off decrees at them was not equivalent to actual destruction of their autonomy. The politicians did not know enough about the scientists' work to rule it directly. But they could recognize clashes and, invoking the principle of "classness" and "partyness," they could arbitrate.

Consider the case of fruit growing. In July 1930 the Council of People's Commissars suddenly realized that fruit had been neglected in the frantic struggle for grain. The commercial orchards of the kulaks were disappearing as those of the nobility had disappeared in 1917–18, without adequate replacement by collective- or state-farm orchards. So the Council of Commissars ordered the Commissar of Agriculture to organize a large expansion of fruit growing.[27] Naturally enough, the Commissar of Agriculture handed the scientific aspects of the job to the Lenin Academy of Agricultural Sciences, which in turn delegated responsibility to the appropriate section of VIR (the All-Union Institute of Plant Industry in Leningrad), and to the All-Union Institute of Fruit Growing in Kiev.[28] Thus the government decree resulted in a sensible task being assigned to an autonomous group of competent specialists. They decided that their first job was to establish a list of varieties of fruit and berries recommended for various regions. According to a leading pomologist, 10,000 varieties of apple and pear were being grown in the Soviet Union,[29] and even allowing for exaggeration (he considered each clone a variety), it is obvious that a real problem existed. The commercial fruit grower in a modern economy produces a uniform product for mass marketing, and Soviet fruit specialists were working out a sensible coordinated program to achieve such modernity. The trouble was that Michurin's group would not fit in.

By accident or because he was tipped off Michurin opened an offensive in the middle of 1930, just before the Council of People's Commissars issued its decree on the necessity of improving fruit growing. Once again he used a sensational journalist to throw his charges, but this time he touched off serious explosions. The literary magazine *October* serialized Michurin's familiar legend, punctuated with direct quotations explicitly authorized by the hero.

> "Not only is it impossible [in plant breeding] to apply any calculation in accordance with Mendel's law, but it is quite impossible to do any strictly precise work in accordance with a plan worked out in advance. . . ." These angry Michurinist lines are appearing in printer's ink for the first time. They were not included in the printed volume of his works.[30]

Michurin emerged as a coarse old man, trembling with anti-intellectual rage. " 'You listen here,' " he exclaimed, using the insulting *ty;* " 'you've got to grab the problem by the tail, and give fools, and wise men too, a poke in the eye.' "[31] The only way to breed improved varieties was to follow Michurin's special methods, which could not be learned from books but only from experience, under the supervision of the rough genius or his disciples. Nevertheless his writings were vital; the long delay in their publication was intolerable: " 'To call this foolishness is not enough; this is a flagrant crime.' "[32] The final appeal was to " 'Michurinize' " the country:

> to knock out sleepiness with punches, with demands, with insistence, with daring. With daring to master and transform the earth, nature, fruit. Is it not daring to drive the grape into the tundra! Drive! Drive! Drive! Into the furrows, into the gardens, into the orchards, into the machines of jelly factories. . . . Faster and faster, . . . faster comrade agronomists![33]

For a time it seemed that this outburst was going to be ignored. In December 1930, at a conference for planning the expansion of fruit growing, the All-Union Institute of Fruit Growing in Kiev was, and showed all signs of continuing to be, the directing center of fruit breeding and testing.[34] VIR was the center of basic research on relevant scientific problems; and Michurin's station in Kozlov was one of many subordinate nurseries, however prized by journalists. In 1931 a preliminary list of standard varieties, published by VIR, included *one* of Michurin's creations, at a time when his press agents were claiming over three hundred.[35] There the matter would have rested, if the normal autonomy of scientists and technicians had been respected. But this was the time when the specialist's autonomy was under attack as a vestige of rotten liberalism, serving only to protect anti-Soviet specialists in their sneaky efforts to wreck the construction of socialism. The chief agency for inspection and discipline investigated the work of fruit specialists, very likely acting on the complaint of Michurin or his associates. This chief agency was the Party's Central Control Commission and the government's Commissariat of Workers' and Peasants' Inspection, which usually acted jointly. In 1931 they were directed by A. A. Andreev, who had been one of the organizers of the grain factories and would become Stalin's chief lieutenant for agriculture after 1937, when the terror would cut off Ia. A. Iakovlev.

The investigation of fruit breeding produced a decree on May 13, 1931.[36] The Lenin Academy of Agricultural Sciences and "individual links of the agricultural apparatus" were reprimanded for giving insufficient support to Michurin. Detailed orders were given for expanding his experiment station and boosting his varieties and breeding methods. A. I. Luss, the chief fruit specialist in VIR, was singled out for individual rebuke for his "tactless and intolerable attacks on Michurin and on his achievements." Luss had formerly worked in Michurin's station, where he had been privately rebuked by Michurin in 1927 for presuming to do a scientific test of Michurin's special breeding methods ("mentors" and "vegetative blending").[37] In effect, the Control Commission and Commissariat of Inspection were taking Michurin's side against his scientific critics, turning a private quarrel into a political conflict.

In effect — that must be stressed. The decree did not specify which methods or achievements of Michurin were being shielded from what kinds of criticism. The explicit order to honor Michurin and expand his domain could be obeyed without capitulating to his attack on scientific breeding. Michurin's station was elevated into the Institute of Northern Fruit Growing, supervising a network of stations in central and northern Russia, independent of the Kiev Institute, which was restricted to supervision of stations in the south.[38] Luss was removed from his chairmanship of the fruit section at VIR (the All-Union Institute of Plant Industry in Leningrad).[39] But he and other specialists refused to believe that they had been ordered to give up science for Michurinism. The result was a nasty wrangle at the end of 1931, when another conference met in Kiev to draw up the list of standard varieties of fruit. Eight of Michurin's creations were certified, which was seven more than VIR had recommended before the official censure of Luss, but still not enough to satisfy Michurin's disciples. They were especially outraged by the refusal of the conference to pass a resolution endorsing Michurin's methods of breeding. When the Michurinists invoked the government's decrees — there had been several following the basic statement of policy on May 13 — one specialist had the audacity to challenge the government's authority: "It is the job of the Soviet government to give awards to Michurin and others; it is our job to recognize his methods and achievements, or not to recognize them." [40] V. L. Simirenko, the leading specialist at the conference, was subsequently assailed in the main journal of horticulture. He replied with a wishy-washy apology for the tenor of the conference, still refraining, however, from an endorsement of Michurin's special methods.[41]

For a time the political authorities seemed to withdraw from further intervention in fruit breeding. In fact, the scientific authorities reasserted their powers at the beginning of 1933. Deciding that the sudden expansion of Michurin's Institute had resulted in confusion and nonfulfillment of assigned tasks, the presidium of the Lenin Academy of Agricultural Sciences ordered a sharp cutback. The staff was reduced from 150 to 67 people, along with a corresponding slash in budget and assignments.[42] Whatever the private reaction at the Institute may have been, in public, moderation was the rule. Thus the 1933 guidebook of the Institute explained that Michurin's methods had

> received a complete scientific foundation in Germany, in the works of Professor Renner. Michurin's opinion concerning the hereditary transmission of characters acquired by an organism under the influence of the external environment has found its reflection in a whole series of works by such well-known professors as Kammerer, Muller, and Bogdanov.[43]

This was a renewal of the confused effort to reconcile Michurinism and science, which had marked Michurinist publications in the 1920s. The author of the guidebook, A. N. Bakharev, Michurin's press agent, may have been unaware that Renner and Muller were standard geneticists who had never examined or commented on Michurin's work and would have been utterly unwilling to endorse his views. Kammerer was a martyr of the dying Lamarckist opposition to genetics (he had shot himself when it transpired that some of his crucial experiments were faked), and E. A. Bogdanov was a livestock specialist who had tried to effect a compromise between Mendelian and Lamarckist principles. It seems more likely that Bakharev was aware of these differences without quite understanding them. He probably chose to ignore them because he was trying to reestablish a live-and-let-live relationship with the scientists at the Lenin Academy. Michurin for his part abstained from further offensives. In fact he passed the word through the journalistic megaphone that the president of the Academy, N. I. Vavilov, was a good friend and supporter of his work.[44]

In 1933 the terror helped Michurin's cause, as it did Williams', by attaching the suspicion of "wrecking" to their opponents. In March there was a highly publicized strike at agricultural specialists, recalling the pogrom of 1929. The press reported that thirty-five specialists were shot and forty jailed, for trying "to create a condition of famine in the country." [45] The most prominent victim was M. M. Vol'f (or Wolf) an economist in charge of the agricultural section of the State Planning Commission. He was charged with malicious opposition to the expansion of acreage, favoring instead an intensification of production, as though the socialist agricultural system, the best in the world, could not accomplish both expansion and intensification at the same time.[46] Most of the other evidences of "wrecking" given out in the press related to similar issues, suggesting that once again as in 1929 agricultural economists were the chief victims. There was no suggestion of a connection between "wrecking" and criticism of Michurin until November 1933, when the main journal of horticulture revealed that V. L. Simirenko had been arrested as a "wrecker." His refusal to endorse the Michurinist resolution at the conference of December 1931 was cited as evidence of his "wrecking." [47] Immediately following this revelation, the director of Michurin's Institute, V. A. Odintsev, was glad to report the removal of the restrictions imposed on the Institute "by the former scientific and technical sector of the RSFSR Commissariat of Agriculture and the Lenin Academy of Agricultural Sciences." [48] The supervisory power of Michurin's Institute, Odintsev explained, had just been extended to fruit experiment stations as far south as Sochi and the Crimea; the sphere of the Kiev Institute of Fruit Growing, where Simirenko had been a major official,[49] had been correspondingly reduced.

Clearly the Michurinists profited from the terror; but it must be noted that their public declarations to that effect were rare, brief, and gingerly. If the Michurinists had special influence in the apparatus of terror, they were not eager to boast about it, and they were afraid to use it very much. Most of their strongest critics remained at large.[50] That is one basic fact to bear in mind when assessing the influence of the terror. Another is the fact that Odintsev himself, director of Michurin's Institute, was arrested a few years after Simirenko, without any setback to the Michurinist cause.[51] Since the most sensitive archives are closed, one cannot answer the question: In each act of terror, who put the finger on whom, with what allegation of crime, and with what real motives? The bureaucratic indifference to individuals that marked the administration of terror is passed on to the historian. Individuals are simply cases, helping to establish a pattern. An important feature of the pattern is the arrest of far less Michurinists than non-Michurinists. That fact will be given due consideration later on.[52] But already, without studying numbers, a basic anomaly should be clear. When the terror struck scientific critics of Michurinism, their cause suffered; when the terror struck Michurinists, their cause did not suffer. In the contest between science and agrobiology terror was not the main arbiter. Something more important was pushing the scientists to defeat, whether or not they were arrested. Something was pushing Odintsev's cause to victory in spite of the terror that snatched him away.

In fact, during the last year of Michurin's life, his cause was being pushed to victory on a far broader front in a much larger conflict than he and his immediate disciples had chosen. Buoyed on a flood of public adulation, Michurin seemed to be content with dominion over fruit-breeding stations and circles of amateur Michurinists. He had made a curmudgeonly peace with science. "'Man's science is lazy,'" he told a journalist biographer at the end of 1934:

> "in many respects it is still aristocratic and lily-fingered. Our scientists have irritated me a great deal in their time with their arrogance and conservatism.
>
> "And how is it now, Ivan Vladimirovich; do you get along with the Academy?
>
> "Now it seems okay," he answers. "At last it seems they understand me."[53]

That peaceful view was not endorsed by Iakovlev, the chief of agricultural journalism who had become Commissar of Agriculture at the time of collectivization and was now moved up to a superior position as Head of the Agricultural Section of the Party's Central Committee.

During the enormous celebration of Michurin's anniversary in September, 1934 Iakovlev's public letter of tribute made a special contrast between Michurin and

> representatives of bourgeois science, [who] have sharpened all their wits to prove the *impossibility* of interspecific crosses and the changelessness of the genotype in an endless series of generations. You, moving from experiment to experiment, have proved in practice both the possibility not only of inter-specific but even of intergeneric crosses, and the possibility, by the influence of external factors, of changing the plants' characters that are transmitted in heredity.[54]

Iakovlev was a politician with little scientific training. He may have been unaware of the falsehood in this statement. The possibility of interspecific and even intergeneric crosses was common knowledge long before Michurin; geneticists and breeders had long argued about the emphasis that should be given to such "distant crosses" in a sensible breeding program.[55] As for changing genotypes through "external influences," farmers and empirical breeders like Michurin had been doing that for centuries, by selection, by crossing, by blind trial and error. Simply moving plants from one ecological setting to another subjects them to altered selective pressures which, over a series of generations, can have a very marked effect on their hereditary nature. Geneticists were just learning how to improve genotypes more efficiently, in the first place by insisting on a rigorous distinction between hereditary and nonhereditary changes, a distinction that Michurin handled very crudely. Iakovlev may have been unaware of these facts, but he was almost certainly aware that his contrast between Michurin and "representatives of bourgeois science" was an implicit endorsement of Lysenko's crusade against genetics, which was just beginning in 1934. How else could Iakovlev explain the publication of his tribute only in the local paper of Michurin's town, while the central papers were filled with bland tributes to Michurin from lesser agricultural officials?

Whether Michurin was quietly opposed to Lysenko's crusade or indifferent, or gratified, is not clear. He did not comment on Iakovlev's letter, and he ignored repeated approaches from Lysenko.[56] He kept a genuine scientist in charge of his Institute's department of cytogenetics, and he and Vavilov fostered the story of cordial relations between themselves.[57] If Michurin did not protest as publicity men transformed him from an unlearned fruit breeder into an epoch-making scientist, neither did he protest when Vavilov published a bland, essentially meaningless interpretation of this new inflation of the legend.[58] A week before Michurin died of cancer in June 1935, he was elected an honorary member of the Academy of Sciences, on a motion presented, *inter alios,* by Vavilov.[59] No convincing description of Michurin's contribution to science could be given then or since, for he had made no contribution to science. By trial and error and much toil he had created a rich stock of breeding material and a few varieties of commercial value. His most remarkable achievement was in the field of public relations. He had projected the image — let Americans wince at their own jargon, their own way of life — of a Soviet Union inundated with fruit, created by uneducated rural folk like himself, in defiance of harsh climate and learned pessimists.

After Michurin was dead, Lysenko made him the patron saint of a crusade against genetics and its applications in breeding. Vavilov tried to prove that Michurin was really a friend of genetics, but his argument was pretty weak.[60] The most one can say for the old man is that, once he had his way in fruit breeding, he was willing to live and let live. He simply did not understand the science of genetics. If men like Luss and Simirenko got out of his way, and administrators like Vavilov let him rule fruit breeding, he did not much care how the geneticists wasted their time.[61]

SYSTEMATIC CONFLICT

Vavilov's praise of Michurin as a scientist seems pretty clearly to have been an insincere political maneuver, forced upon him by his position. Chief organizer and first president of the Lenin Academy of Agricultural Sciences, he was a model "non-Party Bolshevik," that is, a person who shows complete solidarity with the Party without becoming a formal member. At major Party gatherings of the early 1930s he and Tulaikov, who became a formal member, were the usual speakers on the scientific and technical aspects of the agricultural revolution. They were quite willing to use the Bolshevik rhetoric, and in most cases we have no reason to doubt their sincerity. When President Vavilov described the Lenin Academy as "the academy of the general staff of the agricultural revolution," [62] he probably was expressing his heartfelt conviction. So also in his repeated declarations that the collectivization of agriculture would allow the Soviet Union to apply scientific methods more rapidly than any capitalist country.[63] He proved the sincerity of his dedication to scientific socialist agriculture by resisting political authority when it turned against the scientific part of the program.

Vavilov's first test came at the beginning of 1931, when the influential newspaper *Economic Life* — published by the Council of Labor and Defense, a sort of supercabinet — attacked his ambitious plan of assembling the world's greatest collection of plant breeding material. It was denounced as "reactionary botanical studies," the result of Vavilov's "separation from practice," founded on his law of homologous series and his theory of centers of origin of cultivated plants. The author of the attack, A. K. Kol', called attention to the chaotic state of the seed business; let Vavilov put his army of specialists to work on that practical problem. Kol' was not opposed to all plant hunting, only to the collection and maintenance of exotica that might prove useless to the breeder. Vavilov, he argued, should seek out plants that could be introduced directly into farm production.[64] The newspaper went out of its way to endorse this attack, but Vavilov disdained the apology or "self-criticism" that was expected in such situations. He conceded that the Lenin Academy might be intelligently criticized for the enormous sweep (*razmakh*) of its undertakings, but insisted that Kol' was masking a personal issue in public slogans. Vavilov had put him in charge of the Bureau of Plant Introduction at VIR (the All Union Institute of Plant Industry), where Kol' had done such a poor job that his tasks had to be shunted to other bureaus. Now he was trying to get even. Vavilov insisted that the enormous scope of VIR's work was necessitated by the enormous demands of socialist reconstruction of a backward system of agriculture in an extremely cold and dry country. In the long run it would prove very shortsighted, not practical at all, to cut back basic research designed to serve the needs of tomorrow's socialist farmers.[65]

That conflict seemed to blow over quickly, causing no serious disturbance in Vavilov's large and still expanding enterprise. In March 1931 he was an honored speaker at the Congress of Soviets, expressing his satisfaction at the recent doubling of research institutions in the agricultural sciences. (The peak would be reached in 1932–33: almost 1300 institutions, ranging from small experiment stations to major institutes, employing something like 26,000 specialists.)[66] Vavilov thought he saw the end of the shortsighted view "that we should close scientific institutions for a while and devote ourselves to propaganda of what is already known. . . ."[67] At last everyone realized that the new agricultural system was raising such gigantic problems that old science was inadequate. A revolutionized science was required by a revolutionized system of agriculture. Thus Vavilov turned to his advantage the noisy campaign then under way to revolutionize the natural sciences.

That particular campaign petered out, but the pattern of argument between Kol' and Vavilov persisted. Instead of arguing over the relative amounts that a poor country should spend on basic as against applied science, the disputants came close to insisting on all or nothing for both, refusing to distinguish between pure and applied science, reading major theoretical significance into disputes over practical matters and immediate practical significance into theoretical disputes. Either Vavilov's plan for the world's biggest collection of living plant stock was justified by its practical utility *and* by its underlying theoretical principles, or he was wasting money and talent on bourgeois scholasticism. Some will see here a long-standing Russian tendency to think in extremes, interlocked with a habit of confusing practical and theoretical issues. Others will connect these habits with Russia's backwardness, and with the Stalinist method of fighting backwardness by refusing to see it. The best agricultural system in the world required the best agricultural science, nothing less. Talk about modest goals that could be achieved with the modest means available to a backward country had long irritated many Russians. After 1929 such talk became virtual treason. It was regarded as psychic disarmament, surrender to defeatist notions about the tyranny of nature and history. "If there is a passionate desire to do so," Stalin declared in 1931, "every goal can be reached, every obstacle overcome." [68] Of course he was indulging in rhetorical exaggeration. Realistic calculation was not entirely banned; the faith that willing makes things so did not enjoy a total victory. On different sectors of their campaign against backwardness the Bolsheviks give different emphases to these conflicting styles of thought. On the agricultural front they came close to literal acceptance of Stalin's words, perilously close to a murderous and suicidal extreme of political voluntarism. And agricultural specialists were expected to justify such willfulness scientifically.

For most of the new farms, calculating prudence was not only officially despised as a "wrecking" style of thought; it was almost impossible. Just to get the crops planted at something like the proper time had become a major crisis, requiring a "sowing campaign" each spring and fall. That was especially true for spring grains, which demanded a maximum of labor in a minimum of time. Russia was still close to grain monoculture — and most of it was spring grain — the growing season was still short, machines and draught animals few, and rural labor had ceased to be self-driven during the precious days when weather and soil were just right for plowing and seeding. The sowing campaigns were therefore very much like military campaigns, with "mobilization of columns," heroic exhortations to patriots, and quieter but deadly serious threats of punishment for slackers and deserters. A Ukrainian report of 1934 took comfort from the fact that shooting had stopped.

> The fall sowing campaign of 1933 was carried out in a period when the class enemy in the Ukraine was already basically beaten, and therefore severe repression was not applied on a mass scale. There were no sentences to the highest measure of punishment [shooting] for crimes connected with the fall sowing campaign. 13.3 percent of all convicted kulaks were sentenced to ten years, while no middle or poor peasants received such sentences.[69]

In such an atmosphere, trial of a new technique tended to fuse with the elemental trial of political strength between officials and peasants. If failure resulted, it was hard to decide which was at fault, the experimental technique or the peasants, who could not be counted on to do the most time-honored operations properly and on time. Technical experiments became an element of the sowing campaigns, a test of the officials' ability to prod the peasantry into activity that they hoped would be useful.

The major experimental campaign during the early 1930s was devoted to "superearly sowing." In effect, this meant getting peasants out to broadcast seed in the spring mud, which was supposed to give the seed a head start against drought and to relieve the peak work load that would come when the earth grew dry enough to plow properly.[70] A similar function was supposed to be served by "sowing close to winter" (*posev pod zimu*) or "winterization" (*ozimizatsiia*) — planting spring grain in late fall, just before the hard freeze, so that it would lie in the earth without sprouting until the spring thaw.[71] Nobody claimed that these techniques were genuine innovations; peasants had toyed with them for a long time. There was even a rhyming proverb that could be quoted on behalf of superearly sowing — *Sei v griaz' budesh' kniaz'* (sow in the mud and you'll be a prince) — though it may be just another way of saying early bird gets the worm.[72] What was new was government campaigns for such a dubious technique, which, according to a decree of the Commissariat of Agriculture, was supposed to be applied on 2.4 million hectares (almost 6 million acres) in 1933.[73] Glowing press reports made these enforced experiments seem completely successful, if the reader overlooked the scantiness and extreme selectivity of the data offered as proof. In reality it seems that four years of repeated failure on an increasingly large scale caused the agricultural authorities to conclude that the experimental techniques were at fault, not just the peasants who were being pushed to try them. In 1935 Iakovlev, then chief agricultural official, criticized excessive enthusiasm for superearly sowing, and in February 1936 a joint decree of the Party's Central Committee and the Council of People's Commissars completely banned it.[74] The obedient specialists who had published one-sided praise based on arbitrarily selected data were now expected to publish obedient criticisms of themselves.[75]

Dr. Spielmann's magic formula for the "biontization" of seed and Professor Chizhevskii's miraculous "ionization" were also boosted by central newspapers and political authorities in the early 1930s, only to be rejected a few years later. Those wasteful mistakes were blamed on the cranks themselves, who were driven from public view with shouts of "fascist bastard" (*svoloch'*) and "crook" (*zhulik*).[76] The waste in those ventures was largely of research funds; talk of "biontization" and "ionization" in actual production was little more than talk. So also for the lesser stars that flashed through the sensationalist press in the early 1930s. A physicist in Ashkhabad who made rain with electrified smoke was taken up by *Economic Life,* the same paper that sponsored the first attack on Vavilov. When last heard from, the rainmaker was the central hero in a celebration of socialist science by the philosopher I. I. Prezent, who would shortly become Lysenko's chief theorist:

> We are carrying out the grandest task, planned alteration of the climate. . . . A special grand institute for making and stopping rain is being organized. . . . The grandest, unheard of projects are now being worked out, in actual working plans with concrete economic calculation, for the irrigation of dry regions and an all-out assault on the desert. We are solving the problem of *heating Siberia.*[77]

With this grand exception, all the scientific miracles of the early 1930s involved seed treatments — superearly sowing, winterization, biontization, ionization, and of course Lysenko's vernalization. To complete the list mention must be made of a railroad worker in Rostov-on-Don who discovered a technique for improving seed by radiation. Local scientists scorned him until the local Party organization "pushed him up" (*vydvinula*) into a scientific institution. We have only one report of his sensational work before he fell into oblivion. On the brink he was proud to say that he was corresponding with Lysenko, who was very sympathetic.[78]

The obvious question is how Lysenko and vernalization avoided the oblivion that engulfed other cranks and their miraculous seed treatments in the mid-1930s. Of course, exceptional success usually owes a great deal to luck. But the best illustration of this rule is not Lysenko as much as N. V. Tsitsin, an agrobiologist who climbed to the top in the mid-1930s and survived all vicissitudes to remain, thirty years later, a big man of Soviet science. He did not have a recipe for an experimental drive among the peasants, and at first he did not have very good luck. A peasant in origin, who grew up in orphanages while his widowed mother worked as a cook, Tsitsin entered the world of science through a *rabfak* or "workmen's faculty," a cram school to get lower-class students into higher education. His first job was a good one: at Saratov's famous All-Union Institute of Grain Culture, where Tulaikov was director and G. K. Meister, the country's most outstanding breeder of wheat, was Tsitsin's boss. But Tsitsin and his boss did not get along. Meister was a great believer in the practicality of distant crosses, but his main concentration was on hybrids of rye and wheat. He was cool toward Tsitsin's suggestion of an all-out effort to achieve a very hardy wheat, perhaps even a perennial, by crossing it with couch grass (also known as quitch or witch grass, a perennial weed of extreme hardiness). Meister and other senior specialists at the Institute had been trying such a cross on a small scale and were inclined to regard it as impractical, a problem mainly of academic interest.[79] Tsitsin called on the Institute's Party organization for support and seems to have won it, but only for a short time. He was sloppy in his work (he blamed inadequate equipment), and when he produced his first hybrid seed, Meister pronounced them no hybrids. Twice, in April 1931 and again in August, there were quarrelsome meetings that dragged out to very late hours, forcing the Party organization to choose between the renowned specialist and the young rebel. They chose Meister, and early in 1932 Tsitsin was asked to resign. He moved to the West Siberian Experiment Station in Omsk, whose director, V. R. Berg, believed in the practicality of crossing wheat with couch grass. He had been working on the problem himself.[80]

A year after Tsitsin's arrival in Omsk, Berg was arrested as a "wrecker," and Tsitsin was made a hero by the mass media, even though (or because) he still complained that "an enormous portion of specialists up to the present are extremely negatively inclined toward our work." [81] A commission of inquiry from the Commissariat of Agriculture, which was itself being purged of "wreckers," decided that Tsitsin's work had enormous promise and merited great support. He was given space in *Pravda* to promise an annual hybrid of wheat and couch grass ready for production testing by the fall of 1935; a perennial would take a year longer.[82] When 1935 came to an end without a hybrid ready for the testing service, Tsitsin received reassurance from the highest authority. Stalin told him, and allowed the awesome words to be inscribed on newsprint: "Experiment more boldly. We will support you." [83] Tsitsin became the director of the Omsk station, which had been promoted into the Siberian Institute of Grain Culture. Specialists began to pay respectful attention to his work.[84]

Obviously Tsitsin had the kind of luck that is called pull. (In Chicago it is called "clout," which is a stronger version of the Russian "push," *tolkat'*. The French combine "push-pull," *pistonner.*) Of course, Tsitsin's luck at clout-push-pull should not blind us to his merits. Dr. Spielmann of "biontization" fame and Professor Chizhevskii of "ionization" were also venturesome cranks with high-placed patrons. They suffered defeat, in part because their projects were merely seed treatments, in part because they did not have enough class. (Spielmann was a middle-aged foreigner; Chizhevskii was a middle-aged Russian *intelligent.*) In some measure Tsitsin earned his success: his promise to create a perfect plant was perfectly unassailable, and he was a young Russian *vydvizhenets,* a person "pushed up" from the lower classes to a responsible position, in a period when the pushing of such people was basic policy. But there was not enough room at the top for every low-class person on the make; witness the fall of the radiating railroad man of Rostov. So we come back to luck as the basic explanation of Tsitsin's success — luck in dream production and clout-push-pull, not in plant breeding. Thirty-odd years after 1935 he still had not produced a really successful hybrid of wheat and couch grass.[85]

Lysenko was another lucky *vydvizhenets,* but not as lucky as Tsitsin. The terror did not snatch any of his major critics between 1930 and 1935, as he was pushing his way to the top, and when he got there, Stalin's public endorsement was comparatively weak. (At a farmers' meeting in 1935, when Lysenko stumbled in his speech and apologized for being a vernalizer rather than an orator, Stalin interjected "Bravo, Comrade Lysenko!")[86] But outstanding ability more than compensated for Lysenko's comparative lack of luck. By 1935 three major achievements won him undisputed leadership of agrobiology. Indeed, agrobiology could easily have been called his creation — the word seems to have been his coinage — for all the other cranks together could not match his record. Lysenko excited agricultural officials into campaigns for *a series of experimental techniques,* devising new ones even before the old ones palled, each of them promising quick returns without any substantial outlay of government funds. He won not only the praise of officials, journalists, and fellow agrobiologists, but also *the praise of his scientific critics,* which reinforced the high opinion of officials, journalists, and fellow agrobiologists. His most important achievement was to *escape from the limited sphere of agricultural techniques,* within which his scientific critics tried to confine him. Officials who believed in the Stalinist fusion of theory and practice were carried with Lysenko from a campaign for vernalization to an attack on the science of plant physiology, from a campaign for wheat improvement to a crusade against genetics, the central discipline of modern biology. In short, Lysenko raised limited conflicts between scientific and crankish agricultural programs to the level of total war between science and agrobiology.

The reader will recall how Lysenko's vernalization first captured the enthusiasm of Ukrainian agricultural officials. In 1929, while they were very worried about winter-killing, he got his peasant father to moisten and chill winter wheat and plant it in the spring. The Ukrainian Commissariat of Agriculture was so impressed by this demonstration of the way to avoid winter-killing, on half a hectare, that it ordered 1000 one-hectare tests for the spring of 1930. That happened to be "the first Bolshevik spring," the time of maximum turmoil in collectivization. Nevertheless Lysenko claimed that 900 farms were trying vernalization on 342 hectares.[87] Granting that unforeseen difficulties had spoiled most of the experiments,[88] he continued a shift that he had begun already in December 1929. Vernalization of winter wheat was relegated to scientific laboratories, and talk of an escape from winter-killing was dropped. Vernalization of *spring* wheat — in Russian the phrase is even more outlandish, *iarovizatsiia iarovykh* — became Lysenko's gift to farmers, and the technique was greatly simplified, until it came to be little more than a hasty moistening of seed before planting. On this basis Lysenko won support at a higher level than the Ukrainian Commissariat of Agriculture. The All-Union Commissariat invited him to give a report in July 1931, and Iakovlev was sufficiently impressed to issue a decree:[89] Mass trials of vernalization of spring wheat were to be carried out in 1932. Agronomists were to do the bidding of Lysenko "without demur" (*bezogovorochno*). Vavilov's chief institute, VIR, was to cooperate in specified ways. Lysenko was to have a journal for himself, *The Bulletin of Vernalization*, to be published by his own division of the Odessa Institute, where he had been causing a turmoil since his arrival in the fall of 1929.[90]

In the second half of 1931 Lysenko enjoyed another boom, bigger than that of 1929. He was lauded not only by the mass media but by conferences of specialists, which dutifully adopted resolutions endorsing the forthcoming campaign for vernalization.[91] The highpoint came at the All-Union Conference on Drought Control in October 1931, when Commissar Iakovlev chided Lysenko for "underestimating up to now the scale of the revolution that his experiments ought to create in agricultural production."[92] Iakovlev looked beyond the tests that had been ordered for 1932 to really mass trials in 1933, "on a scale of at least hundreds of thousands of hectares. Only in that case will the thing be done in a genuinely scientific, a genuinely revolutionary way."[93] The Conference included in its resolution a "warning against drawing hasty negative conclusions from possible individual failures" of vernalization, for "particular failures are possible, indeed unavoidable in . . . [this] as in every experimental search for new pathways."[94]

The warning was heeded. Lysenko's scientific critics concentrated on the confusion he was bringing to plant physiology with his vague concept of vernalization.[95] Almost all of them conceded — at least in print — the practical success of vernalization in boosting yields, even though Lysenko's spotty reporting of results strongly suggested widespread failure. It would have been useless to demand better data, for Iakovlev had explained that "the Soviet concept of [farm] profit (*rentabel'nost'*) does not coincide with the concept of profit among bourgeois economists. When the interests of the Soviet Union as an economic whole demand it, we can have temporary 'losses'. . . ."[96]

Indeed, that was one of the basic principles of Stalin's agricultural revolution. Practical bosses who stood high enough to see the whole picture would determine what was profitable down on the farm and what was not. Nevertheless, however global their outlook, however mistrustful of the peasants, who could be slobs in vernalization as in all collective tasks, the practical bosses must have felt some disappointment at pesky reports of little failures down on individual farms. (They were probably not great failures, for vernalization of spring wheat did not involve as great a risk as winterization or superearly sowing.)[97] Early in 1933 Lysenko's personal journal was allowed to die, and the propaganda boom for vernalization was significantly tuned down.[98]

Lysenko boasted that 200,000 hectares of spring grain were vernalized in 1933, but that was not very impressive by comparison with the 400,000 hectares that enthusiasts for winterization had boasted in 1931, or with the 2.4 *million* hectares on which the Commissariat of Agriculture decreed superearly sowing in 1933.[99] Clearly vernalization was, in the early 1930s, a weak competitor in the contest for crankish solutions to the critical problems of sowing. Official interest came on rather strongly in 1929, slacked off for a year, was revived in 1931 by a new kind of vernalization, and seemed in 1933 to be slacking off from that kind too. Nevertheless, vernalization and its creator rebounded in great triumph in 1935, while winterization and superearly sowing were being utterly rejected by the practical bosses of Soviet agriculture. An important underlying trend was working for Lysenko. Vernalization promised less gain but risked less loss. The shift of official favor to it was one of the first signs of realism encroaching on enthusiasm in the intuitive judgments of agricultural officials.[100]

But Lysenko was not simply a dumb beneficiary of historical trends. He helped earn the success of vernalization by expanding and diversifying it far beyond the limits of a sowing technique, until it became a recipe for solving almost any agricultural problem. For example, he announced that cotton seed could be vernalized by moistening and warming rather than chilling. Thus he promised to breed cotton that would grow in the Ukraine and the North Caucasus, an important official goal of the 1930s, designed to relieve the overstrained railroad link with Central Asia.[101] Far more daringly he announced that potato tubers can be vernalized, though they are not true seed, if they are sprouted in warmth and light. Thus he renamed and advertised as his discovery what was actually an ancient technique of potato growers — sprouting the tubers before planting in order to give the crop a head start. In this case "vernalization" usually does increase the yield, if it is done the ordinary way instead of Lysenko's. He urged farmers to string up the seed tubers on wire or cord, which would have increased labor costs and the danger of infections.[102]

With each new practical success Lysenko expanded his venture into science. He tried to subordinate all of plant physiology to the principle of vernalization, which he enlarged into an abortive theory of stages in plant development.[103] Plant physiologists defended their discipline against the crippling vagueness and confusion that Lysenko pressed upon it, but almost always they weakened their defense by conceding practical utility to Lysenko and confessing a comparative lack of it in their own work. Only a few brave souls dared to suggest, in rather delicate language, that Lysenko's practical successes might be as meretricious as his contributions to science.[104] Some, most notably N. A. Maksimov, developed what may be called the chimera defense: they pasted Lysenkoite appendages onto the scientific body of their writings.[105] Only foreign scientists failed to see that Lysenko's scientific reputation was the product of political pressure. In the Soviet Union the fact was repeatedly cited by *Lysenkoites,* as further proof that ivory-tower scholasticism was giving way to useful science. In 1932, when the Lysenkoite cause was still confined to plant physiology, a young recruit at Moscow University proudly explained the source of the scientists' new deference to Lysenko: "Comrade Lysenko's works have begun to gain recognition only under the pressure of Soviet public opinion (*obshchestvennost'*)."[106] When Lysenko expanded his practical triumphs and scientific theorizing into plant breeding and genetics, "the pressure of public opinion" moved with him.

Lysenko's venture into plant breeding was a rather belated response to the wildest and most self-defeating of all the decrees issued on behalf of scientific agriculture. "On Plant Breeding and Seed Production," published in *Pravda* on August 3, 1931, took the side of no crank. It simply carried the extravagant optimism of men like Tulaikov and Vavilov to a fantastic extreme. It was issued by the same agency that had come to Michurin's aid in April, 1931, the Party's Control Commission and the government's Inspectorate, jointly run by A. A. Andreev. From fruit breeding the investigators moved to the chaotic seed business. Here they found no conflicts that seemed to require their intervention (if they heard of Tsitsin's clash with Meister, they chose to ignore it). They grew intensely annoyed with the whole fraternity of plant breeders and seedmen. Those specialists seemed content to spend decades at the job of substituting improved varieties for the low-yielding seed that peasants were using on the large new farms as they had on their little primitive strips. The bafflement and annoyance of the inspectors are not hard to understand. Since prerevolutionary times specialists had been saying that the use of improved seed was a fairly cheap and easy method of raising yields. (In many other backward countries specialists are still saying so, and still shaking their heads in bewilderment at the failure of most peasants to take such obvious advice.) Collectivization was supposed to rescue farming from such sluggish and incompetent hands.

The joint agency of inspection therefore decreed: For wheat and a few other crops, the Commissariat of Agriculture and the Lenin Academy of Agricultural Sciences were to accomplish the *complete* replacement of ordinary seed by certified varieties *within two years.* Such an astonishing feat was admittedly impossible "in any of the capitalist countries, even the most agriculturally advanced. It can be accomplished in the USSR precisely because of the advantages of big socialist agriculture — because of the state farms and collective farms and the planned system of directing agriculture in the USSR." [107] Tulaikov and Vavilov must have winced when they saw this, one of their favorite themes, used to justify a wildly impossible assignment. But that was not the end. The chief inspectors also decreed: wheat was to be improved so that it could rapidly replace rye in the north and east. Potatoes were to be improved so that degenerative diseases would no longer prevent local production in regions with hot dry summers. To speed these accomplishments the breeding time of improved varieties was to be reduced from ten to twelve years, which specialists regarded as a minimum, to four to five years. Two basically different methods of accomplishing this magic were prescribed. All breeding stations were to apply "new foreign technology and the newest improved methods of breeding (based on genetics). . . ." At the same time breeding was to move out of scientific establishments: "All Party and Soviet organizations and the broad masses of collective farmers and workers on state farms must be drawn into this work." [108]

For more than a year after this startling decree appeared in *Pravda* the only response to it was a stream of pledges by specialists, telling how they would strive to carry out the order, but suggesting that the government might also help — by supplying large heated greenhouses or, better still, the very expensive growth chambers that were being constructed in the West as the ultimate escape from calendar restrictions and the disasterous vagaries of breeding in the open air.[109] Lysenko himself, for all his daring, responded only in 1933, when the fortunes of vernalization as a sowing technique seemed to sag, and then he acted quietly. Not until 1934 did he begin to reveal his newest and boldest project in his usual sensationalist way. Without benefit of growth chambers or any special appropriation, "in five flowerpots in a corner of a crowded greenhouse," he began to breed an improved variety of spring wheat for the Odessa district.[110] He promised to have the new variety ready for production testing in 1935. The total breeding time would be not four or five years, as the inspectors had decreed, but less than three. Of course, it was vernalization that made this miracle possible.

As usual, Lysenko's procedure was quite simple. He picked a spring wheat with a short "stage of vernalization" but a long "light stage," and crossed it with a variety that had a long "stage of vernalization" but a short "light stage." Exactly what he meant by these stages was not made clear, a difficulty that plant physiologists were learning to endure, out of deference to "public opinion." Now Lysenko brought the difficulty to plant breeders and geneticists. By virtue of his stage theory he *knew in advance* that his cross would produce offspring that would ripen sooner *and therefore* be higher yielding than either of its parents. He did not have to test a multitude of plants through many generations. He selected the earliest plants in the first hybrid generation, *knowing in advance* that their earliness *and therefore* their higher yield would be transmitted to their offspring.[111] The italicized phrases dismayed wheat specialists and geneticists, for they revealed utter ignorance, or disregard, of firmly established elementary principles. But journalists and agricultural officials were delighted to hear that plant breeding could be enormously speeded up and cheapened. Lysenko recovered a journal of his own. Naturally enough he called it *Vernalization*, and, in the middle of 1935, on the first page of the first issue, posted the grand announcement:

Confirmation of Theoretical Expectations

To the Central Committee of the CPSU(B), the head of its Agricultural Division, Comrade Ia. A. Iakovlev. To the People's Commissar of Agriculture of the USSR, Comrade M. A. Chernov.

To the Vice-Commissar of Agriculture and President of the Lenin Academy of Agricultural Sciences, Comrade A. I. Muralov.

With your support our promise to breed in two-and-a-half years, by crossing, a variety of spring wheat for the Odessa district that would be earlier and higher yielding than the regional variety Lutescens 062 has been carried out. Four new varieties have been created.[112]

In 1934, as Lysenko began to tell reporters about his forthcoming triumph in wheat breeding, Vavilov got into serious trouble. This time it was not a newspaper attack on his plant collection, or an abusive assault by rebellious underlings during a meeting of VIR, a nasty episode that punctuated his return from the last foreign trip he was allowed to take.[113] In May 1934 Vavilov's corporate boss, the Council of People's Commissars, called him in to tell what the Lenin Academy was doing to overcome its notorious separation from practice, in other words, its failure to help agriculture.[114] Vavilov probably repeated the argument he usually made on this issue. He ran over the rich variety of coordinated research programs on which the Academy's federated institutions were working, pointing to the steeply rising curve of publications as evidence of achievement.[115] He may even have repeated to the Commissars the pathetic tale of the Academy's first effort to bring the lessons of science directly to collective farmers. At the end of 1932 the Presidium of the Academy, while meeting in Leningrad, adjourned to a nearby village for two days of lectures and colloquia on scientific agriculture. In preparation for the visit of the academicians, the peasants of the district were drummed into 80 new collective farms, which were persuaded to bring 1928 wagonloads of flax and other produce to the state collection points. The Party journal that reported this effort to "discuss agricultural problems directly in the thick of the collective farm masses" was quite satisfied with the results. It called for "still greater efforts to bring science closer to the collective farms, . . . to connect science more closely with practical experience."[116] The peasants, we may be sure, saw their days of science in an entirely different light, and the academicians too seem to have felt little enthusiasm. There is no record of further excursions into the thick of the collective farm masses.

In July 1934, two months after Vavilov's report, the Council of People's Commissars expressed its dissatisfaction in a decree. After the customary acknowledgment of achievements, which included both Lysenko's vernalization and Vavilov's collection of plants, the commissars dwelt on the scientists' failure to help agriculture as much as the state helped them. The basic fault was the Lenin Academy's "utterly insufficient scientific generalization of the mass experience of leading (*peredovykh*) state farms, MTS (Machine and Tractor Stations), and collective farms." [117] This had become the favorite Stalinist diagnosis and cure for all the ills of the new agricultural system: find healthy farms and hold them up as examples or "beacon lights" to the rest. In effect, preaching at the sick farms took the place of a serious effort to understand their troubles.

It would be hard to imagine a more effective way to make genuine science useless. What need was there for a lot of experiment stations and basic research institutes, if the scientist's job was to be publicity agent for isolated beacon-light farms? The Council of People's Commissars did not flinch from the surgery indicated by the diagnosis. Fifteen of the major institutes affiliated with the Lenin Academy were closed, along with the experiment stations subordinated to them; twelve major institutes were transferred to such practical masters as the Commissariat of State Farms; and the fourteen left to the Academy had their projects, staffs, and budgets cut back. Vavilov was allowed to remain president of the drastically pruned Academy until June 1935, when he was demoted to one of three vice-presidencies, which were held by genuine scientists.[118] The presidency was conferred upon A. I. Muralov, a senior Party administrator with prerevolutionary experience as an agronomist, who had been Agricultural Commissar of the RSFSR. In his inaugural address, "Do Not Lag Behind Life," President Muralov pointed to Lysenko's Institute in Odessa as a model of close contact between science and the farms.[119]

The Institute of Plant Breeding and Genetics in Odessa had become indisputably Lysenko's domain in 1933, when the former director was called or driven away to Vavilov's Institute of Genetics in Moscow.[120] At the Odessa Institute ivory-tower science — "locked up" (*zamknuta*) was the usual Bolshevik pejorative — was replaced by agrobiology. Contact with collective farms was not established by attempts at scientific discussions with peasants. Lysenko worked with agricultural officials; through them he was able to mobilize agronomists and "peasant scientists" in "hut labs" to carry out campaigns for his experimental techniques. Where enthusiasm was insufficient, he had official documents ordering agronomists to do his bidding "without demur." [121] Thus he won the praise of P. P. Postyshev, the Party chief of the Ukraine, who decided on a campaign to organize hut labs all over, as a way of getting around the shortage of trained people and the irrelevance of the methods developed by ivory-tower (*zamknuty*) experiment stations. With the guidance of down-to-earth scientists like Lysenko, Postyshev expected the hut labs to work out a socialist agronomy suitable to socialist farms. All this was in accord with the decree of August 3, 1931, which had called upon scientists to involve Party and state organizations and the collective farm masses in scientific research. Postyshev found that Lysenko alone was carrying out the directive.[122]

Simultaneously Lysenko was accomplishing other tasks set by the decree. Plant breeders had been told to develop varieties of potatoes immune to the degenerative diseases that inhibited potato growing in regions with hot dry summers and forced upon the overstrained railroads constant shipments of potatoes from north to south. In 1935 Lysenko announced that breeding new varieties was unnecessary. Southern farmers could solve the problem immediately by planting potatoes in the middle of the summer. They would mature in the cool fall, thereby escaping degenerative diseases.[123] With this snap judgment, somehow vaguely connected with his theory of stages, Lysenko fell into conflict with plant pathologists and virologists, who had proved that the degenerative diseases are contagions rather than physiological disorders caused by heat. But the learned specialists could suggest no cheap way to grow potatoes in the south, and Lysenko could.[124]

Lysenko also had a recipe for the immediate improvement of wheat, another task set by the August decree. While waiting for new varieties from breeders, the peasants themselves could "freshen the blood" of existing varieties. To Lysenko it was obvious that the natural inbreeding habit of wheat caused it to deteriorate. The failure of peasants to maintain pure varieties was, for Lysenko, conclusive proof that his intuition was right. He refused to accept the assurances of specialists that pure varieties of wheat would not deteriorate if genuinely pure seed were supplied by a well-run seed business to efficient farms genuinely free of weeds. He insisted that peasants should be armed with tweezers or scissors and sent through the wheat fields opening up the self-pollinating spikelets to cross-pollination by the wind. It was obvious to him that cross-breeding promotes vigor, and I. I. Prezent, his newly acquired theorist, showed him that Darwin agreed.[125]

In September 1935, when a reporter asked Lysenko whether "intravarietal crossing" of wheat should not be tested at research stations before being pressed on farmers by means of a mass campaign, Lysenko recalled that his triumph with vernalization had not been achieved that way.

> "When we put forward a measure that is as yet founded only on theory, such as 'freshening the blood' of varieties of most important agricultural crops [by intravarietal crossing], do we have the right to lose two–three years in preliminary testing of this measure on little plots at several breeding stations? No, we don't have the right to lose a single year," says T. D. Lysenko. "We will proceed in this case the same way we did when we worked out the method of vernalizing wheat and potatoes and at the same time promoted it among the masses." [126]

He was proud to say that vernalization of spring grain would give the country an extra 15,000,000 poods (270,000 tons) of grain in 1935. He knew this without benefit of controlled tests on little plots at experiment stations. He had estimated that vernalization added 7 to 8 poods per hectare, and agricultural officials had decreed vernalization on 2,000,000 hectares. Simple arithmetic yielded that extra 15,000,000 poods.

A less perceptive or less ambitious man might have rested with such triumphs. Lysenko noticed that Bolshevik officials were cutting their outlay for agricultural science, which had disappointed their great expectations of practical utility. He began to demand a complete recasting of scientific theory to rescue it from useless scholasticism, to make it fit his practical successes. He found an eager collaborator in Prezent, the young philosopher who was overcoming the respect for Mendelian genetics that he had learned from senior Marxist-Leninists.[127] Together Lysenko and Prezent created a homemade theory of heredity, which they pasted onto Lysenko's theory of stages, which in turn was pasted onto vernalization. They fashioned a collage of misused scientific terms and ordinary words, strung out in the run-on style of bureaucrats, the sort of language for which the word "gobbledygook" was invented in America. In Russia, where bureaucrats have more respect for theory, the equivalent coinage is *boltologiia,* jabberology.

Admirers of camp may appreciate a couple of samples, though they should be aware that in jabberology, as in poetry, translation cannot do justice to the original.

> The work of the Institute of Plant Breeding and Genetics (Odessa) is based precisely on the established facts of such an absolutely

> definite sequentiality of the connection of the development of the hereditary base in stages, and of the latter in organs and characters. . . .
>
> These solitary bottlenecks [*uzkie mesta,* in this case meaning undesirable characters of plants] will be overcome in the process of segregation of the heterozygote by means of a mutual replacement of the bad index [*pokazatel'*] of one form by the analogous good index of the second, and conversely.[128]

It is painful to note that Vavilov called this drivel "a major world achievement in plant science." [129] He was not a coward or timeserver. He was trying to achieve peaceful coexistence between science and Lysenkoism. But diplomacy could not subdue the increasingly aggressive spirit of Lysenko's theorizing, or remove the lethal untruths that were scattered through his jabberology like explosive shells in a dense fog.

Most specialists simply kept quiet while Lysenko was acclaimed for breeding an improved variety of wheat within two-and-a-half years, and peasants were driven to correct the inbred perversion of wheat by the therapy of intravarietal crossing. Since agricultural officials insisted that they themselves were the best judges of agricultural practice they would have to learn for themselves the practical or hard way, without benefit of honest opinions from the specialists. But insincere tributes to Lysenko and diplomatic silence were not effective defenses against his insistence that all plant breeders must work as quickly as he. He called on them to abandon their scholastic principle that progeny tests beyond the first generation are the only sure way of establishing the hereditary characters of hybrids. He also extended his campaign against inbreeding from peasant wheat fields to scientific breeding stations. There normally cross-pollinating plants, such as maize, rye, and sunflower, were forced to inbreed in order to establish pure lines for future crossing. Geneticists assured the breeders that this was an effective method that would *not* permanently damage the plants. Lysenko condemned the method and the underlying theory.[130] If official support for his practical successes extended to his theorizing, then scientific breeders and geneticists were in for official condemnation. Nothing less than abandonment of their disciplines would save them. Stalin's principle that practice is the criterion of theoretical truth virtually guaranteed such a result, provided only that the political officials who certified Lysenko's practical successes would also certify the theoretical inferences he drew from them.

Toward the end of 1935 official thinking seemed to be moving in this direction. *Izvestiia's* sensationalist expert on agriculture, the ex-Trotskyist Sosnovskii, summed up the prevailing mood in one of those Stalinist metaphors that were no less menacing for being hackneyed: "If science does not knock at the door of the collective farm brigade, then soon the hard hand of the collective farm brigade will knock at the door of the Lenin Academy of Agricultural Sciences." [131] Pravda told how Chairman Molotov, in a three-hour "reception" for members of the Lenin Academy, had praised practical breeders like Lysenko and sneered at scientific breeders.[132] In December 1935, at a convention of beacon-light grain farmers in the Kremlin, Lysenko was featured as the model scientist for his services to agriculture. Iakovlev, the Party's agricultural chief at the time, urged Lysenko to tell the audience which scientists were obstructing his work. He named a few geneticists but pointed out that individuals hardly mattered; the majority of geneticists were against him. (He should have said nearly all.) And he complained that Vavilov, for all his sweet talk, would not approve of selection in the first hybrid generation.[133]

Stalin himself sat on the dais at this meeting. Everybody must have been very anxious to hear what he would say. Only a month before, in November, he had publicly uttered a few words about science, which the press was promoting to the level of a canonical pronouncement. He had told a conference of Stakhanovites, the beacon-light masters of industrial speedup, to use their own experience as a test of recommendations by conservative engineers. They were not to be intimidated by talk of science. "Science is called science just because it does not recognize fetishes, does not fear to raise its hand against the old and dying, and listens carefully to the voice of experience, of practice." [134] Now, at the conference of beacon-light farmers in December, Stalin was given "stormy applause, turning into an ovation," when an eminent specialist called him "the first agronomist of the whole world." [135] Clearly, if Stalin had chosen to say that he was raising his hand "against the old and dying" in agricultural science, resistance to Lysenko would have collapsed. But he chose to be silent, allowing himself only an occasional word of encouragement to shy rural speakers.

A major conflict was breaking out, and Stalin was taking a position of belligerent neutrality. It was clear to all participants that the chiefs of agriculture favored agrobiologists over scientists, and Stalin obviously approved their policy. But he refrained from saying so publicly and explicitly. Care had to be exercised by Lysenkoite writers who quoted Stalin's definition of science as the hand raised to smash, or recalled the "Bravo! Comrade Lysenko," which he had uttered in February 1935. The Lysenkoites were not allowed to end all opposition by the simple statement that Lysenko's views in plant physiology and genetics had the personal endorsement of Joseph Stalin. They could hint as much and did so; but scientists could ignore the hints, and legitimate disagreement could proceed. In this strange way the Bolshevik chiefs were acknowledging a division of opinion, or opening a discussion, as they put it. The outside observer cannot help noting that they had predetermined the outcome of the discussion. The political bosses of agriculture had stipulated that Lysenko's agrobiology was a practical success and standard science was a practical failure. Only a few scientists were brave enough to question these stipulations, and no one could question the basic principle enunciated by Stalin himself, that practical success in agriculture is the ultimate criterion of truth in biological science.[136] Within this frame of reference, scientists were ordered to defend their métier or, rather, to show cause why they should not be forced to give it up for agrobiology. Logically, there was no room for genuine discussion; "old and dying" scientists should have been struck forthwith by Stalin's raised hand. But Bolshevik officials hesitated to follow their logic all the way when it brought them into unexpected conflict with esoteric principles of breeding and genetics.

There were several obvious and conscious reasons for this hesitation. Lenin had taught his followers to respect the esoteric knowledge of autonomous specialists. For almost twenty years the specialists had been encouraged to build a large establishment, which was almost solidly anti-Lysenkoite[137] and still exercised political influence. None of the political bosses who showed public favor to the Lysenkoites before 1936 were "on the scientific front"; all were agricultural officials, or troubleshooting investigators, or general chiefs like Postyshev, Molotov, and Stalin. On the scientific front at least two very important officials, Bukharin and Gorbunov, were publicly cold to Lysenkoism and friendly to the scientific opposition.[138] Many more must have been so in private, for the professional ideologists or philosophers, whose avowed function was to justify and elaborate the official mentality, held back from condemnation of genetics. They had worked out a delicately balanced official endorsement of genetics only a few years before Lysenko began to attack it.[139] Prezent was exceptional in his readiness to break away from this official position. Perhaps more important than the principles that held his colleagues to it, were the emotions they shared with most Soviet intellectuals. The old habit of looking to the advanced Western countries as models of scientific development was far from dead. Nativistic yahoos did not begin a vigorous attack on it until 1936, when they mounted a campaign of exemplary abuse against an eminent mathematician who had maintained foreign contacts after he was told to break them.[140] The grand old faith in a natural alliance between international science and international Communism was still a part of the official outlook. The main journal of the philosophical establishment proudly contrasted the flourishing of genetics in the USSR with its crisis in Germany, where racist ideology was distorting science. Geneticists of the world, the author claimed, regarded the Soviet Union as one of the two most important centers of their science, on a level with the United States.[141]

For all those conscious reasons the Bolshevik leaders hesitated when Stalinist logic brought them to the verge of condemning the sciences that had disappointed their hopes of agricultural utility. But a deeper, subconscious function was served by this hesitation, indeed by an extended prolongation of the angry "discussion" between aggressive agrobiologists and cowed scientists. It was a neurotic substitute for a genuine discussion of the real reasons why science was largely inapplicable in Soviet agriculture. A genuine discussion was impossible. In 1929 Stalin had decreed the end of agricultural economics as an autonomous discipline, because it pointed to intolerable socioeconomic limits on the official will to modernize agriculture instantly. Now the agrobiologists invited him to decree the end of natural sciences that pointed to biological limits. Stalin and his lieutenants — except for those on the scientific front — were strongly tempted; but they held back, asking for "discussion." "It is better to know less," Lysenko sneered at his learned critics, "but to know just what is necessary for practice." [142] Between scientists and people who held such a belief there could be no meaningful discussion, only an acutely confused, intensely angry quarrel. The underlying issue, which no one could state clearly, was the choice between Stalinist willfulness and scientific realism. Blindly, piecemeal, over a long period of time, Bolshevik officials were going to work their way toward the conscious recognition that they faced such a choice. In the meantime it eased their frustration, their disappointment at the irrelevance of science in their new agricultural system, to hear the blame heaped upon bourgeois science. Until they could endure a genuine discussion of real issues in agriculture, they encouraged a "discussion" of fictive issues in natural science.

5

Stalinist Self-Defeat, 1936–1950

"DISCUSSION"

Practical achievement was the avowed purpose of the Stalinist system, not talk. "Discussions" were interruptions, occasionally desirable in order to rally everyone to the single viewpoint necessary for collective action, but a nuisance all the same. A typical "discussion" was quickly pushed to its foregone conclusion by displays of official support for "correct" views and official irritation with "mistaken" ones. (The quote marks are untidy but unavoidable. In ordinary language *discussion, correct,* and *mistaken,* imply a standard of judgment that is impersonal or equalitarian or conventional. In Stalinist discourse, the boss' will, *quod principi placuit,* was the implicit standard.) At first, the "discussion" of breeding and genetics seemed to be no exception. Formally opened in 1936, it was hurried to a conclusion in December of the same year, when a large conference was called to resolve the issues in a very literal sense, by unanimous approval of official resolutions. But unity was not achieved.

In breeding and genetics, as in many other areas of "discussion," the Stalinists found themselves embroiled in something that would not conform to the prescribed pattern. The official point of view, duly approved at the conference of December 1936, failed to bring concerted action for the improvement of agriculture. In plant physiology and soil science there was a show of unity with subdued bickering, and breeders were inclined to accept a similar arrangement, but geneticists would not concede any significant point to the Lysenkoites, who began to demand total abolition of genetics. At another convention, in 1939, another official resolution was approved, somewhat closer to the Lysenkoite goal than that of 1936, but still not quite there. It was a compromise, however one-sided, and it proved as impossible to maintain as the earlier one. In 1948 a third "discussion" was staged, leading so quickly to a resolution so violently uncompromising that further "discussion" seemed utterly impossible. Genetics was to be totally suppressed; Lysenko was to be the undisputed boss of all biology and all its agricultural application. That was the ultimate victory of the practical Stalinist approach to scientific agriculture, the final arrival at those windy heights from which the nature of things can be revolutionized by a flick of willful intuition. It was also ultimate self-defeat. Within a few years, the Stalinists would find it necessary to dislodge Lysenko by opening a genuine discussion.

Soon after the conference of beacon-light grain farmers in December 1935 gave maximum publicity to the fact that Lysenko enjoyed the support of the highest chiefs, a large conference of beacon-light stockbreeders gathered in the Kremlin. Though Lysenko had not yet moved from the plant world to the animal kingdom, he profited from the renewed display of official insistence that science must be the servant of agriculture. The newspapers repeated Molotov's sneer at scientists who studied absurdly academic problems like "the domestication of the fox." [1] When the director of a stockbreeding institute tried to defend experimental crosses of domestic and wild animals, Stalin himself cut in with sarcastic heckling, culminating in the flat declaration: "You've fallen for exotica, while we need an institute that serves production." [2] If anyone wondered how broadly Stalin and his lieutenants conceived service of production, the answer was given by Iakovlev, then Party chief of agriculture. A favorite young stockbreeder, telling how he had transported bull sperm 8 kilometers to inseminate 800 cows, caused Iakovlev to ejaculate: "Now that's real science!" [3]

About the same time, in February 1936, the chief of the Central Committee's Division of Press and Publishing told a gathering of Marxist-Leninist philosophers that Lysenko deserved their special attention as the man who brought together Darwinian theory and the agricultural version of the Stakhanov movement.[4] The Central Committee and the Council of Commissars jointly decreed that almost 5,000,000 hectares had to be sowed with vernalized seed in the spring.[5] *Pravda* attacked the collective magnum opus of Vavilov's

Institute, *The Theoretical Foundations of Plant Breeding*, as "The Teamwork of Reactionary Botanists." [6] And *Pravda* hailed Lysenko's publications for "discovering the real laws of plant life, which make it possible to solve a number of the most important scientific and practical problems of agriculture quickly, on the basis of new principles." [7] The signals could hardly have been clearer. Breeders and geneticists were to "discuss" their disciplines in terms that would confirm the official judgment that Lysenko was right, his critics wrong.[8]

That hard and fast rule tended to dissolve when pushed beyond personalities, into the abstract issues of science. Officials could not specify wherein Lysenko was right, his critics wrong. After all, the officials knew nothing about Mendelian ratios of combination and segregation, about inbreeding and crossbreeding, chromosomes and genes. Lysenko himself would not — and probably could not — provide specifications. To be precise and rigorous, which was the main demand that critics made upon him, was to fall into impractical scholasticism, the very bog from which he hoped to rescue biological science. The specialists in the besieged disciplines would not help prove that Lysenko was right. They reacted in several different ways, none of them relevant to the Party chiefs' wishes. Most tried simply to keep out of trouble by keeping out of the "discussion," while they continued their usual publications. Only a very small number of full-fledged specialists strongly endorsed Lysenko's position, and they could not explain it any better than he.[9] Specialists in training began to show considerable interest, especially when the Party's Central Committee ordered that groups of them be sent from all over the country to Odessa for brief training periods.[10] But they could hardly clarify what was obscure in Lysenko's mind. Administrators of science — Vavilov, most notably — continued their efforts to appease Lysenko by praising him, while offering scientific hypotheses they hoped might explain his practical triumphs in agriculture.[11] It was a useless tactic. Lysenko and his disciples accepted the praise without thanks, and steadfastly ignored the explanatory hypotheses, which would have confined their intuition within the constraining rules of science. They preferred the freedom of their own makeshift explanations, though they achieved that freedom by flouting science. Only a small number of militant scientists dared to say so vigorously, without debilitating concessions and restraint. In short, the "discussion" was a conflict between two little bands of militants. The bulk of the scientific community simply watched in silence.

At first sight the little band of anti-Lysenkoite militants seems a strange lot. They were mainly Party members or "non-Party Bolsheviks," whose obvious duty, it would seem, was to fall in line behind the leader endorsed by their Party chiefs. Equally strange was the predominance of pure scientists rather than agricultural specialists, and of zoologists rather than botanists, though the objects of controversy were entirely agricultural plants. The chief agrobiologist, elevated by Party chiefs to rule over applied plant science, was opposed for the most part by pure scientists with strong ties to the Party.

These anomalies become comprehensible when examined in historical context. Out-and-out "bourgeois" specialists could hardly criticize Lysenko, for the Party had decreed the extinction of "bourgeois" specialists during the great break, 1929 to 1932. The community of learned people with an outlook independent of the Party — that is the essential meaning of the epithet, "bourgeois" specialists — did become extinct. Or maybe they only learned to keep their independent thoughts to themselves. It was hard to tell which was the case, once the punishment of "bourgeois" specialists was made extremely great — that was the essential meaning of their decreed extinction. In short, a natural suspicion of masked enmity hung over the bulk of full-fledged specialists, who had never exerted themselves to show devotion to Bolshevism until they were forced to do so.[12] Such people could hardly produce uninhibited criticism of the agrobiologist acclaimed by Party chiefs.

On the other hand, debate about Party issues came naturally to the minority of specialists who had been voluntarily trying to develop a Communist position in natural science before the great break made such an effort mandatory. They had been rebuked by Party leaders in 1931, for imagining that they were autonomous creators of Communist thought, independent of control by the political leaders.[13] But they had been allowed to continue their theoretical discussions, clustered about the Timiriazev Biological Institute of the Communist Academy. 1936 brought another and ruder shock to these people. By government decree their center was dissolved. Gestures were made toward continuation of the Timiriazev Biological Institute in another context than the Communist Academy, but they came to nothing in the end. *Timir-In* and *Kom-Akademiia* disappeared together.[14] The chiefs' obsession with the practical problems of boosting yields was obviously making them impatient with the philosophical discussions that had prevailed at the Communist Academy. It took courage to shut one's eyes to this fact and to carry the old philosophical viewpoint into the "discussion" of agrobiology. Such courage was shown by A. S. Serebrovskii, M. M. Zavadovskii, N. P. Dubinin, M. S. Navashin, and A. R. Zhebrak, young biologists who were Party members or active participants in the discussions at the defunct Communist Academy. In 1936 they became the outstanding spokesmen for the scientific opposition to Lysenko, still concentrating on theoretical issues, conceding the practical utility of his agricultural recipes.[15]

Only three or four specialists in applied science had the boldness to question the practicality of Lysenko's agricultural recipes in public, and they did so very inconsistently, granting Lysenko's practicality in general while questioning it in particulars.[16] One man, the plant breeder P. I. Lisitsyn, emboldened perhaps by the common knowledge that he had been Lenin's favorite agricultural specialist, had the nerve to be completely unequivocal in stating what must have been obvious to all full-fledged specialists: The practicality of vernalization and the other Lysenkoite recipes was not proved by Lysenko's little lists of farms that had allegedly used them with success. The similarity with advertisers' proof by testimonials was quite obvious. Lisitsyn preferred a different comparison. Roman priests proved the efficacy of offerings to their favorite gods by posting sailors' thanks for safe return from voyages that had begun with offerings to the gods in question. Where, asked Lisitsyn, was news of the sailors who failed to return? Similarly, where was Lysenko's analysis of the failures of vernalization?[17] Such a question was breathtaking in its exceptional boldness, for it challenged the judgment of the agricultural bosses, who were pushing vernalization on a mass scale. Most agricultural specialists took refuge in silence. The minority who participated in the "discussion" produced, in all but the few cases mentioned, the wishy-washy comment whose only meaning is leave-me-be.[18]

Scientists with theoretical preoccupations could afford to be a little braver than applied scientists, because they could engage Lysenko on a level where Party chiefs had not unanswerably endorsed him. Nevertheless, most pure biologists avoided public dispute with Lysenkoism as most applied ones did. The minority who entered the arena were disproportionately numerous and outspoken because applied specialists were disproportionately few and reserved. The disproportionate number of animal geneticists in a quarrel that began over plant breeding is a simple corollary: the focal discipline for theoretically minded biologists was genetics, and animals like fruit flies and mice were their preferred experimental objects. Of course there were plant geneticists in the Soviet Union, and a few of them were outspoken critics of Lysenko. But on the whole, plant geneticists tended to behave like plant breeders; they could easily prove their practical merit and avoid a dangerous quarrel by moving entirely into breeding. A few ceremonial bows to Lysenko seemed enough to get them out of contested territory.[19] Thus the Party chiefs achieved the opposite of what they intended. Insisting on a "discussion" that would turn all biologists to practical work on the model of Lysenko's, they simply frightened applied specialists into silence or evasive chatter and provoked a quarrel on a purely theoretical level. The demand for a thoroughly practical biology produced little more than a purely theoretical dispute.

Political authorities were being dragged into a vicious circle by their customary assumption that problems are solved by arranging people in hierarchies. Most specialists complied very readily; they would not quarrel with the political elevation of Lysenko. But their compliance was going to be meaningless, unless more political power were applied — unless breeders and geneticists were subjected to the kind of orders that Lysenko and the agricultural officials used on agronomists and peasants. Without an actual decree against inbreeding, without actual inspectors to enforce selection in the first hybrid generation, or "vegetative hybridization," or "additional pollination," these Lysenkoite prescriptions would simply be ignored by well-trained plant breeders. In the late 1930s such extreme forms of political intrusion into plant breeding were out of the question.

Before such political intervention could develop, something had to be done about the geneticists who refused even lip service to Lysenko. Their outspoken leaders had the nerve to attack his prescriptions, though he was officially credited with giving the Soviet state millions of extra poods of grain, while the geneticists studied the inheritance of eye color in fruit flies. At one of the first important clashes, in the Moscow House of Scholars, Lysenko drew the obvious conclusion: there was no serious justification for the existence of genetics. It was something like football or chess, a game for gentlemen of leisure. Forthright Serebrovskii made the equally obvious rejoinder. He shouted "*Mrakobes!*" * [20]

Early in September 1936, K. Ia. Bauman, the chief of the Central Committee's Division of Science published in *Pravda* the official reaction to the irreconcilable conflict. He tried to reconcile the combatants, invoking the rule that Lysenko was to be followed because he was superbly successful at the fusion of theory and practice. Bauman recalled such Lysenkoite triumphs as vernalization, the theory of stages, and the summer planting of potatoes. "In addition," he noted, "Comrade Lysenko does not agree with the point of view that is dominant among geneticists on particular questions." And that is all he wrote on that burning issue, adding only a warning to geneticists: They should not dismiss Lysenko's ideas out of hand; they should listen to the man and reason carefully with him.[21] Militant geneticists hastened to take advantage of the chief's indecision. They published a little flurry of hard-hitting polemics, which provoked Lysenko and Prezent to greater heat in their denunciations of genetics.

At the climactic conference in December 1936 the official goal was still the accommodation of genetics within the dominant agrobiology of Lysenko.[22] The conference revealed the impossible inconsistency of that goal. Vavilov was finally pushed from compromising praise of Lysenko to direct criticism.[23] The speakers selected to give the geneticists' point of view were the uncompromising Serebrovskii and the even more uncompromising American, H. J. Muller. (His acceptance of a permanent post at the Soviet Academy of Sciences had been a special point of pride in the Soviet press a few years earlier.)[24] They could find nothing but ignorance and confusion in Lysenko's ideas about heredity. In the debate that followed the main speeches, several people tried to blame Prezent for this discord and thus to lure Lysenko into an acceptance of the basic concepts and methods of genetics. But Lysenko responded with an explicit endorsement of Prezent, and with maddening obscurantism on the concepts of genetics.[25] Even the plant physiologist Krenke, who made one of the most elaborate efforts to achieve sweetness and light by compromising science and nonsense, almost lost his temper:

* "Obscurantist," an accurate but bookish translation, does not have the proper force, for *mrakobes* is crude and obvious Russian, a standard term of Communist abuse. "Demon of darkness" is the literal meaning.

> Trofim Denisovich, let's speak simply, amicably, without irritation. . . . From your seat you have confirmed your recognition that the quantitative ratios in the segregation of hybrids are determined by reduction division [meiosis].
>
> LYSENKO: Correct.
>
> KRENKE: Then write that way so everyone will understand.
>
> LYSENKO: Quote what I have written.
>
> KRENKE: No, I don't want to. As I see it, that's tactless, that prosecutor's form of conversation. You say we misunderstand you. But look how many misunderstand your article. . . . We're not fools, you know. You ought to explain yourself in a friendly way.[26]

Krenke and the other conciliators were wasting their time. The farthest Lysenko would go toward acknowledging the essentials of genetics was the following declaration: "Academician A. S. Serebrovskii is wrong when he says that Lysenko denies the existence of genes. Neither Lysenko nor Prezent has ever denied the existence of genes."[27] In the next breath he retracted even this politically motivated, essentially ambiguous concession:

> We deny little pieces, corpuscles of heredity. But if a man denies little pieces of temperature, denies the existence of a specific substance of temperature, does that mean that he is denying the existence of temperature as one of the properties of the condition of matters' [*odnogo iz svoistv sostoianiia materii*]? We deny corpuscles, molecules of some special "substance of heredity," and at the same time we not only recognize but, in our view, incomparably better than you geneticists, we understand the hereditary nature, the hereditary basis of plant forms.[28]

As always, Lysenko took his stand on the superiority of vague intuition to precise science. One geneticist told the dreadful truth when he compared the Lysenkoites with people who believe in the possibility of a perpetual motion machine. There was simply no basis for genuine compromise or even for mutually understandable talk between such people and scientists. Far from bringing unity to Soviet biologists, the "discussion" of 1936 simply put a sharper point on the underlying question: How far were political authorities willing to go in their support of Lysenkoism?

The answer depended in large part on the area of Soviet life over which the politicians were trying to exercise authority. Chiefs of agriculture unreservedly supported agrobiology. Chiefs of science, education, and ideology were sympathetic but reserved, conceding dominance to Lysenko in the agricultural field but trying to maintain an academic existence for genuine biology. The highest chief, Stalin, was subtly evasive, though pro-Lysenkoite. On May 17, 1938, at a reception for officials (*rabotniki*) of higher learning, he proposed a toast, which was quickly transformed by the mass media into another canonical ruling on science:

> To the flourishing of science, of that science which does not allow its old and recognized leaders to draw complacently into their shell as priests of science, the shell of monopolists of science; to that science which recognizes that the future belongs to the youth of science.
>
> To the flourishing of science, of that science whose people . . . do not want to be slaves of tradition, who have the boldness, the resolution to smash old traditions, norms, positions. . . .[29]

The implicit blessing on Lysenko was obvious—indeed, it was made explicit in a *Pravda* editorial[30]—but *Stalin* did not make it explicit. He refrained from a forthright statement that would have ended all disagreement and mobilized all officialdom in support of agrobiology. As a result, the chiefs of higher learning and ideology kept on fumbling for a compromise. They were caught between the anti-Lysenkoite pressure of their scientist constituents and the pro-Lysenkoite pressure of the agricultural bosses. Stalin maintained the tension by withholding a public blessing on Lysenkoism, though he allowed everyone to believe that he supported it.

In the year preceding Stalin's toast to the kind of science that smashes, terror had removed two successive presidents of the Lenin Academy of Agricultural Sciences, and Lysenko himself had succeeded to that command post in the campaign for scientific agriculture.[31] He could not immediately suppress all non-Lysenkoite work in the federation of research institutions over which he now presided—he never succeeded in complete suppression—but he was in a position to begin. In the central organization of pure science, the Academy of Sciences of the USSR, he did not have such luck, though unanswerable force seemed to be mobilized on his behalf. The case is worth examining in detail, for it reveals the great power of resistance inherent in the scientific community, and it strongly suggests that this power was quietly acknowledged at the highest level of government.

On May 8, 1938 Lysenko was invited to address a meeting of the Council of People's Commissars, with Chairman Molotov presiding. Five and a half hours were devoted to criticism of the Academy of

Sciences. At the end the Academy's "plan" was sent back to be revised in light of the criticism.[32] The Presidium of the Academy called a meeting, demanded a report from Vavilov, director of the Institute of Genetics, and rebuked him for "isolating the Institute from the trend of Academician T. D. Lysenko's scientific works." The Institute was ordered to attack "class-hostile positions on the theoretical front," to concentrate on economically important problems and solve them quickly, and "to place at the basis of its work the task of transforming the nature of the plant by means of re-education." [33] (The language suggests either a Schweikish caricature of Lysenkoite jargon or swift approval of a Lysenkoite draft without any effort to improve it — which once again suggests the mentality of good soldier Schweik.) The Presidium thanked Lysenko for his readiness to organize his kind of work at the Institute of Genetics, and called upon Marxist philosophers to abandon neutrality and join the struggle for the new biology.[34] The Council of People's Commissars was not satisfied. On July 26 it complained that the Academy was still not studying practical problems and was still not fighting "anti-Darwinist theories in biology." [35] (Anti-Darwinism was a favorite euphemism for anti-Lysenkoite.) In September, the Presidium of the Academy called another meeting, for another round of criticism and self-criticism, leading once again to "full and complete endorsement" of the Commissars' directive.[36]

Meantime the second most important center of genetical research was being driven toward destruction. This was the Institute of Experimental Biology, which had been founded in the first years of the Soviet regime by the cooperation of the first Commissar of Health and N. K. Kol'tsov, with research in eugenics as one of its major tasks.[37] In the 1930s the Bolsheviks decided that eugenics was hopelessly fascist, but Kol'tsov would not criticize himself for failing to see the new truth before it was announced. Even after *Pravda* carried a denunciation of his "fascist ideology," signed by nine biologists — four were geneticists and cytologists in the process of converting to Lysenkoism[38] — Kol'tsov refused to bow his head. He faced a commission of inquiry with his customary polite arrogance, saying that everything he had ever written was "historically correct." [39] The Institute, which had been transferred to the Academy of Sciences, was taken from his direction and reorganized; its new, cumbersome but explicit title — Institute of Cytology, Histology, and Embryology — called attention to the absence of genetics.[40]

When we note that the new President of the Academy, elected just after the conference of December 1936, was V. L. Komarov, an elderly botanist with Lamarckist sentiments and a reputation for political cooperation with the Bolshevik chiefs,[41] it would seem that biologists in the Academy of Sciences had no choice but to submit to Lysenkoism.

Yet most of them did not submit, or they made gestures of submission while standing pat. Komarov's Lamarckist sentiments did not prevent him from showing in print, in 1938, his genuine respect for the science of genetics and for Vavilov's theories of plant evolution.[42] He did not lose his reputation for cooperation with the Bolsheviks by his obvious though silent resistance to Lysenkoism. The two decrees that the Council of People's Commissars sent to the Academy in 1938 resulted in nothing more than self-critical meetings and resolutions and talk of a Lysenkoite laboratory in the Institute of Genetics — until August 1940, when Vavilov was arrested. Then Lysenko became director, most of the staff was pushed out, and the Institute became the country's main center of Lysenkoism. But within the Academy it was an isolated center, at odds with the other biological institutes, which adhered to genuine science. Even the institute that had been taken from Kol'tsov and thoroughly reorganized did not turn Lysenkoite, though its new director was a man who would prove to be a complete opportunist when subjected to real pressure. Until 1948 the new director was spared that trial; he even sheltered some of the geneticists who were driven from the Institute of Genetics following Vavilov's arrest and Lysenko's takeover.[43] The Academy's Institute of Plant Physiology also continued to annoy the Lysenkoites by publishing work that went against their pet notions, though they managed in 1938 to force out the director by attacking him in *Pravda*.[44] There was, to be sure, a sharp reduction in the boldest kind of advanced research. Population genetics, for example, withered away in the Soviet Union, though the pioneering paper in that field had come from Kol'tsov's Institute.[45] One of the world's major centers of advanced biological research, which the USSR had become by the early 1930s, was returned to the backwater where it had been before the Revolution. Lysenko succeeded that far. He could harass and frustrate the community of biologists, confining them to old-fashioned lines of research. He could not turn them into a community of agrobiologists.

Similarly at other institutions of higher learning. In the universities very few professors turned Lysenkoite. Most were evasively resistant, and a minority were actively opposed, with the result that the banner of Lysenkoism had to be carried by student militants. Their agitation, spiked with insinuations that the learned opposition was subversive, scared students away from genetics and forced the establishment of a special course called "Darwinism."[46] But the militants could not win control of the biological faculties, with the result that the course in Darwinism was usually an old-fashioned review of evolution satisfying neither Lysenkoites nor geneticists.[47] Within the research and educational institutions that were subject to agricultural authorities, Lysenko had more success. But even here his growing army of followers were mostly obscure people in obscure places — "all the practical people in the privinces (*na mestakh*)" was the way a Lysenkoite described them in 1939.[48] (Nine years later the Minister of Higher Education gave the same analysis.[49]) At VIR (All-Union Institute of Plant Industry), which was the central institution of plant science, Director Vavilov turned from conciliation of the Lysenkoites to vigorous opposition. "We will go to the stake," he exclaimed at a violent debate in March, 1939, "we will burn, but we will not renounce our convictions."[50] Of course, he could not stop some of the staff from scrambling onto the Lysenkoite bandwagon, and when he was arrested in the summer of 1940, one of them was installed in the director's office.[51]

The agricultural schools were one area of higher learning that the Lysenkoites regarded as unquestionably theirs, but even here they faced resistance. They were especially outraged by a new textbook of breeding and genetics published in 1938. The two men commissioned to do the job (Delone or Delaunay and Grishko) turned out a standard Mendelian handbook, inconsistently sprinkled with ludicrous Lysenkoite passages.[52] The Lysenkoites responded with a noisy attack on authors and publishers who would not help to "DRIVE FORMAL GENETICS FROM HIGHER EDUCATION."[53] That was the headline the chief agricultural newspaper placed over a Lysenkoite blast, but the newspaper was spongy in its support. It also printed — with apologetic disclaimers — defenses of genetics by Delone and Grishko, by Vavilov, and by Zhebrak, a Party member of peasant extraction, who was still teaching genetics at the Timiriazev Academy, the chief institution of higher education in the agricultural sciences.[54]

Agricultural authorities, headed by A. A. Andreev, had got involved in a disagreement with cultural authorities, headed by A. A. Zhdanov. Was the science and education front to be brought in line with the agricultural front; if so, was it to be by further compromise or by the total victory of agrobiology? The Central Committee's Division of Science, which was under Zhdanov's supervision, appointed "a group of comrades" to check on Mendelian ratios of segregation. (They were confirmed.)[55] Aside from that scrap of information, the public record reveals nothing about the role of the Central Committee in the late 1930s. Very likely the main functions of "this areopagus," as Stalin had described it in 1930, were now concentrated in a kitchen cabinet of himself and a few associates, which probably included both Andreev and Zhdanov. Andreev and his subordinates took Lysenko's side publicly and strongly.[56] Zhdanov was steadily silent in public, which gave rise to persistent rumors that he favored the geneticists.[57] A more reliable indication of his position can be found in the publications of leading "philosophers," that is, the official interpreters of theoretical ideology. It is foolish to suppose that every word of the philosophers was a direct emanation of Zhdanov's mind, but it is no less foolish to suppose that they would take a stand on a major issue without some kind of authorization from their chief. When they were silent or evasive, as they were until the fall of 1939, it is fair to assume that they were getting that kind of signal from the summit of the Communist olympus.[58]

Finally, in October 1939, the editors of *Under the Banner of Marxism,* the main organ of the ideological establishment, assembled a conference and gave out the official view. (Back in 1936 the Lenin Academy of Agricultural Sciences had been the sponsor of the decisive conference. Now agrobiology was being formally raised to the level of theoretical ideology.) Vavilov saw the 1939 meeting as an effort to "mutate" scientists into Lysenkoites,[59] but the official summation by M. B. Mitin, high priest of Stalinist philosophy, indicated a different purpose. Compromise was once again the object, with dominance conceded to Lysenkoism. Scientists were directed to recognize the supreme practical importance of agrobiology, to abandon their "seignorial" (*barskoe*) disdain for it, and find a place for it within biology. The Lysenkoites were asked to desist from anti-intellectual — *makhaevskie** was Mitin's actual word — efforts

* The Polish radical Machajski argued that the proletarian movement should get rid of intellectuals because they were trying to become a new ruling class. When

to suppress genetics.[60] This time compromise proved to be even more elusive than in 1936, for both sides had hardened their positions. Once again the impossibility of meaningful discourse between Lysenkoites and scientists was painfully revealed. The geneticist Kerkis, for example, wondered how Lysenko could have said, "In order to obtain a certain result, you must want to obtain precisely that result; if you want to obtain a certain result, you will obtain it. . . . I need only such people as will obtain the results that I need." Lysenko called out, "I spoke correctly!" and Kerkis could only gasp, "We cannot understand how a scientist can obtain in such disputed problems what he needs. I cannot understand that. It just doesn't fit in my head." [61]

In the aftermath of that conference, polemical exchanges were greatly reduced. An effort was made to adhere to the official compromise, but nothing was settled. The crucial textbook of plant breeding for institutions of agricultural education was rewritten by a diverse group, who turned out a mishmash of Lysenkoite and scientific passages.[62] Many chimeras appeared, as we may call scientific works with Lysenkoite appendages. The more academic a biological publication, that is, the further from agricultural interests, the less likely was it to show Lysenkoite influence. Biologists were even able to start a new journal in 1940, in which they were free to criticize Lysenkoism, usually without mentioning names.[63] The closer to agriculture, the greater the Lysenkoite influence. When a journal of plant breeding carried an article defending Mendelian ratios of segregation, the Lysenkoites fumed with indignation. But they felt obliged to drop a parenthetical concession: "No one is demanding that articles of formal geneticists must not be printed." [64] In short, there was a compromised deadlock, with science entrenched in academic institutions of higher learning and aggressive Lysenkoism trying to expand from its agricultural base. The long run seemed to favor the Lysenkoites, for biological education was severely mutilated. In secondary schools it became a mixture of natural history, old-fashioned Darwinism, and meaningless chatter about Michurinism.[65] In higher education everything depended on the balance of forces at particular institutions, with Lysenkoites tending to win in the agricultural sector and scientists tending to

---

Stalin wished to ease up on the *intelligentsia,* as he did in 1939, he denounced *makhaevshchina* and *spetseedstvo* (literally, "eating specialists").

hold their own in the academic. But students were being scared away from real biology in all institutions.

Of course, in the very long run there would be generations of students who might have genuine biological education restored to them. Everything depended on Lysenko's ability to keep political support by maintaining his reputation as the great benefactor of agriculture. He continued to hatch new schemes for cheap, rapid improvements. He started a campaign for millet, to rescue it from its reputation as a primitive cereal unworthy of advanced socialist agriculture.[66] He won Khrushchev, Postyshev's successor as chief of the Ukraine, to a mobilization of chickens for war on the weevil afflicting sugar beets.[67] Lysenko's most important practical accomplishment in the late 1930s was a loosening of standards for the certification of pure seed.[68] But it seems pretty clear in retrospect that his innovative power was faltering on the eve of the German attack. The great boom for vernalization nearly disappeared from the press, as did the campaigns for intravarietal crossing of wheat, summer planting of potatoes, stripping (*chekanka*) of cotton plants, and additional pollination of maize. By contrast the mobilization of Ukrainian chickens, the relaxation of seed standards, and the campaign for millet were fairly feeble substitutes. Lysenko's emphasis was shifting from recipes that promised immediate gain to research that promised future benefit. Most notably he promised to breed a winter wheat for Siberia within three or four years.[69]

That was a dangerous shift of emphasis. The postponement of practical benefit to agriculture had caused the Bolsheviks to grow disillusioned with genuine science. Yet how could Lysenko avoid the same pitfall? Agricultural officials seemed to be growing a bit weary of campaigns, and he had become president of the Lenin Academy of Agricultural Sciences, a federation of research institutions, not farms. Many of the research institutions ran experimental or model farms, but they merely intensified Lysenko's dilemma. If the model farms continued to be impossibly superior to the average farm, Lysenko would be condemned, as his predecessors had been, for developing a hothouse science of no use to ordinary farms. If he and his disciples brought the model farms down to the average level, he would be condemned for incompetence. The dilemma would become apparent only in the long run. At the moment, in the late 1930s, agricultural officials took Lysenko's usefulness for granted.

TERROR*

Generalizations about the beliefs and policies of Soviet authorities in the late 1930s have probably provoked the knowledgeable reader to exclaim: Who *were* the authorities in that time of wholesale terror? Could Stalin himself be sure that he would not be transformed from god to sacrificial victim? The frightful answer seems to be that individuals hardly mattered. All three patrons to whom Lysenko gave public thanks in 1935 were gone by 1939,[70] but Lysenko enjoyed unbroken support from the replaceable parts that took their place in the agricultural hierarchy. Individuals came and went, to sit in upholstered offices or lie in unmarked graves, but the offices they transiently occupied followed steady policies. The historian can hardly avoid the occupational disease of the administrators of terror, a callous indifference to individuals. (The compassionate reader and the Kremlinologist will find in the appendices a list of high officials who were involved in the Lysenko affair. Those who suffered repression are duly noted.) The impersonal fact is the crucial one: No matter who wore the regalia, the chiefs of agriculture supported agrobiology, while the chiefs of higher learning and ideology temporized.

Continuity in policies while politicians were at an acme of discontinuity was one of the most striking features of the time of terror in the late 1930s. We should not be surprised. Lesser creatures than Soviet officials are not deterred from their usual activities by the fact that predators grab some of them. Of course, a change in predation pressure can change the nature of the species. Seen from that angle, the influence of terror is no different. The most obvious effect was to suppress a breed of official that was already extremely

* Terror is used in the Stalinist sense to mean the repression of "enemies." There was no distinction between enemy and potential enemy. The single word "enemy" stood for both and will be used in that sense here, with warning quote marks. Repression (*repressiia*), which meant simply punishment for crime, has acquired, in post-Stalin Russia, the connotation of extralegal punishment for nonexistent crime. Here it will not matter which way repression is understood. When the distinction between real and potential crime disappears, so does the distinction between deserved and undeserved punishment. Repression entailed arrest plus one of the following: execution; internment in jail or concentration camp; or one of the three forms of internal exile — *ssylka* or exile to a specific place with corrective labor, *ssylka* without corrective labor, and *vysylka* or banishment from a given city or list of cities. Simple dismissal from a job is not considered repression.

scarce when the great terror struck: the creative type, the official who tries to improve his job in both senses, as a task as well as a position. By 1939 he was either extinct or carefully concealed behind a mask of conformity to established rules. One is tempted to formulate a rule: The greater the insecurity of officials, the greater their mechanical adherence to approved policies (the sort of behavior that Soviet Russians call *rutinerstvo*). Thus the terror reinforced policies that were already in operation when it began to strike Soviet officialdom.

But there were other, special relationships between terror and the particular policy we are examining here — the government's support of agrobiology. To discover them it is necessary to examine the fragmentary evidence very carefully. Some facts suggest that the NKVD consciously worked for Lysenko. Some facts suggest that it did not. Put together, the inconsistent evidence suggests bloody chaos. Then, if we look at the chaos long and hard enough, certain patterns do emerge.

Toward the end of 1936 the Lysenkoite crusade received its first[71] major assist from the terror, by the arrest of three Communists who were not taking part in the "discussion" of Lysenkoism. Solomon Levit's case was the most noteworthy. He was the respected director of the main institute studying human heredity until November 13, 1936, when the Party boss of Moscow's scientists, Ernst Kol'man, staged a public meeting to denounce Levit as an abettor of Nazi doctrines. Levit was subsequently arrested without publicity — he died in prison — and his institute was closed in silence, the news transpiring by rumor and belated casual references in the press.[72] In the same way the public was allowed to learn that Israel Agol and Max Levin had been "exposed" as "enemies." They were known as outstanding Communist theorists or philosophical interpreters of biology; Agol, indeed, had grown sufficiently involved to become a full-fledged geneticist. The brief printed references to their crimes were nothing more than hints, vaguely suggesting connections between their "Menshevizing idealist" deviation from "correct" philosophy and their association with Trotskyite conspirators.[73] The fact that all three "enemies" were Jewish was never explicitly mentioned in print, though it was obvious to all from their names. Nothing was made specific beyond the brute fact of arrest and condemnation. Chances are that little more will be found in the files of the terrorist apparatus. To try and specify what exactly was criminal in the behavior of the condemned would have been to try and turn extralegal justice into legal justice, terror into law.

When the *New York Times* reported that Vavilov had also been arrested and Soviet genetics was on the verge of destruction by Lysenkoism, Vavilov wrote a stiff letter to the *Times,* also printed in *Izvestiia,* proving that he was still at large, and asserting that the science of genetics was not threatened by the arrest of criminals.[74] If he had been free to discuss the terror frankly — in that case, once again, it would not have been terror but law — he might have made a plausible argument for his assertion. At any rate the retrospective outside observer cannot see the arrest of Levit, Agol, and Levin as conclusive evidence that the administrators of terror were purposefully working with Lysenko to destroy the science of genetics. "Exposures" of "enemies" were coming thick and fast toward the end of 1936, in other fields of intellectual endeavor than biology, and even more in government. A few arrests were publicized but most were not. The administrators of the terror were giving out only cryptic hints of the intuitive methods by which they were culling subversives from the mass of loyal folk. The public record is therefore quite fragmentary; without access to the archives it is impossible to know exactly how many people were arrested in various disciplines, and exactly what the connection was, if any, between their work and their supposed crime. But there are enough fragments in the public record to make certain patterns fairly clear and to suggest others as hypotheses, to be checked some day in the archives of terror.

At the outset one must realize the inadequacy of explanations that simply point to Lysenko's malevolence, or to Stalin's, or to some abstract nonhistorical principle of totalitarianism. We are obviously dealing with an historical process, recurrent but changing spasms in an evolving social organism. The first, or Cheka, peak of terror, between 1918 and 1921, was connected with the Civil War. People with the wrong politics or the wrong "class origin" — well-bred or non-Bolshevik people — were the typical victims. The second, or OGPU, peak of terror, between 1929 and 1933, was connected with the forced collectivization of the peasants and the forced conversion of the intelligentsia from "bourgeois" to "red" specialists. In number of victims and degree of violence, that second peak probably exceeded all others, but obscure peasants were the principal victims, so the public imagination, both Soviet and Western, has mistakenly pictured the next spasm, "the blood purge" of 1936 to 1939, as the grandest one of all. In that third, or NKVD, peak of terror, the intelligentsia was mauled once again, worse than ever; and, to everyone's amazement, the Bolshevik leadership was decimated. Of course, the rare spectacle of a ruling class tearing at itself is amazing, but it is not the only puzzle or even the most important one. The most important puzzle is the intensification of terror in the absence of any Civil War or collectivization of peasants or conversion of the intelligentsia. No major change in social organization or in government institutions or even in government policy accompanied the great terror of the late 1930s. It seems utterly without purpose, and the same eerie pointlessness characterizes the fourth and last — and much lower — peak of terror in the late 1940s and early 1950s, the final years of Stalin's reign.[75]

After the mauling of anti-Bolsheviks and well-bred people between 1918 and 1921, and the wild assault on the peasantry from 1929 to 1933, both spasms helping to force major political and social change, the random bludgeoning of officialdom and the intelligentsia came on strong between 1936 and 1939. It then declined in diminishing tremors until Stalin's successors stopped it altogether by dismantling the apparatus of terror. (There is still some residual trembling in the body politic, but it is qualitively different from the earlier massive fits.) Irrationality, in this case meaning the lack of intelligible purpose, seems to be the hallmark of late Stalinist terror. Indeed it may remain so when all the archives have been opened and studied. The apparatus of terror seems, after 1936, to have ground out its quota of victims as aimlessly as automobiles and cancer and bombing planes grind out theirs. Even blindly operating causes may prove quite elusive. But certain functional — or dysfunctional — correlations will probably be established, as patterns of cancer or bombing are described by specialists who do not know the basic causes and send us to the clergy or the public relations men if we ask the reason why.

One such correlation is fairly clear already, from a study of the public record. When the terror intruded in a scientific "discussion," it tended to favor the crude, anti-intellectual, nativist side, but did *not* suffice to give it victory. In physics, to take a notable example, the leading Communist, B. M. Gessen or Hessen, was condemned toward the end of 1936, about the same time as Levit, Agol, and Levin. Hessen had been a philosophical defender of modern physics; and, as chief administrator of physics at Moscow University, he had been a patron of genuine scientists. The obscurantists who were attacking modern physical theories — a tiny group of old-fashioned physicists surrounded by a crowd of semiliterate ideologists — seized upon the condemnation of Hessen as an occasion for intensified attack.[76] In the late 1930s, the apparatus of terror arrested at least twenty-one other physicists and philosophers who defended modern physics. That is the number revealed in the public record, which shows no arrests of obscurantists.[77] Yet physicists stubbornly continued to defend their science, and successfully; it came through repeated "discussions" in flourishing condition. Geneticists and philosophical defenders of genetics, who lost at least twenty-two of their colleagues to the terror, stood up for their science no less stubbornly, but in vain. Their scientific work was sharply restricted and finally, in 1948, it was altogether prohibited.

If we add to the number of arrested geneticists and philosophical defenders of genetics all the other non-Lysenkoite biologists and agricultural specialists who are known to have suffered repression, the total rises to seventy-seven, as against twenty-two repressed physicists, and six repressed Lysenkoites.[78] The difference in numbers has become significant, but it still does not support the conclusion that terror determined the contrasting fates of genuine biology, physics, and Lysenkoism. Some of that difference is undoubtedly due to the fact that I searched the public record more intensively for biologists and agricultural specialists than for physicists, while I collected no figures at all for research engineers, the industrial equivalent of the people I call agricultural specialists. Some of the difference may be evidence that the apparatus of terror struck biologists and agricultural specialists more frequently than physicists (and engineers?), because the former were held responsible for chronic agricultural disappointments, while the latter were given credit for industrial success. In any case, the difference between the number of arrested biologists and the number of arrested physicists loses much of its significance when one thinks of how many were *not* arrested. Only small fractions of the respective scientific communities suffered repression.

I will be very much surprised if the final archival count shows that the number of arrested biologists and agricultural specialists constituted more than 5 percent of the entire group. I will be surprised if less than 2 percent of all physicists suffered repression.[79]

The difference between 5 percent and 2 percent, if it is real, suggests that administrators of terror had different feelings about biologists and physicists. It does not suggest that biologists and physicists had different feelings about the terror. People in mortal danger do not measure it so finely. The community of physicists was probably no less scared than the biologists. They could nevertheless continue their work because their centers of education and research were continued — indeed, they were expanded. The community of geneticists was checked and finally prohibited from its work, not by the apparatus of terror but by the administrators of agriculture, of education, and finally of science. To put the matter as simply as possible: By 1948 terror had helped convert to Lysenkoism a bare handful of geneticists — something like 5.[80] It had repressed a few — something like 22 (including philosophical defenders of genetics). All the rest — something like 300 (philosophers *not* included) — were forced into other types of work by other officials than the administrators of terror.[81]

Terror was so vivid a reality, yet so elusive in its import for people still at large, that further detail cannot be avoided, if we wish to grasp its strange self-defeating role in the development of Soviet society. On the eve of the conference of December 1936, official articles and speeches associated the "pseudoscientific" work of Levit with the ideology of the foreign racists and fascists who were threatening the Soviet Union, but the science of genetics was not specifically blamed.[82] At the conference itself, the chief of the Central Committee's Division of Science let it be known that the issue of human genetics was not to be raised, and three of the four major speakers (Vavilov, Lysenko, and Serebrovskii) complied.[83] The fourth, H. J. Muller, American, defiantly concluded his speech with an argument that *Lysenko's* views on heredity were a logical basis for racism and fascism.[84] (If training alters heredity, then centuries of wretchedness have made the less fortunate classes and races genetically inferior to the fortunate classes and races.) As a result of that minor scandal, seven of the people who commented on those speeches touched on human heredity. One man, who was trying to achieve a compromise between Lysenkoism and genetics, provoked cries of "Correct!" by declaring that Muller did not understand what the "discussion" was about.[85] All the anti-Lysenkoite speakers ignored the issue, except for the geneticist Dubinin, who staunchly defended Muller's point of view.[86] Most of the Lysenkoite speakers also ignored the subject, but four of them insinuated that there was more than an accidental connection between the science of genetics and the ideology of racism. (They did not actually reply to Muller's argument, perhaps because it stumped them.[87]) The plant breeder G. K. Meister, who gave the official summation of the conference, rebuked all the efforts to inject an irrelevant issue — and inconsistently opined that Lysenko's views had less affinity with racism than the belief in stable human genotypes.[88]

The pattern here is elusive because men of authority were elusive, though violent. But a pattern is discernible. The arrest of Levit, the virtual suppression of research in human heredity, the hints of an affinity between the science of genetics and enemy ideology — all associated phenomena — frightened Lysenko's opponents. If the majority of geneticists and breeders nevertheless went on with their non-Lysenkoite work, while a brave minority spoke out against agrobiology, nearly all of them complying with the ban on human genetics, they were able to do so because the science of genetics and its applications in breeding were under *suspicion* of subversive connections. They were not *prohibited* as a criminal pursuit, which would have been an act of law rather than terror. When such an act finally came, in 1948, Lysenko's critics became law-abiding by falling silent. Geneticists who would not turn into Lysenkoites — and most of them would not — were obliged to take jobs outside their banned discipline. Only two or three were arrested following the edict of 1948.[89]

It is widely believed that the Lysenkoites had a direct line to the apparatus of terror and deliberately used it to get rid of their opponents. Insinuations to that effect have even been printed in the Soviet Union, though the evidence offered has been extremely weak.[90] One is struck, for example, by the meager quotation of Lysenkoite aspersions on the loyalty of their scientific opponents. If one goes back to the record of the controversy, one begins to understand: in the major publications such aspersions are notable by their scantiness. Prezent will serve as a good example. He is widely credited with heavy use of subversive charges, but I have found only four instances in the things he published before 1948, while the terror was working for agrobiology rather than law; and two of the four do not concern genetics or Lysenkoism. More significant than numbers is the way he used the label of "wrecker" or "enemy," pinning it only on people who had already been pinned by the apparatus of terror. He did not in public initiate accusations.[91]

Calling attention to condemned opponents might have been a tempting polemical device, a way of insinuating that other opponents and their doctrines were subversive.[92] But the Lysenkoites used this device only occasionally and gingerly before 1948. The few cases of such public aspersions are worth digging out, for they shed a little light in a very dark place. They suggest what may emerge when secret denunciations are turned up in the archives. At the conference of October 1939, G. N. Shlykov, a nominal subordinate of Vavilov's who made a specialty of attacking his superior, was so abusive that the shocked chairman rebuked Shlykov, and his charges were omitted from the printed record.[93] We can guess what they were, when we learn that the "wall newspaper" at VIR (the All-Union Institute of Plant Industry), where Vavilov was director and Shlykov his nominal subordinate, charged Vavilov with protecting "wreckers" and "enemies."[94] In 1940 he was arrested. The directorship was given to a subordinate who had turned Lysenkoite. The directorship of the Institute of Genetics at the Academy of Sciences, which had also been Vavilov's, was given to Lysenko, who had already in 1938 taken the presidency of the Lenin Academy of Agricultural Sciences, still another post made vacant by terror. Those facts are widely regarded as more than circumstantial evidence, as virtually conclusive proof that the apparatus of terror consciously cooperated with the Lysenkoites to destroy the scientific opposition and transfer its institutions to Lysenkoite control. People who feel that way usually shrug off the news that *Shlykov* did not get Vavilov's job at VIR, nor did another subordinate who distinguished himself by public attacks on Vavilov. They both followed the man they had defamed to jail.[95] A shrug may be the appropriate reaction to the news, though poetic justice is spoiled by Shlykov's survival to work again at VIR, while Vavilov died in jail, of malnutrition. Justice aside, it would seem that those who lifted the sword of terror were wielding a two-edged weapon. At any rate, a general rule cannot be made of Vavilov's history until analogous cases have been examined.

At the conference of December 1936 only one Lysenkoite had the nerve to impugn the loyalty of the scientific opposition. The eccentric Lamarckist and incompetent biochemist, S. S. Perov, did the job very deftly: "Levin is arrested, Levit too, Serebrovskii has not produced a single achievement in practice, and up to now has not dissociated himself from Menshevizing idealism."[96] The audience knew that Levin and Levit had been Serebrovskii's colleagues and close associates at Moscow University and at the defunct Communist Academy, and that "Menshevizing idealism" was their philosophical deviation, which was now treated as virtual treason. Neither Serebrovskii nor anyone else commented on Perov's obvious insinuation. Subsequently, the newspaper of Moscow University published analogous aspersions on the loyalty of Serebrovskii, over the signatures of students and junior faculty who have never distinguished themselves in any other way.[97] The newspaper, an organ of university yahoos, gave the same treatment to a professor of plant physiology who was also anti-Lysenkoite, though not as militant as Serebrovskii.[98] It also charged a professor of physical anthropology with evading the race question, which "enemy Levin used for his fascist purposes."[99] None of the smeared professors was arrested or even pushed out of his professorship. Instead, without preliminary public denunciations, the terror snatched the young dean of the biology faculty, a *vydvizhenka,* only recently "pushed up" to her administrative post.[100] An upheaval followed in the Party organization of that faculty, but the Lysenkoites failed to turn it to their advantage. They had on their side only some students and a professor of botany, who was expelled from the Party for past association with the Socialist Revolutionaries.[101] (He was subsequently reinstated.) As the turmoil quieted in 1938, a young geneticist emerged as a chief of the biology faculty at Moscow University, and was still holding the fort against Lysenkoite attack in 1948, when the formal condemnation of genetics forced him out, but not to jail.[102]

The newspaper of Leningrad University was more successful with its denunciations. It published aspersions on the loyalty of four anti-Lysenkoites, and two of them were finally arrested (the geneticist Karpechenko, the cytologist Levitskii).[103] Even so, the Lysenkoites failed to get control of the biology faculty. Prezent's course, Darwinism, was restored and expanded into a small department or *kabinet.* (It had been abolished in 1935, the newspaper pointed out, when the University was run by administrators who were subsequently "exposed" as "enemies."[104]) But Prezent remained disgruntled, for two of the biologists whose loyalty was impugned in the newspaper did more than escape arrest; they gained control of the biology faculty. Expelled from the *komsomol* in 1937 for consorting with "enemies," the physiologist Airapet'iants returned in 1938 as one of the local Party chiefs.[105] The young geneticist Lobashev, who was placed "under surveillance" for the same reason, declared at a stormy meeting that he could not in good conscience teach any Lysenkoite doctrines.[106] Nevertheless, when Karpechenko, chairman of the genetics department of Leningrad University, was arrested in 1941, Lobashev succeeded him, and was still at the post in 1948 — indeed he was dean of the biology faculty — when the formal condemnation of genetics finally gave the Lysenkoites the power to replace him, but not to repress him. He marked time in another job, and kowtowed to the Lysenkoites, until a department of genuine genetics was restored at Leningrad University and Lobashev, once again an intransigent scientist, resumed the chairmanship.[107] Institutions of higher learning, it will be noted once again, were very resistant to Lysenkoism, even when terror removed key geneticists.

In institutions run by agricultural authorities, Lysenkoites tended to win important posts with or without the aid of terror. But even in such cases the pattern was hardly uniform. After terror removed the rector of the Timiriazev Academy — to be precise two were removed, one right after the other[108] — the job was finally given to an agricultural statistician named Nemchinov. He tried to keep the academic peace by maintaining an internal balance of Lysenkoite and non-Lysenkoite forces. In 1948, when he was ordered to fire the geneticist still teaching at the Academy, he created a sensation by a well-staged public refusal. "You have to resign!" the Lysenkoites cried. "I may," Nemchinov replied, and he did, without suffering repression.[109] Of course, he was an exceptionally brave man. The important point is that such people existed even in agricultural institutions, and they were not automatically arrested.

Many scientists and technical specialists had their loyalty publicly impugned during the third peak of terror, 1936 to 1939, usually by undistinguished junior colleagues or students speaking at meetings convened for that purpose — "the liquidation of the consequences of wrecking." Some of the slanders were published, usually in little local newspapers, which are the best source for students of the terror who cannot get into the archives. Most of the slandered people were in fields far removed from Lysenko's, for it was official policy during those years to harass all communities of professional people (that is what *intelligentsia* had come to mean in Soviet Russia) with insinuations of disloyalty. It is artificial to study cases of published slander only in fields affected by Lysenko's crusade, but this artificial limitation confers an important advantage. It helps one see the terror in its social context, in its relations with other Soviet institutions. Limiting the count to biologists and agricultural specialists whose loyalty was publicly impugned, *and whose subsequent fate can be discovered,*[110] discloses fifteen cases. Only four of the fifteen were arrested. Of the four positions vacated in that way, only two were given to Lysenkoites.[111] If we were to add the accusations of disloyalty that the Lysenkoites hurled at their opponents in 1948, their rate of success would decline still further, for none of the 1948 accusations was followed by repression.[112]

Secret tips to the NKVD were probably much more common than public denunciations, but also, one is tempted to guess, even less sure of success.[113] How often the Lysenkoites sent such tips, with what rate of success, can be determined only by searching archives that are firmly closed. How often the poisoned pen was used against *them* is another question that must be left hanging, though it is significant that the public record reveals no case of a scientist impugning the loyalty of a Lysenkoite. But the public record does show that at least six Lysenkoites suffered arrest.[114] It also shows that most of the most vigorous opponents of Lysenkoism were not arrested.[115] Any way one searches it, the public record simply will not support the common belief that the apparatus of terror consciously and consistently worked with the Lysenkoites to promote their cause.

Whoever is puzzled by the apparent inconsistency between this conclusion and the earlier statement, that the terror did help Lysenkoism, should remind himself that we are studying the work of terror, not of law. The difference lies precisely in the commitment of law to an ideal of rationality and consistency in the state's use of force. Terror was an evasion of that ideal. The administrators of terror wished to defend the state by punishing its potential enemies, that is, by repressing criminals before they became criminals. The insane logic of that attitude, as many critics have pointed out, would make everyone a potential enemy of the state, including its highest officials, including even the administrators of terror. The conscious goal, it is often said, systematically pursued by Stalin, was the atomization of society so that one man might be omnipotent. Perhaps that was Stalin's goal, though it must be noted, if we are to consider his mind the engine of Soviet history, that it resembled other minds in having many different, frequently inconsistent purposes. But reading Stalin's mind is not nearly as important as analyzing social reality. His presumed goal was impossible to realize. If, between 1936 and 1939, his regime was indeed striving to atomize society, it was striving for chaos.

A close examination of the terror at work in actual social groups — Moscow and Leningrad Universities, the Timiriazev and Lenin Academies, the Academy of Sciences, the All-Union Institute of Grain Culture in Saratov — discloses a reality much more complex than the favorite schemes of political theorists. Those social groups were far tougher than the speculations of, say, Brzezinski would lead one to expect.[116] It should also be apparent, with respect to the ruler's purposes, that terror was not a variety of consistency and rationality in the state's use of force, not even a lunatic variety. It was a series of convulsive efforts to evade those lawyerish ideals. It was a recurrent preference for vague and intuitive judgments of criminal behavior, for jailing and shooting while putting off the framing of rules for jailing and shooting. Many of the criteria for the judgments of the NKVD were probably never reduced to the repugnant precision of written language. Criteria existed, if only in the administrators' neurons; they are implicit in the list of the repressed. As lopsided regularities manifest themselves in the random violence of mob outbursts or military bombardments, so they do in the repression of specialists by administrators of terror.

The appendices list eighty-three repressed biologists and agricultural specialists, with information on their ages, specialties, posts, and stand taken in the controversy over agrobiology. Most of them did not take an active stand. That is the most striking feature of the list. Nevertheless, all but six were *non*-Lysenkoites, though they were not active *anti*-Lysenkoites. A few tried appeasement. Most tried to pay no attention to agrobiology, while they went on with their own work. In other words, most of the repressed people were genuine scientists and technicians, professionals with a strong commitment to their disciplines. On the most conservative estimate, such non-Lysenkoites were ten or twelve times as numerous as Lysenkoites among the victims of Stalinist terror. Three inferences can be drawn from this lopsided pattern. The first is adequately confirmed by the public record; the last two will probably be confirmed by archival research. (1) There were far less Lysenkoites than non-Lysenkoites when the terror was at its height among specialists, for Lysenkoism was just beginning to win a large following. (2) Other things being equal, the administrators of terror regarded a specialist's enthusiasm for "our native agrobiological science"[117] as a sign of loyalty to the Soviet Union. (3) The *type of person* likely to become a Lysenkoite was also likely to arouse feelings of trust in the administrator of terror, while the *type of person* likely to be a non-Lysenkoite was also likely to provoke suspicion.

It is tempting, considering the fact that most of the most vigorous opponents of Lysenkoism were not arrested, to add a fourth inference: Other things being equal, the apparatus of terror regarded forthright opponents of Lysenkoism with less suspicion than compromisers and specialists who avoided comment. One might call this the Prianishnikov effect, to commemorate the boldest man of honor in the Soviet scientific community. He was the elected director of the Timiriazev or Petrovskaia Academy who resigned when the Bolsheviks came to power. Subsequently he made peace with the Soviet regime and became its leading authority on agricultural chemistry. Twice in the 1930s he made shocking displays of disobedience in public. In Stalin's presence he refused to confess that science was responsible for the low productivity of the new farming system. It was "organizational measures," he said, that kept the farms from applying the sound recommendations of scientists.[118] In 1937, when he was made chairman of a meeting that was supposed to condemn recently arrested soil scientists, he ruled such condemnation out of order. He silenced any speaker who departed from scientific and technical issues, saying "You are not the procurator or the NKVD." *Pravda* denounced him as a protector of "enemies," but he was not arrested.[119] Indeed, when his former student Vavilov was arrested in 1940, Prianishnikov sent a long appeal to Beria, which is supposed to have saved Vavilov from the firing squad.[120] Whether it was better to die in jail of malnutrition is beside the present point, which is the possibility that the administrators of terror may have felt less suspicious of audacious types like Prianishnikov than they did of types like Vavilov or Tulaikov, whose uncompromising principles were concealed beneath a layer of complaisance. There were other people like Prianishnikov — Serebrovskii, Rapoport, the physicist Tamm — and I have found no cases of audacious public disobedience followed by arrest. But there undoubtedly are such cases in the archives. In short, to establish the Prianishnikov effect requires finer reasoning than the coarse and meager data in the public record will allow, even in the tentative way that the second and third inferences are offered. Let us return to them.

In the late 1930s Stalinist xenophobia was not yet an explicit, theoretically justified ideology, as it would be in the late 1940s. But it was very much in evidence, partly as an outgrowth of the Stalinist method of fighting backwardness by refusing to recognize it. Gradually this gave rise to perverse self-congratulation for many of the backward features of Soviet life. In part Stalinist xenophobia was the result of a deliberate suppression of contact with foreigners, as a method of combating subversion. The Lysenkoites fit in perfectly on both counts. They proposed primitive agricultural recipes so scientifically advanced as to arouse the hostility of "bourgeois" specialists; that was the essence of their appeal to Bolshevik leaders. As for the unity of world science, while they were not yet prepared to denounce it, they were headed in that direction. Lysenko might declare, in 1936, "We are not against the use of the factual material of world science." [121] The significant fact is that he had earned the accusation; he proudly ignored and flouted world science. The terrorists who broke up the "nest of enemies" at the Saratov Institute of Grain Culture in 1937 must have felt a close affinity to the crude nativism of the Lysenkoites.[122] The published announcement of that pogrom — perhaps the worst that any scientific institution suffered, except for the Kharkov Physics Institute, which seems to have been hit even harder — was menacingly vague on crimes and punishments, but the terrorist's mentality was revealed clearly enough, in characteristically sloppy language:

> The land of socialism prides itself on progressive Soviet science, on the works of Michurin, Lysenko, Williams, the continuers of Darwin's great work. People in agronomic science should constantly have before themselves, and take examples of creative work, from these remarkable scientists.
>
> To liquidate the consequences of wrecking in agronomic science means to totally destroy, to root out false "theories" in agronomy, and to raise agronomic science to the level of our socialist agriculture, the most progressive in the world.[123]

Except for the growing obsession with the difference between "ours" and "theirs," the administrators of terror showed no interest in the substantive issues of the quarrel over agrobiology. Yet they tended, when they culled "enemies" from the community of biologists and agricultural specialists, to pick an overwhelming majority of non-Lysenkoites. The third inference is especially important for an understanding of this unplanned regularity. The *type of person* likely to be a non-Lysenkoite specialist was also the type likely to provoke suspicion and hostility in the apparatus of terror. He was likely to be *older* than the average Lysenkoite, *better educated,* with a record of *prerevolutionary training and experience,* of *less desirable "class origin,"* in a *higher position,* with a *better reputation* among fellow specialists, and more *involved with foreigners.* Those characteristics were judged very harshly between 1936 and 1939. It is noteworthy that three of the six Lysenkoites who suffered repression were directors of important institutes, and two others were obliged by their positions to have extensive dealings with foreigners (they were officials in plant introduction).[124] In both respects those four Lysenkoites were not typical of the breed, which tended at that time to be *young, poorly educated, subordinate, anti-intellectual nativists.*[125] When they could they boasted of their peasant origin, and in any case their chief contention was that their ways of thought were more congenial to Soviet farmers than to "bourgeois" scientists in foreign countries.

At times the distinction between the two types of person came so close to explicit formulation as to approach an edict outlawing the agricultural scientist in favor of the agrobiologist. In the fall of 1937, for example, when the president of the Lenin Academy of Sciences was "exposed" as an "enemy" and the authorities were looking about for a reliable successor, the central agricultural newspaper boosted Lysenko as "the distinguished son of the collective farm peasantry," the man whose practical recipes had been "furiously opposed" by people now exposed as "enemies." "Relying on the collective farm masses, Academician T. D. Lysenko has completely defeated *enemy* efforts to put the brakes on the growth of Soviet agrobiological science."[126] (Italics added.) Thus a newspaper saw the connection between Lysenko's crusade in agricultural science and the war on "enemies." Very likely the administrators of terror and the Party chiefs were inclined to agree, but they would not come right out and say so until 1948. With the formal announcement in 1948 that Lysenkoism was the patriot's only choice, they would move, as far as agricultural scientists were concerned, from the reign of terror to the rule of law. The period of terror was their period of deliberation, leading to a clearcut distinction between the true servants of the state and its enemies. Jailing specialists who seemed subversive was an extended trial by ordeal, helping the chiefs make up their minds on a precise definition of subversion in this area.

Between 1930 and 1933, when the terror first raked the specialists (relatively lightly, as compared with the spasm in the late 1930s), the declared purpose was to turn all "bourgeois" specialists into "red" specialists, unreservedly dedicated to the service of the Party chiefs. The autonomy of scientists and technical specialists was sternly declared to be a bourgeois illusion, which would henceforth be replaced by the honest, open submission of specialists to political authorities. That talk was little more than bluster, though punctuated by gunshots. Political authorities were still largely incapable of judging technical service.[127] Even the condemnation of specialists for "wrecking" turns out, when inspected closely, to have been a haphazard selection of sacrificial victims, a way of "explaining" the disappointments of collectivization and thus of "encouraging the others," to borrow the euphemism of French officers. In plain English, political authorities were trying to scare technical specialists into reliable performance of their esoteric services.[128] The administrators of terror did not become expert judges of those services. They simply found exemplary criminals to blame for such serious problems as "economic crisis, . . . dissatisfaction in broad strata of the laboring masses, disruption of the alliance between the proletariat and the middle peasants, . . . kulak risings against the Soviet regime."[129] Imaginary criminals were punished for very real troubles: for breakdowns in machinery, for fires, for winter-killing of grain, for "intentionally choking fields with weeds, . . . [for] disorganizing sowing, harvesting, and thrashing, with the aim of undermining the material condition of the peasantry and creating a condition of famine in the country."[130]

"Reliable service" is actually a very ambiguous name for the thing that Soviet authorities were trying to scare out of agricultural specialists. Disappointed and angry, ignorant of science and technology, the authorities were demanding instant solution of problems that had resisted solution for many decades, problems that they, the political authorities, had greatly exacerbated by forced collectivization. They were mistaken or dishonest in their belief that conservative old specialists were responsible for the lack of instant solutions. They were very much mistaken in their belief that terror would help them escape from dependence on unreliable, autonomous specialists.

Indeed, the habit of "encouraging the others" by shooting or jailing a few miscreants had the effect of encouraging underground types of autonomy.

The truly red specialist was urged to follow the example of collective farm "shock brigadiers" and pledge to keep an eye on ten colleagues, reporting any wrongdoing to the proper authorities.[131] And vice versa: when the apparatus of terror "exposed" an "enemy," his colleagues were asked to explain why they had not exposed him on their own. The Bolshevik authorities were thus reviving the *krugovaia poruka* — the "circular guarantee" or "group bond" — that had been Peter the Great's device for overcoming *his* shortage of reliable personnel. It was the rule that made every member of a group responsible for the work of every other member. The expectation was that reliable people would see to the improvement of their unreliable colleagues or report them for repression. The result was usually different. The disloyal and the disobedient, the crooks and the bunglers — all are "unreliable" in the language of bosses — established some underground type of autonomy. The very term, "*krugovaia poruka,*" came to mean the mutual cover-up by which subordinates protect each other against their superiors. In the 1930s within each scientific or technical institution specialists and police agents and halfbreeds worked out a variety of relationships. Sometimes, as in the case of Moscow and Leningrad Universities, the specialists wound up with effective control. Sometimes, as in the case of VIR (the All-Union Institute of Plant Industry) blackmailers, incompetents, and complete opportunists came to power. Sometimes a complex fragile balance was achieved. My guess is that in most cases genuine specialists managed to keep a considerable measure of subterranean autonomy. Very few were openly disobedient, like Prianishnikov. Most were pliable men of principle; their basic principle was that somehow or other, within the limits of the possible, they had to get on with their work.

Between 1936 and 1939, when the flailing of the intelligentsia reached its highest peak, the functions of terror changed somewhat. The scapegoat function was much diminished. "Enemies" were still blamed for confusion in the "coupling campaign" (*sluchnaia kampaniia*) — witness the fact that some Yaroslav cows were covered by an Ostfries bull.[132] But such charges were infrequent and subdued by comparison with the terror between 1929 and 1933.[133] Vigilantism, the "circular guarantee," on the other hand, received greatly increased emphasis. The apparatus of terror snatched "enemies" from virtually every institution, and the remaining staff were assembled to discuss "the liquidation of the consequences of wrecking." They were expected to reproach each other for failing to anticipate the arrests, and thus to purge themselves of the stupidity and "rotten liberalism" that had caused them to assume that the vanished colleagues were loyal specialists like themselves. Such meetings produced occasional charges of responsibility for real troubles, as in the case of the poorly serviced cows, and sometimes they gave agrobiologists an opportunity to denounce genuine science.[134] But mostly they produced a staggering amount of trivial accusations: this or that person was especially close to the arrested person, or maintained correspondence with an émigré, or failed to provide an adequate and clean bathhouse, or was responsible for the intolerable food and long lines in the cafeteria, or misappropriated funds to hire an unauthorized secretary, or gave his students too much work, or failed to prepare his lectures, or gave dull ones, or had not written good popular accounts of his subject.[135] Studied close up, the terror of the late 1930s provokes the same overwhelming sense of pointlessness as it does from a bird's-eye view. The people directly involved seem to have struck at each other more and more aimlessly, moved more and more by simple confusion over the distinction between "us," who are members of the community and deserve fair treatment, and "them," who are not and do not.

Perhaps we have stumbled on the most important function of late Stalinist terror — a latent function, unrecognized by the people involved. Haphazardly, by random violence, terror was being pushed toward self-defeat; a political community, to be governed by the rule of law, was being formed. In agriculture the political chiefs were drifting toward agreement on a clearcut definition of the reliable specialist: He would be "our" kind of person, anti-intellectual, authoritarian, nativist, ever ready to promise that agricultural problems would be solved cheaply within "the next two-three years." The administrators of terror aided this drift toward consensus by repressing specialists who rubbed them the wrong way, that is, specialists who seemed to be *ne nashi,* "not ours." The administrators of agriculture helped much more by placing agrobiologists in key positions. Whichever way it was pushed along, this drift toward consensus and the rule of law in favor of agrobiologists had one consequence that would prove fatal to them. Responsibility was coming along with legal authority. Repression made it increasingly difficult to blame the shortcomings of collectivized agriculture on "them," the alien specialists. As Soviet political authorities approached the point where law could take the place of terror because "our" kind of people could be explicitly defined and given formal authority, they were also approaching recognition of a terrible dilemma: "Our" kind of people could not run a modern agricultural system.

THE FINAL "DISCUSSION"

The Second World War gave Lysenko a new chance to show the *deistvennost'* of his agrobiology, its "action quality," as we may translate this favorite term of Bolshevik praise for approved ideas. His chief contribution to the war effort was a campaign among urban gardeners. He urged them to sprout potato stems for planting in place of tubers. Thus the gardener would not have to hold back for planting tubers that were desperately needed for the table. (A similar recipe had been urged on starving city folk during the country's preceding mortal crisis, the Civil War.)[136] To improve the germination of grain seed and thus to increase the yield of those essential crops, Lysenko urged warming of the seed before sowing. He simply ignored the puzzlement of those who remembered his insistence on chilling grain seed as the best way to increase yields.[137] He tried to convince Siberian grain farmers — it was there he spent the war years — that winter wheat could be grown in their fabled climate right away, before his promised new variety was ready. They had only to sow an ordinary variety in the stubble left from spring wheat, which would protect the crop from winter-killing.[138] With the mass media applauding those contributions to the war effort, Lysenko began the most audacious ventures of his career. Toward the end of 1945 he endorsed an attack on cytology, and undertook on his own the subversion of Darwin's theory of natural selection.[139] In short, he expanded his war with twentieth-century biology into an assault on the great achievements of the nineteenth century.

Once again Lysenko was using flights of fancy to justify his practical intuition. Beginning with *kok-saghyz,* a rubber-bearing plant that resembles dandelion, and progressing to oak trees, he convinced himself that the easiest method of cultivation is the best — plant in clusters and leave the clusters to thin themselves. He insisted that those plants knew how to thin themselves to best advantage.[140] Specialists in the creation of shelterbelts for the Russian steppe disagreed. Lysenko, as usual, disdained factual arguments. He backed up his recipe with a homemade theory, which was plainly contrary to the theory of natural selection. Oak seedlings growing in a dense cluster, he argued, did *not* enfeeble each other by competing for moisture, sunlight, and nutrients. Competition occurred only between species. Within a given species individual organisms cooperated for the good of the community. The weaker oak seedlings died out so that the stronger might thrive. In a medieval author this kind of teleological reasoning has a quaint charm. In the twentieth century, among professional biologists, it was a scandal.

With that scandalous theory backing up his practical recipe Lysenko prepared for the last great campaign of his career. He proposed to drive peasants all over the steppe to plant acorns and seedlings in clusters. With a minimum of expense the government would quickly achieve an epoch-making transformation of the climate and a great boost to agricultural yields. Of course shelterbelts were not original to Lysenko; many specialists had long been trying to erect arboreal defenses against the hot dry winds from Central Asia. Even the science-fiction notion of a major transformation of the climate by a suitably large-scale planting of trees was not Lysenko's.[141] He was proud to say that the great plan derived from Williams, who was being posthumously placed beside Michurin and Timiriazev in the pantheon of agrobiological gods. Lysenko's characteristic contribution to shelterbelts was a simple, uniform method admirably suited to another campaign. He urged the use of Soviet power to get all the peasants in the steppe to plant in clusters, and within a few years the elements would also submit to Soviet power. The first step was to silence critics, who cited not only the data of field trials but also the ideas of Karl Marx and Frederick Engels.[142] For more than fifty years Marxists had been taught to venerate the theory of natural selection as "the final blow to teleology" in the life sciences. Now the great plan to transform nature rested on a revival of teleology. In the language of Stalinist ideology this was a conflict between the demands of practice and the demands of theory, and Stalin had decreed (in 1929) that practice must win out in such conflicts.

Lysenko had other incentives for expanding his war with science. He could not win the limited engagement he had been fighting in genetics and plant physiology. His critics repeatedly cornered him on two related issues: Lamarckism, which was officially disapproved because it clashed with natural selection; and cytology, which gave splendid support to genetics. These issues will be examined in detail later on.[143] Suffice it here to say that Lysenko had been constrained, in large part by respect for theoretical ideology, to deny that he was a Lamarckist and to grant that his views on heredity could not be correlated with the facts of cytology. After the war he broke out of these constraints by abandoning the theory of natural selection in favor of outspoken Lamarckism,[144] and by endorsing the work of Olga Lepeshinskaia. She was an old Bolshevik physician who had been denouncing cytology in vain for more than ten years. With suitably sloppy laboratory techniques and reasoning to match she could prove anything she liked about cellular processes and their connections with heredity. She claimed, in fact, that she could create cells out of formless globs of organic matter. By endorsing her work and embracing Lamarckism, the Lysenkoites could brush aside criticisms that had previously seemed unanswerable. It was not even necessary to concede that natural selection was being abandoned. With sufficient political strength to back them up, Darwinism could mean whatever the Lysenkoites wished it to mean or, rather, whatever "practice" required it to mean.

From 1945 to 1947 it was questionable whether they would have sufficient political strength. There was even a little evidence suggesting that agricultural authorities and the biggest political chiefs who stood over them might be wavering in the agrobiological faith, at least in the Lysenkoite part of it. Williams' scheme, the *travopol'naia* or grassland system of crop rotation, enjoyed a boom in the early postwar period, as it had in the late 1930s. Decrees were issued on its behalf, and the agricultural press backed them up with a stream of uniformly favorable propaganda.[145] The same press faltered in its endorsement of Lysenko's recipes. Toward the end of 1945 the chief agricultural newspaper appealed for a discussion of his proposal to sow winter wheat in Siberian stubble; and to show that a genuine discussion was intended, it printed a devastating critique of the proposal.[146] The author, a courageous defender of rational agronomy who had challenged the usefulness of vernalization in the mid-1930s, concluded his thorough, factual rebuttal of Lysenko's new scheme with a general plea for liberation from "the hypnosis of accidental figures," — as neat a characterization of agrobiology as one could wish. For three months the newspaper printed the comments of agricultural specialists, some defending Lysenko's recipe, but many rejecting it.[147]

Early in 1946 the paper veered about, resuming its steady pro-Lysenkoite course, but two important agricultural magazines continued for some months to publish things that enraged the Lysenkoites. (Only one article was an outspoken criticism of Lysenkoism; the others were quietly offensive, taking for granted various scientific principles that the Lysenkoites denied.[148])

A verdict on this furtive dueling within the agricultural establishment might have been expected from the plenary session of the Party's Central Committee that met in February 1947 to discuss the serious agricultural situation. The war's devastations and miseries had been compounded in 1946 by one of the periodic droughts that used to threaten Russia with famine. No figures were given out; the government was satisfied to report that grain production had fallen below the level of 1945, "but not as low as 1921." [149] When the minutes of the February 1947 meetings are published, they will probably reveal arguments between aggressive agrobiologists and very circumspect critics, under the terrifying gaze of a noncommittal Stalin.[150] The published resolutions and subsequent commentary indicate a decision to muddle along as before. There were points for Lysenko in the resolutions: vernalization of grain and the summer planting of potatoes were to be reintroduced in areas where they had been successful before the war, and the Central Committee noted with approval the campaigns to grow millet and to mobilize Ukrainian chickens for war on the sugar beet weevil.[151] But the resolutions also noted that "the collective farms of Siberia still do not have winter-hardy varieties of winter wheat." [152] Names were unnecessary; this was an obvious reproach to Lysenko — and to Tsitsin, who had promised a *perennial* wheat for Siberia back in 1935. Perhaps the greatest encouragement to genuine science was the omission of the usual charge of impracticality against it. Indeed, there was an implicit recognition of its practicality in the call for a great expansion of hybrid corn production, a program that the Lysenkoites had severely curtailed in the 1930s. The resolutions did not specify whether the Central Committee wanted old-fashioned varietal hybrids, which Lysenko approved, or the modern hybrids of inbred lines, which he condemned. But it is clear from subsequent comment that work was desired on both types. Lysenko himself, spelling out the Central Committee's wishes in a speech to the Lenin Academy, blandly endorsed the modern type of hybrid corn.[153]

These glimmers of possible division within the agricultural establishment seem feeble in retrospect. To Soviet biologists of the time they must have seemed the first light of dawn, encouraging genuine scientists to rise and get moving again. They not only printed anti-Lysenkoite articles — by themselves such polemics would have indicated merely the beginning of another "discussion." The scientists also undertook measures of an organizational nature, to borrow a favorite Soviet phrase. An eminent professor of botany at the Timiriazev Academy published an appeal for a return to genetics. He reminded the world of higher learning that the subject had not been outlawed, though many students and editors seemed to think so.[154] The response from the summit of power must have been at least an encouraging silence, for the Ministry of Higher Education at the end of 1946 opened a genuine discussion of the course called Darwinism.[155] Decreed in the 1930s by "the former leadership" (that is, by officials who had subsequently fallen victim to the terror), the course had languished. Where it was offered at all it was merely an elective, Mendelian in a few places, evasive in others, militantly Lysenkoite almost nowhere, though it was originally devised as a way to bring Lysenkoism into institutions of higher education. Zhebrak, who taught genetics at the Timiriazev Academy, argued that Darwinism as a separate course should be abolished; its scientific content should be incorporated in expanded courses of genetics.[156] Schmalhausen, the famous student of evolutionary morphology who headed the Department of Darwinism at Moscow University, thought the course should be preserved but placed in the control of genuine scientists.[157] Other people had other proposals; the upshot was the common bureaucratic standoff. Nothing was changed. But everyone must have noticed the absence of Lysenkoite opinions in the discussion. The only Lysenkoite who published a brief comment was vague and evasive.[158] In short, it may be said that Minister of Higher Education Kaftanov earned the reproaches that Lysenkoites would hurl at him in 1948. In 1946–47 he showed some favor to the scientists in his constituency. But the credit or blame should not be his alone; he was a subordinate of the Party's chief of science and culture, A. A. Zhdanov.

Within the Academy of Sciences, Zhebrak and Dubinin agitated for a new institute of genetics to replace the center of their science that had fallen to Lysenko in 1940, when Vavilov was arrested. The president of the Academy — Vavilov's brother Sergei, a physicist — seems to have received a reassuring nod from above, for he got the Presidium of the Academy to give its formal approval and propose to the government the creation of an institute of cytogenetics.[159] Even Oparin, Lysenko's lone partisan among the Academy's distinguished biologists, voted for the proposal.[160] In preparation for the expected approval, the Academy added staff to its existing laboratory of genetics and elected Dubinin to corresponding membership. (He was slated to be the director.) In December 1946 Lysenko wrote a letter of protest: Dubinin had no practical or scientific achievements to his credit, and he was the chief (*vozhak*) of the anti-Michurinists, "who represent in our genetical science the ideology of foreign biologists, who are conservative and even reactionarily inclined in the ideological respect." [161]

Lysenko was obviously trying to hitch his wagon to the "anticosmopolitan" juggernaut that A. A. Zhdanov was then setting in motion. Art and the humanities, Zhdanov decreed, were to become aggressively chauvinist — Russian chauvinist, of course; the other Soviet nationalities were formally demoted to the status of junior brothers. Outside of the humanities and the arts, Zhdanov dropped a fleeting hint that the physicists might be chastised for preaching foreign ideology.[162] He said nothing about biology. In that field his son, Iurii, who was head of the Central Committee's Division of Science, was still trying to maintain the precarious compromise of 1936 and 1939.[163] To add anticosmopolitan charges to the burden already carried by genetics would have been to crush it altogether.

Zhebrak and Dubinin felt encouraged not only to press for an expansion of their science within the Academy of Sciences but even to pursue a colloquy with scientists abroad. It began in the friendly days of 1945, when Western scientists discreetly inquired into the possibility of holding the next international congress of genetics in Moscow.[164] (One of the ugly incidents marking the rise of Lysenkoism had been the abrupt cancellation of the 1937 congress scheduled for Moscow. It had finally convened in Edinburgh in 1939 with a long list of Soviet delegates on the program and not one in attendance.)[165] In 1945 Serebrovskii, Zhebrak, and Dubinin not only expressed delight at the suggestion of their foreign colleagues; the last two moved from private conversation to publication in American learned journals. They wished to show that genetics was flourishing in the Soviet Union. Zhebrak could list only eight laboratories, but the prospects, he felt, were bright.[166] For two years this daring venture in foreign contacts[167] provoked no comment from Soviet authorities. Then, in September 1947, *Pravda* let everyone know that the anticosmopolitan campaign was coming to biology. It published a vehement attack on Zhebrak and Dubinin, written by a leading agricultural economist. He explicitly denied the unity of world science and pictured Lysenkoism as the only biology for Soviet patriots.[168]

Resistance still seemed possible. At a meeting convened for the special purpose of chastising Zhebrak, the victim was defiant, saying, if we can trust the minutes, "that Lysenko's doctrine is incorrect, and that several years hence Lysenko will be a blank space in science, while the doctrine that he, Zhebrak, serves will go on developing. His, Zhebrak's, theories and Lysenko's will always remain poles apart." [169] Zhebrak, a Party member, may have told himself that defiance was still possible because the elder Zhdanov was still silent on the issue of biology, while the younger Zhdanov continued to seek a compromise.[170] Other scientists shared Zhebrak's myopic courage in the latter part of 1947 and early months of 1948. At Moscow University the faculty council (*uchenyi sovet*), with the rector's support, reorganized the department of dialectical materialism in such a way that a Lysenkoite seminar for biology students was dropped.[171] The rector even published a strange article at the end of 1947, arguing that *support* of genetics was the proper stance for anticosmopolitan patriots, because genetics had been developed by Soviet scientists.[172]

Another "discussion" had obviously been opened, and courageous optimists assumed that it would lead to much the same conclusion as the earlier ones: The Lysenkoites would be blessed for their contributions to agriculture, the academic biologists would be rebuked for opposing or ignoring Lysenkoism while they themselves gave nothing to farmers, and the ideological theorists would stand firmly in their perpetual straddle. With such expectations, the scientists organized their defense. It may seem bold and strong in light of the total condemnation that was being prepared for genetics and cytology, but, compared to the defense of 1939 or 1936, it was fatally weak. Early in 1948 the Institute of Plant Physiology at the Academy of Sciences organized a meeting to defend work on plant hormones.[173] In February 1948 a group of biologists struck a spongy blow at the weakest point in the theoretical ideology of the Lysenkoites. Gathering at Moscow University they read papers and adopted resolutions in support of natural selection, without mentioning Lysenko's attack on it.[174] As late as July 7 organized resistance still seemed feasible, if only in cytology. On that date, the newspaper of the Ministry of Health, which had previously ignored Lepeshinskaia's anticytology, published a criticism of it, signed by thirteen Leningrad scientists.[175]

As luck would have it, July 1948 was probably the time when the highest chiefs decided to give Lysenko full and uncontested control of the biological sciences, as a prelude to "the great Stalin plan for the transformation of nature," and to a strong anticosmopolitan drive in the natural sciences. Without access to the archives it is futile to guess precisely which political chiefs played precisely what parts in the decision. The public record reveals these suggestive facts. Iurii Zhdanov, who was head of the Central Committee's Division of Science and therefore was the subordinate as well as the son of Andrei Zhdanov, who was supreme chief under Stalin of the entire cultural front — Iurii Zhdanov, so fathered and so situated, wrote a letter to Stalin apologizing for his previous efforts to preserve the science of genetics, and pledging to join in the effort at total suppression, which was now planned. The letter was dated July 10, 1948.[176] Events then moved very quickly. On July 15 the Council of Ministers decreed that the Lenin Academy of Agricultural Sciences would seat thirty-five new academicians, most of them Lysenkoites or collaborators. The Council instructed Lysenko to give the Academy a report "On the Situation in Biological Science," and the campaign for massive afforestation was launched.[177] On July 31 the Academy assembled for its famous August Session to hear Lysenko's report, with the Minister of State Farms acting as chairman. Lysenko announced the total suppression of genetics, the total victory of

> Michurinist, Soviet biological science, fundamentally new, our own, engendered by socialist agriculture, by the system of collective and state farms, and developing in close unity with agricultural practice as an agronomic biology.[178]

The record does not indicate the presence of either Zhdanov. Zhdanov senior died at the end of August and was given a most illustrious funeral. Junior soldiered on as head of the Central Committee's Division of Science, and husband of Stalin's daughter. Whatever the archives may ultimately reveal about the intricacies of the Zhdanovs' situation in 1948, the most important fact is the institutional one, which is in the public record for all to see. Agricultural chiefs presided at the climactic triumph of agrobiology, while the heads of the cultural and scientific establishment still showed signs of doubt and lingering reservations.

The obvious explanation is that the Party's chiefs of culture and science were not only rulers but also, in some measure, representatives of their constituencies. That will sound absurd to those who think of Zhdanov senior as the personification of Stalinist anti-intellectualism. Beyond doubt he was that, but he was also the ultimate chief under Stalin of a large bureaucracy of higher learning, which had been putting up strong resistance to Lysenkoism. Indeed, the resistance was manifest at the August Session, mostly in the refusal of key scientists to speak. The chief of biological sciences at the Academy of Sciences walked out after Lysenko's speech and would not return. When pressure was put on him, he pleaded incapacity for his administrative job and resigned.[179] The director of the Timiriazev Academy, the central institution of higher education in agriculture, also tried to avoid speaking. When pressure was put on *him,* he mounted the platform and threw the meeting into an uproar by a hard-hitting defense of Zhebrak's right to teach genetics at the Timiriazev Academy. To shouts of "Resign!" he shouted back "I may!" and did so, without "self-criticism." [180] A few other men spurned silent resistance in order to place on record the fact that angry willpower can support the scientist's commitment to reason no less than the crank's adherence to his private whim. Most notably the geneticist Joseph Rapoport, a short, slight war hero with an eye patch, defied the chiefs of his own Party with a blunt warning that "the transformation of animals and plants cannot be achieved merely as a result of our wish." [181] He was so outspoken and unbending that he was expelled from the Party.

Young Zhdanov's letter was published on the last day of the meeting, in part to show such recalcitrants that they were defying Stalin himself. Three of them immediately apologized for their recalcitrance, noting, however, that the letter still allowed a little room for criticism of Lysenko.[182] And strangely enough, it did. Zhdanov wrote that he still could not agree with Lysenko's "underestimation of the internal specificity of the organism," or with his "denial of intraspecific struggle and mutual aid." [183] The two phrases were very clumsy, but their meaning was clear enough in the context of the "discussion." The first referred to Lysenko's views on cytology, the second to his denial of natural selection within a species. On these issues he might still be criticized. Thus young Zhdanov, in the very act of surrendering biological science to Lysenko, tried to specify limits to the surrender. He tried to continue, if only in an extremely attenuated form, the sort of compromise that had marked the settlements of 1936 and 1939. It is difficult to think that he was especially courageous or intelligent; he later reneged this attenuated compromise, and he continued to write Lysenkoite nonsense even when he had no official obligation to do so.[184] His writings reveal a clumsy mediocre mind, even within the area of his special training at the Institute of Philosophy.[185] He seems to have acted as a normal bureaucrat, trying to keep the peace in his realm and to protect it against depredations from without — always within the limits of orders from above.

Young Zhdanov also noted that "Lysenko has not bred any significant varieties of agricultural plants," a fault attributed to his "weak use of the treasury of Michurinist doctrine."[186] Whether Zhdanov was here recording or trying to provoke the first glimmer of disenchantment among the chiefs of agriculture is not clear. The same question is raised by his observation that Lysenko had allowed the Lenin Academy, which he had directed since 1938, to become weak in stockbreeding, agricultural economics, and agricultural chemistry.

Taken at face value, these comments, coming from the head of the Central Committee's Division of Science, would seem to have allowed a considerable right of continued disagreement with Lysenko. In fact no one dared to take advantage of this right until specifically authorized to do so, beginning in 1951. And small wonder. Zhdanov's little demurrer was drowned by the roar of the mass media, turned up full blast to deliver the message of the August Session. A crude farm chairman stated it most appropriately with a threatening Lamarckist joke:

> We must now finally and irreversibly dethrone this antiscientific and reactionary theory [genetics], and unless we strengthen our "external influences" on the minds of our opponents, and create for them "appropriate environmental conditions," we will not transform them.[187]

The ultimate "external influence" was brought to bear on the last day of the conference. "The Party's Central Committee," Lysenko was authorized to say, "reviewed my report and approved it."[188] Everyone rose at this invocation of the highest authority under Stalin and gave Lysenko "stormy applause turning into an ovation," which was usually reserved for Stalin.

It seemed quite clear that absolute power had been delegated to Lysenko. Young Zhdanov's qualifications, which appeared that same day in *Pravda,* must have seemed as academic as the doubt whether absolute power can be delegated, or the related doubt whether political power in a large complex state can be literally absolute. Nobody expressed such doubts in Stalin's Russia. But they were as unavoidably fixed in the nature of political things as Mendelian ratios and plant hormones were fixed in the nature of biological things, and they made themselves felt in the rush to establish the absolute power of Lysenko. To whom could he delegate that power without losing it? Unbending foes were, of course, discharged from research and teaching posts, and the Minister of Agriculture warned that the same would be done to those who tried *"perekraska pod Michurina,"* or "Michurinist camouflage." [189] He was unwittingly acknowledging the law of enforced belief, as we may call the rule that hypocrisy springs up where free thought is cut down. But how were the authorities to distinguish the true convert from the hypocrite biding his time?

N. A. Maksimov may serve as a fine specimen of the second type. As the country's leading plant physiologist he had been one of the first to feel the crushing political force of Lysenkoism. In the early 1930s he bowed before it, inserting Lysenkoite passages in his world famous textbook and desisting from printed criticism of Lysenko.[190] Thus, after a six-year exile in Saratov, where he survived the 1937 pogrom, he returned to a capital (Moscow, this time) and became director of the Institute of Plant Physiology at the Academy of Sciences. The reader will recall that this Institute early in 1948 adopted a resolution condemning the Lysenkoite condemnation of research in plant hormones. After the August Session, Maksimov appeared before a meeting of the Academy to apologize and swear his limitless devotion to Lysenkoism.[191] He was obviously a master in the art of *perekraska.* While he denounced plant hormones he defended growth substances, which, he recalled, the Council of Ministers had directed his Institute to study for their possible use in agriculture. (Maksimov conveniently ignored the fact that the Council had formerly ordered the widespread study of genetics for the same reason.) He apologized for the scientists in his Institute who had bad records of public disapproval of Lysenkoism. This one should have been ordered to renounce his views — no doubt Maksimov said this without a trace of sarcasm — that one had surely changed his mind on his own, and in general the Institute would henceforth be purely loyal to the Lysenkoite school.[192] Thus Maksimov kept the directorship, until death took it from him in 1952. Ten years later an old acquaintance still remembered his "headwaiter's smile."

Maksimov was succeeded in 1952 by a younger version of himself, the pliable man of principle.[193] In the long run, however, the supply of such people was seriously threatened by the educational consequences of the August Session. Biological curricula and textbooks from gradeschool through university were rewritten to eliminate the patchwork of Lysenkoism and genuine science that had been their chief characteristics since the late 1930s. Now a militant, uncompromising Lysenkoism was written into biological education at every level.[194] Undoubtedly many teachers and professors were principled hypocrites like Maksimov; in any educational system what is in the textbooks is not identical with what is actually taught. Especially at the higher levels of the Soviet educational system circumspect transmission of genuine biological science was continued even in the worst years, 1948 to 1951. But it was so sharply reduced, and the reproduction of Lysenkoites was so sharply increased, that genuine biologists might well have become extinct within a generation or two, if the government had seen fit to maintain for so long a period the "unlimited rule of Michurinist biology."[195] (The phrase is that of the repentant Minister of Higher Education, promising to overcome his previous toleration of genetics.)

Everything depended on Lysenko's ability to produce great agricultural benefit now that opponents and skeptics were forbidden to hinder him. Beyond any doubt he was now the top specialist responsible for the "Great Stalin Plan for the Transformation of Nature." In a painting of Generalissimo Stalin unrolling great maps of shelterbelts and forest zones before his radiant Politburo, Lysenko is not to be found.[196] After all, he was not a member of the Politburo — indeed he was never a Party member — and did not enjoy their godlike responsibility only for good. He could be blamed for evil, if the gods should ever wish it. Between 1948 and 1951 they wished him only good. His cheap method of planting trees in clusters and then letting them thin themselves was transformed in the usual way from a mass experiment to a mass prescription. The Great Stalin Plan forecast five million hectares of trees to be planted by 1955, almost twelve and a half million acres within seven years.[197] It was the grand finale of the Stalinist agricultural campaigns; in one huge rush cold dry Russia was to be made a land of mild moist weather. Periodic crop failures, like that of 1946, were to become a dim memory. Shostakovich put the whisper of life-giving leaves in a choral symphony, "The Song of the Forests." Leonid Leonov and many lesser storytellers triumphed over nature by rustling those leaves in their fictions.[198] Why should they have doubted when the most eminent scientists expressed unqualified approval? Sukachev, a world renowned ecologist who must have known that the Great Plan and the Lysenkoite method of planting were a noisy haste-makes-waste, pledged his support nevertheless.[199] Thus he kept his crucial posts as head of the Forestry Institute and chief editor of the *Botanical Journal.* How much was spent on the Great Plan, whether by the state or by the farms, has never been revealed. We have only scattered facts, such as the planting of almost one and a half million hectares of trees on collective farms from 1948 through 1951,[200] when the campaign faltered and Sukachev got the signal he had obviously been waiting for. In the *Botanical Journal* he was allowed to open a discussion of Lysenko's views on natural selection; that is, he was allowed to attack the faith that justified the cluster planting of trees.[201]

Between 1948 and 1952, regardless of young Zhdanov's initial note of reservation, Lysenko's hostility to natural selection and to cytology were official doctrine. Indeed, young Zhdanov himself, still head of the Central Committee's Division of Science, published his endorsement of these ultimate absurdities.[202] The Lysenkoites, who seemed to be released from any possibility of criticism, gave completely free rein to their fancies. Not only was the gene proscribed and the chromosome under suspicion; Lysenko's more extreme partisans denounced plant hormones and viruses as metaphysical vaporings of the bourgeois mind.[203] The world of living things became magic putty in their hands; they could effect any transformation they wished by manipulation of the environment. Lysenko had written that "any character can be transferred from one breed to another by grafting as well as by crossing." [204] He had also shown that varieties of wheat can be transformed without the burdensome necessity of grafting, by simple exposure to the climate in which he wanted them to live.[205] Now he turned his omnipotent mind to the old problem of wheat fields choked with weeds, still the curse of peasant agriculture, in collective farms as in the old communes. Lysenko found the explanation very easily. The environment transformed wheat into weeds:

> Under the influence of changing conditions, which become unfavorable for the nature (the heredity) of the organisms of the species of plants growing here, in the body of the organisms of these species are engendered, are formed, rudiments [*zachatki;* elsewhere he calls them *krupinki,* granules] of the body of other species which are more suitable to the changing conditions of the external environment.[206]

That discovery was not only applauded in the central newspapers, it was immediately enshrined in the *Great Soviet Encyclopedia,* in the article, "Species." [207] It is as if one said that lap dogs turned loose in the forest will give birth to foxes and wolves, because the wilderness is a suitable environment for wild animals.

Let us not ask what use those peasant fancies might have been in the struggle against weeds, or how many trees grew in the steppe as well as the novel and the concert hall, or whether Soviet peasants enjoyed the "limitless growth in harvests" that Lysenko forecast on New Year's Day 1949.[208] To ask such questions is to take the first steps from the total victory of agrobiology toward its defeat; to be precise, it is to follow the Stalinist chiefs of agriculture in their first steps from self-defeating willfulness to self-conquering objectivity.

6

Self-Conquest, 1950–1965

When Stalin died in March 1953, his chief subordinates put his mummy in a glass case and moved quickly to reform the system they had helped him to establish. Before the year was over, they quietly dismantled the apparatus of terror, publicly confessed the shocking stagnation of agriculture, and in all sectors began to attack the ancient Russian malady that had reached nearly mortal extremes during Stalin's reign: the concentration of power without responsibility, the diffusion of responsibility without power. All three lines of reform pointed to the end of Lysenkoism, yet the end was delayed more than eleven years, for Lysenkoism was the product of a system; and a system is not quickly changed by the subtraction of an individual, not even an individual as mighty as Stalin. His death accelerated changes that were beginning while he lived. Of course, the beginnings were small when measured against the need for change, but the same can be said of the accelerated reform that followed his death. It has not been as great or rapid as outside critics and inside victims have wished. One must always remember that changes in the system have so far been enacted by its masters, who have frequently bowed to superior force but have never yet lost control. Before or after Stalin's death their minds have been fixed on the best interest of the Soviet state *as they explained it to themselves.* The arguments of outside critics and the complaints of inside victims have influenced them, but they have strenuously insisted on maintaining their "leading role," that is, their monopoly of political power. In short, Stalinist bosses have been struggling to extricate themselves from the self-defeating vices of their own system, their own mentality.

Self-Conquest, 1950–1965

UNDER STALIN

A favorite form of Stalinist self-correction was silent abandonment of measures that persistently failed to achieve anticipated results. Perhaps this neurotic habit should not be called reform. It rests on a great reluctance to acknowledge shortcomings, which is a block to genuine reform. The decline of terror, for example, following its two peaks in the 1930s, was an unacknowledged retreat from a method of administering justice that threatened to destroy the state it was supposed to defend. Following Stalin's death the retreat was greatly hastened by the abolition of the secret tribunals that had the power to arrest and punish without anything like due process. That crucial turn toward legality was taken in 1953, but quietly. It was not announced to the public until 1956, when the Communist chiefs seemed ready to turn their surreptitious retreat into a program of genuine reform.[1] But then they grew alarmed and quickly tried to stamp finished on this bill, as on many others. Perfunctory insistence that any "mistakes" in the administration of justice have been corrected, that any necessary change in the system has already been effected — this has been the dominant attitude, and it has been a major block to the further growth of genuine legality. As late as 1964 a Lysenkoite president of the Lenin Academy of Agricultural Sciences called for criminal prosecution of a biochemist who had written a critical history of Lysenkoism.[2] To be sure, the Lysenkoite president invoked the law of libel rather than the defunct "law" concerning "enemies of the people." And the case, if it had come to court, would have come to a court, rather than the defunct secret tribunals that had no obligation to observe any kind of due process. It is unlikely that such a politically inspired case would have been tried in strict accordance with the law, and the law itself, especially in the matter of "antistate" publications, is badly in need of greater precision.[3] The biochemist's critical history of Lysenkoism is still unpublished in the Soviet Union, though manuscript copies circulate, along with other works that Soviet publishing officials are afraid to put out. When bold authors have published abroad, kangaroo trials have sent some of them to prison, while others are precariously free. The fact that courageous intellectuals are pressing for genuinely legal guarantees of genuinely free expression shows how far the political bosses have retreated from terror. The fact that courage is required simply to raise the issue shows how far the bosses are from submitting to strict legality.

Just as the surreptitious retreat from terror began under Stalin at the end of the 1930s, so did the abandonment of Lysenkoism. Lysenkoite agricultural recipes continued to be advertised as great practical successes long after officials had dropped the campaigns on their behalf. As late as 1947 the Party's Central Committee included in a batch of agricultural resolutions a call for renewed campaigns on behalf of vernalization and summer planting of potatoes,[4] but the campaigns were not actually renewed, and no significant number of farmers revived these nostrums. The officials were preaching Lysenkoism while abandoning it in practice, just as American politicians used to preach the virtue of balanced budgets while moving ever deeper into deficit spending. Pledging allegiance to Lysenkoism, Soviet officials were reassuring themselves that they were still as wise as they had been in the 1930s, while in fact they were relaxing their efforts to press this wisdom upon farmers. Standoffs of that kind were accumulating in the late Stalinist system, as willful bosses succumbed to refractory facts of life without acknowledging that they were doing so.

If the Soviet agricultural system had corrected itself only in this way, by a widening breach between the central bosses' word and the local farmers' deed, long-run stagnation would probably have been the result. Whether, as in the 1930s, a vigorous campaign obliged farmers to vernalize their grain seed, or, as in the late forties, the Central Committee vainly wished that they would resume vernalizing, either way, grain yields hovered around an intolerably low level.[5] Agricultural officials were extremely reluctant to ask the reason why — which is another way of saying that their words were increasingly unrelated to the facts of agricultural life. They clung to the belief that campaigns are a cheap way to raise yields, shutting their eyes to contrary reports (or shutting the mouths of those who dared to give such reports). As old campaigns petered out, they repressed their disappointment and looked for new ones. The reader has already seen how, in the late 1940s, they rallied their faltering enthusiasm on behalf of a grandiose final solution of the agricultural problem, the "Great Stalin Plan for the Transformation of Nature." That was a joint campaign for the massive planting of trees and for the grassland (*travopol'naia*) system of crop rotation. The cropping system had been pressed on Soviet farmers in the

1930s, but not so forcefully as in the late 1940s, when its disciples were given the power to apply the whole scheme of their crackpot master Williams (who died in 1939), including even his insistence on spring rather than winter grain. But with this crotchet Williams' followers provoked immediate resistance among agricultural officials. Long before the biggest bosses could appraise the global gains and losses of the Great Stalin Plan, whether in their usual sloppy, intuitive way or with some newfound precision, the order to substitute spring for winter grain provoked immediate hostility in the Ukraine. Agricultural officials there knew that they stood to lose by substituting spring for winter grain. The simple fact is that, in southern regions where the winter is not too harsh, winter grain returns a much greater yield than spring grain. Williams' disciples were demanding that this advantage be sacrificed on the expectation that soil structure would sooner or later be sufficiently improved to reverse the ratio, that is, to make spring grain outproduce winter varieties.[6] Here was a choice that could not be evaded, with or without nostalgic pledges of devotion to the agrobiological wisdom of the 1930s.

In 1949 about five hundred specialists and officials, gathered in Kiev to settle the contested issue, heard Nikita Sergeevich Khrushchev scold Williams' disciples with a coarse loquacity that would soon become world famous:

> . . . If you went to a noble landlord [*pomeshchik*] and offered him your services as a learned agronomist, he would certainly ask you, What kind of returns do you promise me from my farm? You would say what you have said here at the conference: For the first year of spring grain I make no promise, but I will run the farm in a scientific way, according to Williams' system. Maybe even the second year will bring nothing, and for the third I make no promise, but I believe that sometime good seeds will be grown and spring wheat will beat winter wheat. What would this landlord do? He would either call a doctor, because he would think that you, as they say, aren't quite right in the head, or, if he was a landlord of the Sobakevich type, he would simply set the dog on you: You [*ty*], what did you say? You want to ruin me? You promise no harvest, you'll leave me without pants![7]

Khrushchev's unembarrassed analogy between *pomeshchik* and Communist management of agriculture shows how far the Communists had moved from their original utopianism. In the 1920s and early 1930s such analogies were considered subversive, in part because they suggested continued exploitation of the peasantry, but also because the Communist leaders at that time felt emancipated from the *pomeshchik's* nearsighted understanding of profit. The major Soviet dictionary defined "profit" by quoting Stalin's sneer at the "shopkeeper" (*torgasheskii*) view of it, his insistence that profit must be seen from "the viewpoint of the whole national economy on a scale of several years." [8] But Khrushchev, speaking for Ukrainian officials in 1949, was unwilling to force "their" farms into another losing venture on the hope that the whole national economy might profit. The longer collectivized agriculture endured, the more difficult it grew to think that the state might gain from the farmer's loss. The farmer was increasingly a Communist manager of peasant labor.

It is also noteworthy that Khrushchev was unwilling to get involved in the scientific aspects of the quarrel over spring and winter grain. Indeed, twelve years after 1949, when he would finally denounce the *travopol'naia* system altogether, he would still refrain from pronouncing judgment on the crackpot theory of soil structure that underlay it. He confessed that he did not understand science and insisted that he judged rival agricultural proposals in a purely practical way.[9] Paradoxical as it may seem, this refusal to intervene in science was one of the main reasons why he was extremely slow in breaking free of faith in the agrobiologists. He might reject one or another of their practical recommendations, but he followed their basic pattern of thought in doing so. Both he and they were intuitive, dogmatic, and willful in their assessment of agricultural methods. Neither he nor they were willing to make serious, realistic calculations, on the basis of rigorous tests, of relative gains and losses likely to accrue over a significant period of time from the use of this, that, or the other method. The substitution of spring for winter grain was indeed a crackpot proposal, but not for the reason that Khrushchev gave, simply because it entailed a sacrifice of income in the immediate future. It was crackpot because the sacrifice would not be recovered in any time period. That truth could be learned only by someone willing to examine the scientific issues and recognize that the *travopol'shchiki* based their proposal on a travesty of scientific reasoning. No great erudition or special expertise was required, but Khrushchev refused to get involved in such "theoretical" issues. He was completely committed to Stalin's rule that "practical" considerations are the test of truth. He was indeed a prisoner of that rule.

In rejecting a proposal simply because it involved a decline in income for the first two or three years, Khrushchev was also continuing the perverse logic (or illogic) of desperation. That is the tendency to equate savings with losses, or, more precisely, the tendency to regard savings as losses unless they pay off very quickly. Many previous generations of Russian and Ukrainian farmers had ignored genuinely scientific recommendations for improved crop rotations because they involved present sacrifice on the chance of future gain. Of course, Khrushchev and his Bolshevik colleagues were sufficiently modern to press for change and to take chances. But the traditional mentality — and the decision to favor heavy industry — still shaped their impatient lust for immediate large returns in agriculture, their refusal to make painstaking calculations of gradual returns on long-run investments. It was that mentality that had caused their abandonment of genuine agricultural science as soon as it disappointed their utopian expectations. Genuine science proved incompatible with the hope of getting a lot for a little in a very short time. It was that mentality that kept the Communist leaders stuck to agrobiology so long, in spite of repeated disappointments. After Stalin's death as before, they longed for cheap breakthroughs to agricultural abundance "within the next two–three years," and schemes for such breakthroughs were forthcoming from agrobiologists, not from scientists.

Thus Lysenko, though he presided over the *travopol'naia* campaign of the late 1940s, survived recognition of its obvious shortcomings. (The prejudice against winter grain was only one of them, as the reader will see in a later chapter.) Indeed, Lysenko's reputation was enhanced by the reform of *travopol'e.* As the chief agrobiologist, he was given the job of explaining which aspects of Williams' schemes required amendment, and his 1950 explanation — an article in *Pravda,* widely reprinted — was a huge triumph, hailed on all sides without the slightest note of criticism.[10] He seemed to be acquiring the marvelous power of Stalin, who could deny today what he had said yesterday without anyone ever asking whether he had been wrong at least one time. (At worst, a foolish acolyte might humbly ask how the contrary statements might be reconciled.)[11] In this respect also the Bolshevik chiefs were akin to the agrobiologists and alien to the scientists. In their essentially political world reason was subordinate to the boss' will; *quod principi placuit* was a basic criterion of truth in Lysenko's Academy as in Stalin's Party. In both places the dicta that issued from the chief's mouth were hailed by his followers as *scientific truth,* but this bizarre notion seemed to be a traditional survival of antique Marxism, without real meaning in Stalin's Russia.

One can therefore appreciate the shock of amazement in the summer of 1950 when Stalin published a denunciation of despotism in science. ("Arakcheev regime" was his actual term, after the tsarist minister whose name had become an eponym for stupid bureaucratic tyranny.) With the deafening amplification that was used for any canonical pronouncement by Stalin, the mass media broadcast his observation

> that no science can develop and prosper without the clash of opinions, without freedom of criticism. But this generally recognized rule has been ignored and violated in the rudest way. A closed group of infallible leaders has been created, which, insuring itself against any possible criticism, has begun to act in a willful, highhanded manner.[12]

Stalin sent up this rocket while discussing the state of Soviet linguistics, but his choice of words was deliberately universal. The mass media raised a cry for an end to "Arakcheev regimes" in all sciences. Very little was done in Stalin's remaining years to achieve the noble goal, but that little was the beginning of the big push toward freedom of scientific thought which followed his death. And, however small the immediate result, one cannot overlook the enormous anomaly of those liberal words coming from *Stalin,* whose little person was the focus of the greatest concentration of despotic power Russia has ever known. In 1930 he had subjected scientists to a strict new rule of *partiinost';* in 1950 he turned about and opened the door to demands for release from *partiinost'.*

Problems of nationalism seem to have been the main stimulus to the publication of Stalin's famous articles on "Marxism and Problems of Linguistics." He was giving the first sign of retreat from the wild extreme of Russian nationalism to which he had led his multinational state. He was rebuking suggestions that Russian, "the language of socialism," should supercede lesser tongues, which were allegedly products of inferior stages in social evolution. On the practical level he was committing his government to indefinite toleration of minority languages within the Soviet state (except for Yiddish and a few other special cases). On the theoretical level he was dissociating language and the scientific study of it from the superstructures that change as the social base evolves from stage to stage. It was therefore possible for Soviet linguists to forget the distinction between proletarian and bourgeois science. They could rejoin the world community of scientists; indeed, Stalin told them to do so.[13] Those were his main concerns, but he used the occasion to reopen the general question of *partiinost'* or party spirit in science. On the theoretical aspects of this question he was vague, leaving it to the philosophers to decide which parts of which sciences are subject to Party control because they belong to the ideological superstructure, and which parts might be emancipated because they share the classless nature of language and technology.[14] (Soviet philosophers and sociologists were — and still are — extremely reluctant to discuss this problem meaningfully. Their disciplines were the first to feel the rigid rule of *partiinost'*, and the last to enjoy its relaxation.) On the practical level Stalin's vagueness had a very clearcut implication. All the little caesars in charge of various sciences were on warning: they were not to consider themselves immune to criticism. Even if they enjoyed the government's unalloyed support for many years, as the deposed chiefs of linguistics had, they might suddenly be crushed between disavowal from above and emancipated criticism from below.

Those who are inclined to see here merely another manifestation of Stalin's personal dictatorship, his discontent with intermediate bosses standing between himself and "the masses," must explain why this time he did not indulge his discontent as he had in the 1930s, by poking up Communist students and the little men of *agitprop*, inciting them to attack senior scholars. In 1950 Stalin not only removed politically imposed crackpots from control of the community of linguists; he restored genuine self-government to that community. To understand why he did such a thing, why, in fact, he declared it to be a precedent for all the sciences, one must face the terrible problem that he could not avoid: "The Arakcheev regime . . . breeds irresponsibility. . . ."[15] Stalin, the most powerful of bosses, could least escape their constant dilemma: the more power they arrogate to themselves, the less effective responsibility they can lay upon their subordinates. If a boss were a genuinely omnipotent god, he would be completely responsible for his creatures' actions; they could perform only the mechanical motions that he programmed for them.

Though Stalin did not strive for omnipotence as much as Communist and anti-Communist propagandists have often alleged, he did push the centralization of authority to such extremes as to threaten paralysis of the system. Irresponsibility (*bezotvetstvennost'*) does not describe the threat as vividly as the Soviet coinage *obezlichka* or *obezlichenie,* "depersonalization," the disappearance within a group of any person willing to make decisions. Constant denunciations of depersonalization and irresponsibility simply pointed to the problem. A solution required that higher authorities divest themselves of some power, transfer it to their subordinates. To use the Soviet lexicon, a serious attack on depersonalization (*obezlichka*) required an attack on the cult of the person (*kul't lichnosti*). It was Stalin who began this struggle in 1950 in science. His denunciation of the Arakcheev regime, rather than the cult of the person, was merely a matter of word choice. The substantial point was the principle he laid down: The obligation to discover scientific truth resides in autonomous communities of science; it cannot be evaded by reference to the higher authority of political chiefs. He was far from abrogating such authority, and he did not say how it was to be distinguished from the autonomous authority of scientists in various fields. Scientists could still be damned if they asserted their independence. But henceforth they could also be damned if they did not.

When the archives are opened we may discover whether, apart from linguistics, Stalin had in mind specific sciences that he wished to see liberated from Arakcheev regimes. Perhaps he intended to legitimize criticism of *travopol'e.*[16] Perhaps he proclaimed freedom of criticism as an exploratory gesture, to see which groups of scientists might be dissatisfied enough to take him at his word and criticize their particular Arakcheevs. The physicists were among the first to stick their necks out. In 1951 they broke a three-year silence in the face of obscurantist attacks on modern physical theories. Very likely they felt encouraged to resume the philosophical defense of their science because they had just earned the bosses' admiration by creating the first Soviet nuclear weapons. But the crucial fact, which has frequently been overlooked, is that an Arakcheev regime had never been established in physics itself, merely in the philosophical appraisal of physics. It was in this peripheral area that the physicists revived free discussion, beginning in 1951 with publication directly in the chief journal of philosophy.[17] Obviously they had the approval of the ideological establishment. In biology, where an Arakcheev regime had been fastened onto science no less than the philosophical interpretation of science, the bosses of theoretical ideology were given pause by Stalin's proclamation of free discussion, but they did not deviate from the commitment they had made in 1948. In 1951 the chief journal of philosophy sharply reduced its output of articles praising Lysenkoism, but it did not show an inclination to favor genuine biology.

Since the ideological authorities seemed to doubt that Stalin intended the right of free discussion to apply in biology, individual scientists and their organizations began to make only tiny and tentative tests of the right. Within days after the publication of Stalin's articles on linguistics, the Moscow Society of Naturalists sent to the printer a non-Lysenkoite book on plant physiology, which had been sitting on the editor's desk almost three years.[18] In 1951 the *Botanical Journal*, which was published by another learned society, began discreet agitation for a discussion of natural selection and the transformation of species. Though the *Journal's* authors did not explicitly criticize Lysenko's outlandish views on those subjects, the periodical was clearly headed in that direction.[19] But the most momentous signs of the times were subtle indications that anti-Lysenkoite feeling might be developing *within the agricultural establishment.* In October 1951, a major conference on problems of scientific agriculture virtually ignored Lysenko and his recipes.[20] Perhaps more significant was the turn of events at VIR (the leading Institute within the federation called the Lenin Academy of Agricultural Sciences). Lysenko was President of the Academy, but in 1951 the Institute's thoroughly Lysenkoite director was quietly replaced by P. M. Zhukovskii, an eminent plant breeder who had been a principal opponent of Lysenkoism during the "discussion" of 1948.[21] It seems that influential people in the agricultural establishment were beginning to look for an alternative to Lysenkoism.

Here was a strange reversal of roles, which would ultimately prove fatal to Lysenkoism. From its emergence in the early 1930s to the climactic triumph of 1948 Lysenkoism had been carried along by the support of agricultural authorities, while the ideological establishment had temporized, striving for a compromise between agrobiology and science. Now, in 1951, it was the agricultural establishment that began to show faint signs of possible sympathy with anti-Lysenkoite scientists, while the ideological authorities seemed to stand behind the commitment they had made in 1948. Indeed, Iurii Zhdanov, chief of the Central Committee's Division of Science, the very man who in 1948 had stipulated the limited right of scientists to criticize Lysenko, in November 1951 took steps to cut off the incipient criticism. In *Bol'shevik,* chief theoretical organ of the Central Committee, he published an unequivocal endorsement of Lysenko and Lepeshinskaia, explicitly praising the very doctrines that in 1948 he had declared open to criticism (the rejection of natural selection and the subversion of cytology).[22]

The renewal of biological controversy was only slightly delayed by this rallying of the ideological establishment to the Lysenkoite cause, for Lysenko was in agricultural trouble. The "Great Stalin Plan for the Transformation of Nature" was being abandoned in 1951–52. Lysenko had survived many previous surreptitious retreats from his recipes, but there was a most important difference this time. A careful appraisal was ordered of all the trees that had been planted in accordance with standard recommendations and in clusters according to Lysenko's recommendation.[23] A full report has never been published, but already in 1952 shocking figures were passing along the grapevine: Of the millions of trees planted in Lysenko's clusters, more than half had died. By 1956 only 15 percent were still alive, only 4.3 percent flourishing.[24] More astonishing than the figures was the fact that agricultural officials were seeking reliable figures. They were beginning to check up on Lysenko's recipes, and the immediate result was the withdrawal of official support. In April 1952 the Ministry of Agriculture published extensive new directions on afforestation, spelling out in great detail the alternatives to Lysenko's cluster method of planting, and stressing the right and duty of local specialists to choose the most suitable method.[25] Lysenko was not publicly rebuked. On the contrary, he was given space in the main agricultural newspaper to boast once again of the merits of planting trees in clusters and leaving them to thin themselves.[26] But the boasting was obviously empty, now that the Ministry of Agriculture had withdrawn its support.

When the archives are opened it should be possible to discover which agricultural officials at what levels of authority were the prime movers in this turn toward realistic calculation. It may even be possible to discover what their motives were, to what extent they were trying to satisfy disappointment with Lysenko at the highest political level, to what extent they were trying to provoke it. The public record contains only gross indications of the turn within the agricultural establishment; it does not show who initiated it and pushed it along. In March 1952, while the Ministry of Agriculture was preparing its new directive on tree planting, the Central Committee of the Party shifted its public stance. The periodic endorsements of Lysenkoism that had been appearing in *Bol'shevik* stopped.[27] In October, George Malenkov, reading the Central Committee's report to the Nineteenth Party Congress, sounded strangely discordant notes. In the agricultural part of the lengthy report he repeated the complaint that had proved disastrous to genuine science back in the 1930s, but now Lysenko's school was responsible: Though "all anti-scientific, reactionary ideas have been exposed and destroyed in agricultural science, and it is developing now on the only correct, materialist, Michurinist basis, . . . nevertheless it is still lagging behind the requirements of production on the collective and state farms." [28] In the ideological part of his report Malenkov seemed to reaffirm faith in Lysenkoism; he endorsed the obscurantist positions that had been taken in the scientific "discussions" of the late 1940s, explicitly including biology. But then he cast doubt on the endorsement by repeating Stalin's objection to "the monopoly of various groups of scientists, who bar the way to fresh and growing forces, fence themselves off from criticism, and try to settle scientific questions by administrative fiat." [29] Like his master, Malenkov refrained from specifying which groups he had in mind. Scientists who wanted freedom would have to bell their own monopolistic cats.

Confused and indistinct as these signals may seem to an outsider, the behavior of insiders shows that they understood. The incipient critics of Lysenkoism were being told to proceed at their own risk, which would not be great if they observed two restrictions. Detailed public comment on the practical failings of Soviet agriculture was not allowed, and the theoretical decision reached in 1948 must not be called in question. Without these restrictions "disorganization of the cadres" might set in, that is, subordinates might start criticizing their superiors. With these restrictions, it may seem to the outsider that debate was simply impossible. Lysenko's critics were, in fact, being turned loose on two parts of his doctrine: the denial of natural selection within a species, and the claims of sudden transformations from one species to another. It was certainly a torment to criticize the denial of intraspecific competition without citing the failure of the trees planted in clusters, or the sudden transformations without using the genetical concepts that were banned in 1948. The resulting contortions of Lysenko's critics were sometimes distressing to observe. But to see only the restrictions and the contortions is to miss the momentous change that occurred in 1952. The issues that Iurii Zhdanov had declared still open for discussion in 1948 were now open in fact.[30] From August 1948 until 1952 no one had criticized Lysenko in print, and now, with the sanction of the highest authorities, such criticism was being published.

Nor was this a simple reprise of previous "discussions." This time it was not scientists that had provoked official disappointment. It was Lysenko, who had justified the planting of all those dead and dying trees by denying the theory of natural selection within a species, asserting that oak seedlings cooperate rather than compete. This would now be debated — the theory, that is, not the tree-planting campaign. As in the past, controversy over natural science was the authorized substitute for intolerable public discussion of agricultural trouble, but this time scientists were on the offensive. It was the Lysenkoites' turn to invoke the unreserved endorsements that they had formerly received from the highest authorities.[31] In vain. It no longer mattered that Lysenko's basic article on species formation had appeared in *Pravda* as recently as 1950, and was already enshrined in schoolbooks and encyclopedias.[32] Consistency of politicians was not the issue; bosses have a right to change their minds without apology or explanation. Now it was the Lysenkoites' turn to be shocked at sudden apostacy within their ranks. Toward the end of 1952 N. V. Turbin, who had been dean of biology at Leningrad University since 1948 and one of the most virulent Lysenkoites, published one of the first major criticisms of Lysenko's theory of species formation.[33] It would be unfair to say that rats were leaving a sinking ship. No one could be sure in 1952 that the ship was sinking, and most Lysenkoites were not leaving. Turbin was farsighted.

The most significant innovation in 1952 was probably unintended. For the first time since 1929 Lysenkoism was involved in a genuine discussion rather than a "discussion"; the end was not preordained. There were still prohibitions on what might be said, but there were no prescriptions of what must be said. No one with unquestionable authority was trying to impose on the community of biologists an official view of species formation. When the Lysenkoites responded to their critics with savage name calling and demands for suppression, the critics invoked Stalin's words on freedom of criticism, and the Lysenkoites were obliged to calm down.[34] They were as usual quite unable to engage in meaningful scientific discussion, and they no longer had the political authority to shut it off. Thus, the scientists came quickly to dominate the discussion of species formation, raising ever more insistently the question whether political authority would continue to maintain the ban on genetics. Everything depended on the ability of the Lysenkoites to keep their reputation as masters of practical agriculture. Lysenko had come up with two new recipes: increasing the butterfat content of milk by suitable breeding and "training," and stretching mineral fertilizer by composting it with organic wastes and earth.[35] Agricultural publications praised these recipes, but the authorities did not launch a campaign for either one. Indeed, in February 1953 the chief agricultural newspaper allowed a book reviewer to attack the fertilizer scheme.[36]

UNDER KHRUSHCHEV

After Stalin's death in March 1953, the attack on Lysenko's theory of species formation grew sharper and stronger. Many more biologists joined in, and some began to charge deliberate fraud — and prove it, especially in Lysenkoite reports of sudden transformations. A hazelnut limb growing out of a hornbeam was the most notorious case of intentional deception; the rye seeds that grew on wheat plants seem to have been a product of self-deception.[37] By 1955 the editors of the *Botanical Journal,* which continued to be the chief forum for anti-Lysenkoite authors, declared the discussion exhausted, and plainly stated what many of the participants had been hinting at: The time had come to reopen the discussion of genetics.[38] This was a public challenge to the ideological establishment, which, far from reconsidering the ban on genetics, had rallied to the defense of the Lysenkoite theory of species formation.[39] At the same time pressures for freedom of discussion were mounting in other fields blighted by agrobiology — cytology, plant physiology, breeding, and soil science. Indeed, far beyond biology and agricultural science the ideological authorities were besieged on all sides by demands for freedom of expression, in the arts no less than the sciences.

One almost feels sympathy for the ideological authorities. They were trapped in positions to which they had committed themselves when the supreme political bosses were pursuing the mirage of omnipotent *partiinost'*. Now that the bosses were turning away from this self-defeating quest, appealing to *specialists* to tell *them* what was true and useful, the ideological authorities were in a box. To give up the positions they had taken in the various arts and sciences seemed tantamount to giving up any claim of authority over the intelligentsia. Defense of the Party's "leading role" seemed to require defense of such idiocies as rye seeds growing on wheat plants or cuckoos hatching out of warblers' eggs.[40] The supreme bosses were involved in those dilemmas of their ideological bureaucrats, but they were also involved in more urgent problems — the stagnant condition of agriculture, to take a notable example. So they called upon specialists to break free of "bootlicking and servility," to make agricultural science genuinely useful by "creative discussions and free exchange of opinions."[41] If such discussions threatened "the Party's leading role," not just the authority of the ideological bureacracy, they could be stopped. But it worked the other way too. If the biggest chiefs decided that agrobiology was useless, it would be thrown to the scientific wolves. Stalin had set the example. First, he allowed his ideological bureaucrats to raise a school of linguistics above criticism, and then, changing his mind, he forced the bureaucrats to eat crow. That is, after all, one of the main functions of second-rank authorities in a political system, to eat crow when the first rank changes its mind.

It took Stalin's successors eleven years to change their minds completely about the usefulness of agrobiology. A complex struggle effected the change, mostly in private, erupting only occasionally into the public record. But these occasional eruptions are enough to reveal the large trends:

1. By a series of local conflicts many institutions of higher learning were torn from Lysenkoite domination.
2. The ideological officials tried to maintain the uniform rule of a slightly reformed Lysenkoism, but they were obliged to overlook more and more autonomous diversity.
3. The struggle involved not only biologists and agricultural specialists; liberal editors and writers and allies from other scientific disciplines, such as physics, struck occasional blows at Lysenkoism.
4. In the secondary schools and other mass media the uniform rule of Lysenkoism was successfully maintained, until the highest bosses explicitly withdrew their support from it.
5. Agricultural publications kept up a deceptively steady support of Lysenkoism while a covert struggle was proceeding within

the agricultural establishment, steadily sapping Lysenkoite strength.

6. These interacting trends had repercussions at the highest level of politics, leading to the final crash of agrobiology in the fall of 1964.

While Stalin was still alive, Turbin, dean of biology at Leningrad University, defected from the Lysenkoite cause. At first he tried to restrict his defection to the issue of species formation, but within a few years he made the complete move to genetics. Why he also moved physically, from Leningrad University to the Belorussian Academy of Sciences in Minsk, is not revealed in the public record, which does show, however, that his move hastened the resurgence of genuine biology at both institutions. At Leningrad University the chairmanship of a revived Department of Genetics was restored to Lobashev, who had lost the post in 1948, when Turbin replaced him as dean of biology. In the 1950s, in Minsk, the renegade Turbin built a major center of corn genetics.[42] The revival of genetics at Leningrad University should have been checked, one would think, by the Minister of Higher Education of the RSFSR, for he was V. N. Stoletov, one of the original and most virulent Lysenkoites.[43] He did succeed for a long time in checking the revival of genetics at Moscow University, where he continued to be chairman of the Department of Genetics at the same time that he was Minister of Higher Education. But already in the mid-1950s he was unable to suppress genetics at Leningrad University. By the early 1960s Lobashev was able to start a genetics journal and publish a genetics textbook, the first since 1938.[44] The Lysenkoites fumed, but were unable to suppress the publications or the department from which they issued.[45] Perhaps Minister Stoletov attempted suppression and was blocked; perhaps he held his hand, quietly waiting to see which way the highest authorities might move.

Within the Academy of Sciences of the USSR the resurgence of genuine biology was symbolized in 1955 by a change of officers. The biochemist Oparin, who had been serving the Lysenkoite cause as chief — Academic Secretary, to use the proper title — of the Biology Division, gave way to the biochemist Engelhardt, a friend of genetics.[46] Soon afterward the Academy's main periodical in the field of biology was purged of its Lysenkoite editors and restored to genuine scientists.[47] Those were outward signs of much internal conflict. Genuine scientists were freed for combat in 1954, when the highest political authorities endorsed the principle that Academician Lysenko could no longer force his colleagues to sign their approval of dissertations that they actually found unworthy of approval.[48] (Stoletov's dissertation, a massive compilation of nonsense, was not at issue; it had been approved in the good old days.)[49] But this emancipation did not settle the question of who would sit in judgment of which dissertations and, even more basic, who would be in charge of what laboratories and institutes. Efforts to reestablish genetics laboratories excited particular contention, ultimately involving Khrushchev himself. The geneticist Dubinin, who had been bird-watching at a forest station since 1948,[50] returned to Moscow in 1955 and revived the project for a new Institute of Genetics, to compete with the old one that Lysenko had taken in 1940. Dubinin succeeded, but not in the central Moscow complex of the Academy. His new Institute was created in Novosibirsk, where a great new branch of the Academy was growing. By 1960 he was forced out of it, which nevertheless continued to flourish, while he took shelter back in Moscow, in a genetics laboratory within the Institute of Biological Physics.[51]

Other geneticists found equally unlikely places to do their work: in the Institute of Physical Chemistry, within the Academy of Medical Sciences, and at the Botanical Garden, whose director, Tsitsin, Lysenko's onetime rival for the leadership of agrobiology, was completing his long drawn-out defection to genuine science.[52] All over the country analogous compromises were being arranged, and impasses created, as Lysenkoites struggled to beat back the resurgent tide of genetics, cytology, plant physiology, and soil science. In Kiev a genetics laboratory was established by a scientist who truckled to the Lysenkoites, but the teaching of genetics at Kiev University was not revived.[53] At the Ural branch of the Academy of Sciences genuine science seems to have prevailed; in Armenia, Lysenkoism; in Kharkov, genuine science. Important help came to genuine science from foreign countries, including Communist ones. The Lysenkoites were increasingly hard pressed to produce any significant evidence of support outside the Soviet Union — the day had passed when they boasted of their nativist isolation — while the scientists could call to witness the president of the East German Academy of Agricultural Sciences. Since 1949 he had been checking on Lysenko's claims of graft hybrids in tomatoes, which he had found utterly illusory.[54] In general the trend within institutions of pure science and higher education seems to have been increasingly favorable to science. By early 1964 a Lysenkoite was complaining to the Party's Central Committee that Lobashev's textbook of genetics was being used "not only in Leningrad University, but in many biological faculties and in institutes and in teachers' colleges." The Central Committee applauded his demand that Minister Stoletov "should give serious attention to the teaching of biology in higher education."[55] A cynical outsider wonders at the insinuation. Was Stoletov getting ready to switch? And the Central Committee's applause — how sincere was it? They issued no directive to Stoletov.

Only the archives may reveal which people in the hierarchy of science, education, and ideology were taking what positions in this complex struggle. The favored position, very likely, was some kind of evasion, following the example of the man who, we may assume, stood at the peak of the hierarchy, M. A. Suslov. He avoided public comment, while his little men in charge of philosophy followed a modified Lysenkoite line.[56] The obvious inference, that Suslov did too, was greatly strengthened when he accompanied Khrushchev on a well-publicized, admiring visit to Lysenko's experimental farm.[57] But Suslov never came right out and told the world exactly how he appraised the dissension in biology. Similar silence emanated from the Central Committee's Division of Science, though it was headed, during most of the Khrushchev decade, by a genuine scientist. (His field was physical science, but the personal qualities of the bureaucrat mattered less than the institutional setting; in 1965 his place would be taken by a pseudo-historian, who would administer the reestablishment of genuine biology.)[58] As long as they could, militant biologists chose to interpret high-placed silence as tacit permission for increasingly bold attacks on Lysenkoism. They began, in fact, to discuss the agricultural record of Lysenkoism, at which point the bosses cried halt.

In December 1958 an editorial in *Pravda* called for the dismissal of the editors of the *Botanical Journal,* the main organ of militant biologists. *Pravda* insisted that the right of free discussion was not being abrogated. The editors of the *Journal* had abused that right. They had published articles which ignored, distorted, or even denied completely Lysenko's great contributions to Soviet agriculture. For example, one article had recently estimated that Lysenko's cluster method of planting trees had caused a loss of a billion rubles — just after the Soviet government had awarded Lysenko another Order of Lenin in recognition of his great services to agriculture. By casting doubt on the government's practical wisdom, the editors of the *Botanical Journal* had broken the rules of "etiquette and good order" (*vezhlivost' i dobroporiadochnost'*) in discussion. Thus they threatened confusion and disorganization in the great task of mobilizing scientific forces for reaching the goals set by Party and state.[59] Khrushchev endorsed this outburst, and demanded that the editors of the *Journal* be replaced by "Michurinists." [60] They were fired — he was the boss — but it is noteworthy that their replacements did not campaign vigorously for Lysenkoism. They turned the *Botanical Journal* into another of the many biological periodicals that tried to keep out of trouble by avoiding controversial issues. Along with most other specialists in affected fields, the new editors seem to have been looking for a safe position, perhaps a spot from which they might jump to either side in case of unexpected developments.

For five years, 1959 to 1964, public debate was held at a minimum, thereby strengthening the pretense of the ideological officials that differences had been resolved and unity achieved in support of progressive "Michurinism." On the surface it seemed that the Lysenkoite monopoly of 1948–1951 was revived: Lysenkoites were free to criticize science, while scientists were not free to criticize Lysenkoism. But the Lysenkoites were comparatively restrained in their attacks on science, and occasional scientists found outlets for circumspect criticism of Lysenkoism, reminding the world of the continuing covert struggle. It mattered little that official support of "Michurinism," as Lysenko preferred to call his doctrine, was squeezed into the final version of the new Party Program of 1961.[61] The endorsement was brief and vague. Subsequently the Central Committee spelled out its views at much greater length, in a decree "Concerning Measures for the Further Development of Biological Science and the Strengthening of Its Connection with Practice." [62] Nowhere was the evasive skill of the bureaucrat and politician more in evidence than in this long document, which will probably prove to be the Central Committee's last effort to decree truth to natural scientists. The exhaustion of the effort was evident not only in paragraphs of platitudinous twaddle but also in the extremely brief definition of "Michurinism," which was so vague that it could be interpreted as opening the door to genuine biology.[63] Indeed, the decree called for further development of biochemistry, biophysics, and genetics. What kind of genetics the Central Committee had in mind was indicated by simultaneous publication in the central press of a long article by Lysenko, reiterating his fundamental confusions about heredity and its alterations.[64] Real geneticists could take heart from the fact that these confusions were not included in the decree itself.

Perhaps the political and ideological authorities were not simply trying to patch together an expedient compromise without regard to the basic issues. They may have genuinely believed that agrobiology could be reconciled with world science, to the mutual enrichment of both. They were sufficiently ignorant and anti-intellectual to believe this. Khrushchev said plainly that he did not understand the scientific issues, which should in any case be settled "in the fields." [65] The only high-placed official who was bold enough to discuss the scientific issues in public was L. F. Il'ichev, second in command to Suslov, the presumed chief of the cultural front. Maybe Il'ichev was not bold; maybe he was just sticking his neck out on behalf of his chief, as second rankers are supposed to do. It has even been suggested that he was being used to strike at Suslov, his chief competitor. In any case, he tried to sound like a man in the middle. He censured die-hards on both sides, that is, people who utterly denied any merit in the work of their opponents. But he concentrated his fire on the anti-Lysenkoites. They were trying, Il'ichev complained, "to renew an unnecessary discussion, to distract scientific forces from the solution of the problems that are important for science, for the national economy." [66]

The official formula for reconciliation and a united move ahead came to this: The position taken in 1948 was essentially correct; genetics was still banned. There had been mistaken excesses, but now they were happily corrected. Soviet biologists were once again free to learn from the work of foreigners (though the effort to translate work in genetics provoked constant quarrels, and Soviet geneticists were not allowed to travel abroad); cytology was being revived (though it could not be reconciled with Lysenkoism any more than it could be separated from genetics); and Soviet biochemists were permitted to study nucleic acids (though this study also was inseparable from genetics and incompatible with Lysenkoism). The parenthetical discrepancies did not disturb the official belief that a sensible way had been found to keep agrobiology supreme while acquiring such practical marvels as the selective weed killers and the polyploid sugar beets seen in the West by Soviet farm delegations.[67] Il'ichev and his colleagues at the top of the Soviet hierarchy seemed completely blind to the dependence of such wonders on "bourgeois" genetics and plant physiology. They supported Lysenko's abortive "stage theory," though it was essentially incompatible with the biochemical analysis of plant development, which produces the selective weed killers. They supported his insistence that genes are a metaphysical fiction, though they allowed study of nucleic acids, and vigorously promoted a modern program of corn hybridization. It is as if one granted the digestive significance of the alimentary canal and promoted the development of internal medicine, while denying the existence of the stomach and intestines. Only the centralized bureaucracy that ran the country's secondary schools could be counted on to support such nonsense without constant quarrels and evasions. In the Soviet Union as in the United States the educational bureaucracy is submissive to political authority, domineering to teachers and pupils.[68]

Perhaps the strangest situation was to be found in agricultural research and publication. Logically impossible combinations had long been accumulating in this area: chimeras, as we may call scientific works encased within Lysenkoite introductions and conclusions, and monsters, as we may call Lysenkoite works with scientific appendages. Lysenko and his followers were sufficiently indifferent to the requirements of logic to be tolerant of such things, when circumstances required toleration. Inbreeding may serve as an example. The geneticists' theoretical analysis of it was something the Lysenkoites could never tolerate, unless they were to quit being Lysenkoites. But the use of inbreeding among plant breeders was a different matter. It was totally suppressed only at the acme of Lysenkoite power, between 1948 and 1952.[69] Before and afterward the Lysenkoites contented themselves with railing at the technique. And following 1955, when Khrushchev endorsed a modern corn program, which requires inbreeding, the Lysenkoites blandly began to approve the method they had formerly condemned. They reserved their indignation only for those who dared to recall the past or to give the geneticists' analysis of inbreeding.[70] In their final decade of apparent power over agricultural research the Lysenkoites learned to look the other way more and more, as specialists used more and more methods that implicitly clashed with Lysenkoite doctrine. Even verbal efforts to reconcile method and doctrine began to decline. If the agricultural specialist refrained from making an explicit point of his divergence from Lysenkoite doctrine, he would be tolerated. He might even be invited to take part in Lysenkoite symposia.[71] Indeed, when one notices that Kushner, one of Lysenko's closest colleagues, began to write about stockbreeding with hardly a trace of Lysenkoism, assuming genetics as a matter of course,[72] one begins to wonder whether agrobiology was not undergoing a spontaneous transformation from pseudoscience to the real thing.

In fact it was not. Individuals were deserting agrobiology, and some of them, like Kushner, knew how to return to the genuine science they had knowingly deserted when Lysenkoism was on the rise. But such men were a dwindling breed, especially in the applied fields, for a very simple reason: genuine scientists did not control agricultural education; Lysenkoites, ignoramuses, and timeservers did. Thus confusion and ignorance were increasingly the rule. From the late 1930s the basic textbook of agricultural experimentation and plant breeding had been a thoroughly inconsistent jumble of Lysenkoism and science.[73] With each new edition the section devoted to statistical methods dwindled into a more and more incongruous irrelevance. The chief editor of the textbook was a distinguished old plant breeder who probably knew what his book refrained from telling the students, or rather, what his book told in one place and inconsistently subverted in a dozen others. But a growing number of certified specialists, trained with this textbook, did not know the difference between scientific method and a crude travesty of it. The textbook's shrinking section on statistical method was introduced, in each edition, by a quotation from Lenin: "Only after the essence of these forms is clarified . . . does it make sense to illustrate the development of this or that form by means of statistical data. . . ." [74]

No matter whether Lenin really meant that statistical reasoning can only illustrate preconceived notions. That is clearly what he was taken to mean within the context of this Lysenkoite textbook, thereby strengthening Lysenkoite contempt for "mathematical fetishism in agronomic science." [75] A practical man could tell success from failure at a glance; a few figures were useful only to illustrate what his intuition told him. With growing numbers of specialists trained in this spirit, agrobiologists might drift away from Lysenko, but they could hardly change spontaneously into scientists. Increasing toleration of methods borrowed from genuine science merely heightened the confusion, for the oncoming generations of specialists understood neither the scientific bases of such methods nor the proper procedures in using them.

By the early 1960s centers of pure science were escaping from Lysenkoism, but the campaign for scientific agriculture seemed hopelessly entangled in thickets of ignorance, endlessly propagating itself. The men who ran the government seemed invincibly convinced that they could enjoy the practical benefits simultaneously of science and of pseudoscience, with predominance guaranteed to the latter. It seemed almost ominous that they no longer suppressed dissenters altogether; the truth, once too dangerous to be uttered, seemed impotent. Without apparent effect an eminent chemist or physicist occasionally published criticism of agrobiology and a plea for genuine science — once a geneticist and a biochemist did so — and an occasional liberal writer added his voice.[76] A few writers used fiction to publish truths that could only be hinted at in plain exposition. " 'They think,' " said a fictional biologist about fictional Lysenkoites,

> "that a living organism is the same kind of plastic as children use to mold any figure they like. . . . [They don't understand] that we are groping our way only on the first little paths toward the transformation of the nature of plants. I won't even speak of all sorts of falsifications, which are released for careerist purposes. In general that has no connection with science, that is crime. . . ."[77]

A feuilletonist, who explicitly exempted Lysenko from such charges and showed scant sympathy for genetics,[78] can hardly be suspected of partisan exaggeration in painting a bleak picture of agricultural research and education, turning out incompetent bureaucrats who showered stupid orders on hapless farm directors. To be sure, the feuilletonist thought he saw a way out of the mess: protest. Against the weary skepticism of a fictional farm chairman, a scientist hero exclaims: " 'But I will protest. I will try. And I do not think you are right. Answer back, protest, write, complain. Otherwise you are not Communists.' "[79] Outside of fiction the printed record reveals very few protesters against agrobiology. Moreover, the few real protesters were mostly "Varangians," as a journalist named the physical scientists who ventured into the foreign territory of biology in defense of science.[80] Agricultural specialists, if we can trust their publications as a complete expression of their views, were making little or no effort to combat the official doctrine that agrobiology and science were compatible.[81]

Khrushchev appealed, already in 1953, for an end to "bootlicking and servility" among agricultural specialists,[82] but almost none were willing to stick their necks out so far as to publish the plain truth about agrobiology. Even when the *Botanical Journal* and the Moscow Society of Naturalists were leading a spirited attack on Lysenkoism — that is, before the crackdown at the end of 1958 — the attack was mostly confined to pure science and conducted by its adepts. It was unusual for a forestry specialist to recall the waste resulting from Lysenko's cluster method of tree planting,[83] and it was quite exceptional for a polemicist to generalize about the practical origins of agrobiology:

> T. D. Lysenko's doctrine has its source and seeks its support in the past, in the moribund survival, here and there, in our otherwise progressive agriculture, of a negligent attitude toward labor, of low agrotechnics, of impure seed, and of much infestation with weeds.[84]

Most specialists did not have to be told by *Pravda* that such comments were violations of "etiquette and good order."

Khrushchev made this abundantly clear in his frequent speeches to farm officials and specialists. With evident sincerity he solicited honest advice from them, but with equal sincerity he revealed his passionate conviction that he, the chief of "practice," understood the needs of agriculture better than any narrow specialist. He would urge a gathering of specialists to argue freely and disdain to hide behind official formulas,[85] and a bare two days later, speaking at another gathering, he would demand the firing of those who had sent in a letter questioning his rule that corn must become a universal fodder crop:

> People like this could be ignored, they hardly deserve attention. But the trouble is that the education of students is entrusted to them. Imagine how they are educating the future specialists of agriculture. That is intolerable. We cannot permit scientists of this type to pollute the brains of those who will replace us.[86]

This plangent self-assurance among the chiefs of "practice" was the greatest obstacle to the victory of science over agrobiology. They knew what agriculture needed, and they wanted specialists to confirm, amplify, and propagandize their knowledge. At the same time they demanded honest advice from the specialists.

Those who doubt that such a domineering attitude could be sincerely joined with a desire for honest, independent advice, should examine a talk that Khrushchev gave to a meeting of specialists at the end of 1961. "Should I," he asked, and published the question in *Pravda,*

> be the highest authority in scientific problems of agriculture? . . . I can make mistakes, and you, if you are an honest scientist, should say: "Comrade Khrushchev, you don't have a completely correct understanding of this question." If you explain the correct understanding to me I will thank you for that. Let us assume that I have made a mistake. But then you say, Khrushchev has spoken, and I support him. What kind of a scientist is that, comrades? That is bootlicking, timeserving. (*Applause.*) Timeserving in science is intolerable; in general bootlicking is intolerable, and for a scientist it is death. (*Applause.*)[87]

In context, this was far from the self-denying ordinance that it appears to be. Khrushchev was commenting on the effort of P. A. Vlasiuk, president of the Ukrainian Academy of Agricultural Sciences, to explain why he and his colleagues had been preaching *travopol'e,* the grassland system of crop rotation, which Khrushchev had now come to reject *in toto.* Vlasiuk confessed that he had not been sincere when he supported *travopol'e,* but he protested that the highest organs of Party and state had given him and his colleagues no choice. Thereupon Khrushchev sharply rebuked him for trying to transfer his personal responsibility to the Party chiefs, who could err, or to the Party itself, which must never be blamed for the mistakes of its members, not even when the mistakes are included in the most authoritative decrees of the highest Party bodies. He held up the sacred example of Tulaikov, who, Khrushchev implied, had paid with his life for opposing *travopol'e.* "Evidently," he told the live specialist who had supported *travopol'e,* "you don't understand what the Party is, and you have a poor understanding of a Party member's obligations."[88]

Cynical laughter will not dismiss Khrushchev's version of a universal dilemma. It goes without saying that his version was peculiarly Communist, or even Soviet Russian. Where else would a chief of state assume that a specialist had died in a prison camp for opposing official policy on crop rotation? It is beside the present point whether the assumption was correct; Khrushchev, like other worshipers of power, may have been crediting the terror with more rationality than it deserved. But the peculiarly Soviet Russian features of the dilemma were ebbing. Everyone was agreed, for example, that shooting or jailing was no way to deal with a specialist who criticizes his superiors on matters like crop rotation. A handbook for legal officials, *Especially Dangerous State Crimes,* fumbling with the distinction between criminal attacks on the state and legitimate criticism of its officials, cited disapproval of the virgin lands program as an example of permissible criticism, and objection to the suppression of the Hungarian rising as the contrasting example of punishable crime.[89] The Soviet leadership was groping for the kind of distinction that Catholic leaders make between the church militant, which is subject to criticism because it consists of imperfect people, and the church triumphant, which is absolutely perfect, as we will all see if we live long enough.

Perhaps, since no Soviet leader after Bukharin has shown an interest in metaphysics, it would be more appropriate to seek an analogy in the law. The Communist leaders, one may say, have been groping for the kind of distinction that lawyers make between the authority of the state, which cannot be questioned lest sovereignty be impugned, and the deeds of the state's imperfect agents, which can be challenged because these are fallible men who may act *ultra vires.* But no matter what analogies are most appropriate. One should not overlook the reality and seriousness of Khrushchev's problem just because he was self-serving and ridiculously inconsistent in struggling with it. He urged specialists to help him purge agrobiology of its harmful stupidities, such as *travopol'e,* and at the same time he invoked political authority to keep them from telling him that the whole of agrobiology was stupid. "To whom would that be useful? Only to our enemies. I am for criticism, comrades. But criticism must be done in a business-like, principled way; it should be directed against concrete mistakes and shortcomings."[90] However clumsy and inadequate, this was part of the ongoing effort to define the relative powers of scientists and politicians. And that was itself part of a much larger problem: To keep political power from defeating itself it had to be rationally defined, that is, confined, restricted, limited — by Stalin's pupils.

They had been trained to think of themselves as supremely practical men, which meant that they were inclined to brush aside the fact that pure scientists were hostile to agrobiology, as they brushed aside the dissatisfaction of artists with the official school of socialist realism. In neither case would they allow the intelligentsia to tell them, the chiefs of "practice," what the country really needed. They had to learn for themselves, the hard way, as practical men usually do, for their most basic ideological beliefs seem the most obvious common sense and are therefore largely impervious to rational examination.

Stalin's associates must have been changing their intuitive apprehension of agricultural reality while their master was still alive, for they began an impressive series of reforms within months after his death.[91] They said they were correcting his violations of the principle of *zainteresovannost'*, having a material interest in one's work. It would be more accurate to say that they were breaking away from the understanding of this principle which they had shared with Stalin. It was a variant of the traditional attitude toward peasants, similar to Malthus' basic assumption about lower-class people in general: They are dull sluggards who produce only as much as dire need obliges them to. The less that is squeezed out of such people, the less they produce. And conversely, the more they are exploited, the more they produce, up to a limit of physical endurance. Of course, Stalin and his associates never expressed this assumption in public as bluntly as Malthus did in the first edition of his famous *Essay*. In private they seem to have expressed such feelings rather freely, of course without using such words as exploitation or acknowledging their similarity to traditional rulers of peasants and to Malthus.[92] (Even he put a liberal fuzz on his views in later editions of the *Essay*.[93]) But it mattered little what words were used, in private or in public. The vital issue was the agricultural policy and the self-fulfilling prophecy on which it was based. When rulers assume that their subjects are dull sluggards, and treat them that way, the subjects tend to behave that way.[94] Stalin's death eased the process of self-correction. His associates blamed him for the effort to increase the productivity of peasants by forceful exploitation and repression of spontaneity. They began to try and raise productivity by raising the prices paid to peasants, by relaxing the obligations imposed on them, and by increasing the scope for self-direction in the operation of each farm.

Many difficulties were involved in this new approach, not the least being a serious financial shortage. The peasants' immediate gain was the state's immediate loss. Mutual gain could be expected only in the long run. To avoid the immediate shortage Khrushchev and his colleagues decided to launch another campaign, this one to produce cheap grain quickly, by rounding up volunteers to plow the semiarid plains of Siberia and Kazakhstan. Thus, in very act of reforming the Stalinist agricultural system, the Communist leaders used one of Stalin's favorite methods for obtaining a lot of grain in a great hurry with a little outlay: Mobilize tractors and Communists and expand the acreage under cultivation. (In fact, the expansion of the mid-1950s was more spectacular than the one that had accompanied collectivization.)[95] The anomaly deepens as one examines Khrushchev's reasoning in support of the virgin lands program. He recognized that the agricultural problem would not be solved by pushing inefficient production on to more and more marginal hectares. He knew that the only genuine solution was to improve efficiency and raise yields on good land that had a dependable supply of moisture. But when he sent the virgin lands proposal to the Politburo early in 1954, he did not present it as a stopgap, a risky way of getting the state past the initial costs of encouraging productivity on good land. Quite in the Stalinist manner he pictured this new campaign in purely rosy colors. And to prove that respectable stable yields could be expected from the virgin lands, he appended a "scientific" memorandum by Lysenko.[96]

Maybe Khrushchev looked to Lysenko for expert advice simply because Lysenko was president of the Lenin Academy, the formally designated chief of agricultural science. Maybe his reliance on Lysenko was a continuation of a habit he had picked up in the 1930s, when he succeeded Postyshev as chief of the Ukraine and inherited Lysenko, who had been Postyshev's major agricultural advisor.[97] Maybe Khrushchev turned to Lysenko because no genuine specialist on land use would swear that high reliable yields were to be expected from the virgin lands. The first two possibilities can hardly have been as important as the third, but whatever the cause, or combination of causes, the effect was a close relationship between the two men.

For a short time in 1954 it showed signs of strain. Khrushchev took the side of Lysenko's critics in their attack on V. S. Dmitriev, the man who had been in charge of the agricultural section of Gosplan (the State Planning Commission) in 1948, when the Great Stalin Plan was promulgated. When the Plan was abandoned, Dmitriev was dismissed. Khrushchev thought he should have been sent to work on a farm. (For all his talk about getting the ablest people to work on the farms, Khrushchev could not shake the common Soviet habit of regarding farm assignments as suitable punishment for incompetents.) But Lysenko helped Dmitriev into a research position and graduate study at the Academy of Sciences, where he became the center of a public scandal in 1954. His Lysenkoite dissertation on the transformation of species was rejected by his doctoral committee, until Lysenko's vituperation forced reconsideration and won Dmitriev his degree. The scientists protested, and Khrushchev took their side, mainly because he regarded Dmitriev as an egregious example of the *travopol'shchiki* who wanted to replace intensive crops by grass.[98] The scientists regarded the Dmitriev case as an egregious example of the Arakcheev regime in science, and that is the way it was featured in the central press during the first half of 1954.[99] But at the same time that the central press was slapping Lysenko's wrist for his dictatorial behavior toward scientific opponents, it was also applauding him as the Central Committee's chief advisor on the virgin lands campaign.[100] By the end of 1954 Khrushchev had forgiven or forgotten Lysenko's transgressions both against intensive crop schemes and against freedom of scientific criticism. (In fact, Khrushchev's public comments leave room to doubt whether he ever took the second charge very seriously.) For the next ten years, until his sudden dismissal in October 1964, he gave periodic public testimonials of faith in Lysenko's wisdom as a practical master of agricultural science.[101]

Khrushchev was a politician, which means that it is hard to distinguish between what he said he believed and what he really believed. His repeated tributes to Lysenko may have been tributes to the opinion of his colleagues in the Politburo. He may have had no more inkling that they were changing their minds about Lysenko than that they were changing their minds about himself. His published papers include a tantalizing memorandum to the Politburo (August 4, 1962), disputing "the arguments that we are backward in agronomic biology. They are untrue arguments, spread by people who want to defame everything Soviet." [102] Were such people in the Politburo, or was Khrushchev disputing arguments heard only from outsiders? We will not know until the archives are opened. This may have been one of the issues on which Khrushchev and his colleagues nodded smiling agreement with each other, while one or another was privately wondering when and to whom he might safely broach his disillusionment. I feel that Khrushchev was fairly sincere in his repeated endorsements of Lysenko; they were frequent, and they were often spontaneous. In that case his colleagues, especially Dmitrii Polianskii, the chief of agriculture, must be accused of hypocrisy for echoing the boss' words in public while privately disbelieving them.[103]

Whoever was playing the hypocrite at that exalted level — one cannot even rule out the wild possibility that they were all deferring to the Lysenkoite faith in public while they were all abandoning it in their private thoughts — there was no outward hint of disagreement until October 1964, when Khrushchev's colleagues sprang on him and threw him out. Yet all the time that the highest body of rulers were maintaining a public show of perfect unity, the officials just below them felt less and less constrained to show deference. There was a growing rift between the Politburo and the Ministry of Agriculture during the Khrushchev decade, which undoubtedly put increasing strain on the unity of the Politburo. Agricultural specialists, who were not allowed to publish criticism of the Lysenkoite faith, were allowed to subvert it in a quiet stubborn way. Their intermediate bosses — various ministers of agriculture — increasingly took their side in opposition to the ultimate boss or body of bosses at the top. The bureaucratic machine showed a life of its own, thereby helping or causing the supreme boss' colleagues to do the same.

AGAINST KHRUSHCHEV

Speaking to a group of agricultural officials in June 1955, Khrushchev noted the shortage of fertilizer, and urged the widespread use of Lysenko's method for stretching mineral fertilizers by composting them with organic wastes and earth. This method had given good results in many farms of the Moscow district, he said, "but there are still people who are skeptical of . . . [it]. It's time, it seems to me, to move from talk to action. This is a big question, and it should be settled as a matter of state [*po-gosudarstvennomu*]." [104] Not only did the agricultural officials fail to respond. In January 1957, they quietly accepted, from the directors of soil science and chemistry at the Lenin Academy, a resolution condemning Lysenko's recipe.[105] In April, Khrushchev publicly rebuked the specialists in the course of another speech to agricultural officials, and went on to rebuke their superiors:

> I am amazed at the calm of Minister of Agriculture, Comrade Matskevich, and Minister of State Farms, Comrade Benediktov. Like "saints" they fold their hands and do not intervene in this dispute. Ministers must not stand aside. Why do you turn your back on what the people [*narod*] says and knows? [106]

The ministers made no public response, and did nothing to promote the use of Lysenko's recipe. At a plenary meeting of the Central Committee in December 1958, Khrushchev again criticized the opposition to Lysenko, which had just been dealt a serious blow by the dismissal of the editors of the *Botanical Journal.* Lysenko told the Central Committee how grateful he was for the faith shown in his recipes by the Party and by Comrade Khrushchev personally. Yet Matskevich, though he made the ritual confession that his Ministry deserved to be criticized, stubbornly held his tongue on the fertilizer matter.[107] A year later, in December 1959, the Central Committee heard much the same dialogue again, with Lysenko thanking Polianskii and Podgornyi as well as Khrushchev, while Matskevich still refused to comment.[108]

Perhaps this was one of the reasons for the demotion of Matskevich a year later[109] — he was put in charge of a virgin lands district — but it hardly seemed to matter who took his place in charge of the All-Union Ministry. The institution simply would not become an effective agency for propagating Lysenko's schemes, neither the use of his bulls and cows to increase butterfat nor the composting of mineral fertilizer with earth and dung. Even when Lysenko's old comrade Ol'shanskii was in nominal command,[110] Khrushchev continued to complain that the Ministry was not effectively promoting widespread adoption of these recipes.[111] This standoff between the chief and his agricultural bureaucracy was one of the issues in February 1964, when the Central Committee met to discuss the renewed crisis in agriculture. Yields had shown impressive gains from 1953 to 1958, but they had stagnated afterwards, and an especially bad harvest in 1963 necessitated the costly humiliation of importing grain from the capitalist countries that Khrushchev had recently sworn to overtake in per capita output. The meeting began with a long, wide-ranging report by the current Minister of Agriculture, I. P. Volovchenko, who simply ignored Lysenko's recipes. Perhaps he meant to suggest that they were "unrealistic schemes" [*prozhekterskie plany*], which he considered a major cause of the current agricultural crisis.[112]

When Lysenko got a chance to speak, he complained that Volovchenko's silence was indicative of the Ministry's obstructive attitude.[113] Another Lysenkoite speaker, Ol'shanskii, pointed the finger at the country's two leading institutes for research in fertilizer: for ten years they had refused to recommend Lysenko's recipe.[114] A director of one of the institutes replied with tables of data showing that standard methods of fertilizing were better than Lysenko's.[115] Another Lysenkoite gave a rejoinder, and finally it was Khrushchev's turn, while summing up the results of the meeting, to comment on "the silent, covert dispute" that had been going on over fertilizer methods. He took Lysenko's side once again, and criticized Minister Volovchenko as, seven years earlier, he had criticized Ministers Matskevich and Benediktov. He insisted that he was not trying to prescribe truth to scientists. But they should realize that they only seemed to be arguing about scientific data; what they were really debating was methods of fertilizing crops, getting the most effective use of the limited supply of fertilizers. If the scientists who disagreed with Lysenko could get 30 to 35 centners of grain from a hectare with the same small amount of mineral fertilizer as Lysenko used, Khrushchev promised support to *them.*[116]

The bewildered reader is left wondering how Khrushchev could have been so inconsistent. In alternate paragraphs he brushed aside as irrelevant the specialists' reasoning with figures, and then adduced his own crude version of the same thing. He urged that the Ministry of Agriculture see to the application of Lysenko's recipe all over, and then denied that he was prescribing a single method to all farms and all specialists. At either end of these oscillations he seemed to speak with equal sincerity. Indeed, he was almost eloquent in his insistence on the practical necessity of freedom for specialists.

> The Central Committee of the Party is prohibited from saying that only one method must be used. And I, as First Secretary of the Central Committee and Chairman of the Council of Ministers, I do not want to say that, and I cannot say it. Why? Because there must be a competition of minds, of ideas, of scientists. What scientists propose must be verified in practice, and the best should be applied in the collective and state farms.
>
> By the way, one sometimes hears in orators' speeches, and not only in speeches, such expressions as, "This is done by order of the Central Committee," or "This is done by order of Comrade So-and-So." And sometimes it happens this way: a man crooks his finger, pointing to heaven, to say that there's been a directive "from above." . . . Comrades of this type seem to be disciplined executives carrying out instructions. But none of us demands thoughtless "fulfillment of orders," and a concrete culprit cannot justify poorly managed business or the ruin of this or that enterprise by any kind of pointing to authority. (*Applause.*) . . .
>
> When the Party decisively condemned the ways of the period of the cult of personality, it condemned ways according to which one man was supposed to see all and one could supposedly decide all. Our Party came out for . . . collective leadership, heightening

> the responsibility of each man for the work entrusted to him. Let us, comrades, firmly hold to this wise Leninist principle. (*Prolonged applause.*)[117]

Khrushchev was elaborating the plain truth that Stalin had proclaimed in 1950—"an Arakcheev regime breeds irresponsibility"—but, like Stalin, he was reluctant to follow his own advice to its logical conclusion. His colleagues in the Politburo finally helped him along. In October 1964, they suddenly convened the Central Committee and got it to fire him. They blamed him for the "unrealistic schemes"—the very words that Volovchenko had used in February—that were a major cause of the agricultural crisis.[118]

That meeting at the end of October 1964 marked the first time in Soviet history that a supreme boss was voted out of office in something like a constitutional manner. "Etiquette and good order" were undoubtedly violated, so the minutes have not been published and it is not possible to prove that the Central Committee heard a lively expansion of the February debate, with Khrushchev trying once again to endorse both Lysenko's recipes and the freedom of science. But indirect evidence makes it fairly clear that this did occur; Lysenkoism was almost certainly one of the issues at the October meeting. The major charge against Khrushchev was the stagnation of yields since 1958; his support of Lysenkoism and his other repeated interferences with sound agronomy were probably cited as contributing causes. Subsequent publications indicate that the Central Committee resolved to practice what Khrushchev and Stalin had preached: no political intervention in the autonomous judgments of scientific specialists. Of course they did not acknowledge their debt to Stalin and Khrushchev for this principle, any more than they acknowledged the responsibility they shared for violations of it. The record will probably show that Polianskii, chief of agriculture, and Suslov, chief of the cultural front, blamed Khrushchev for subordinating science to agrobiology, ignoring or glossing over the fact that they had helped maintain that wasteful tyranny. But the turnabout of particular bosses is hardly as important as the pattern of institutional change that preceded, and presumably determined, their shift. The men in the Politburo turned away from Lysenkoism after, and presumably because, their subordinate bureaucracy began to turn away from it. Not the bureaucracy that supervised the cultural front. In that priestly establishment bureaucrats were extremely conservative, cautiously clinging to the old testament till a clear new revelation might come from on high. Until 1939 they had preached faith in science, then an inconsistent mixture of science and agrobiology; in 1948 they were ordered to uphold agrobiology alone, and they did so until ordered back to science at the end of 1964.[119] It was the agricultural establishment that pulled the highest political chiefs away from agrobiology in the 1960s, just as, in the 1930s, it had pulled them to it.

The attentive reader will have noted an inconsistency. Earlier the agricultural establishment was said to have been hopelessly entangled in thickets of self-perpetuating ignorance; now it is described as the agency that pulled the biggest political chiefs away from agrobiology, back to genuine science. The trouble is that different parts of the evidence point in different directions. Though the administrators of agriculture refused to campaign for any more Lysenkoite recipes after the crash of the Great Stalin Plan in 1951–52, neither would they allow the publication of explictly anti-Lysenkoite material. It is true that the flow of explicitly Lysenkoite material diminished steadily during the Khrushchev decade, which indicated a drift away from Lysenko's version of agrobiology, but only a minority of senior specialists knew how to find their way back to genuine science. The puzzling question is how this minority managed to win the support of the men in charge of the agricultural establishment, against the will of the Politburo. It may seem more exciting to ask a different question, how Khrushchev's colleagues in the Politburo finally came to appreciate the significance of this impasse, and how they worked out the exceedingly delicate problem of dumping and blaming the boss, so that specialists might once again be autonomous masters of their particular domains. It may seem more exciting to ask this question: How did light finally penetrate the frightful self-assurance of the ultimate bosses? But the necessary archives are closed, and anyway it is more puzzling, and more important, to ask the prior question: How did intermediate bosses manage to see the light first, through the double thickness of their own self-assurance and their masters'?

Many students of Soviet history think they have explained this and other triumphs of rationality when they note that the bosses of agriculture were increasingly graduates of technical institutes, as their predecessors of the 1920s and 1930s had not been. But this simple faith in the power of education is subverted by too many facts to be maintained. Item: Though the typical agricultural commissar of the 1920s had little or no technical education, he was an ardent supporter of genuine autonomous science — until the 1930s, when he veered about to support agrobiology.[120] Item: The climax of terror in the 1930s hastened the replacement of such old-fashioned, uneducated officials by new men with irreproachable diplomas — who carried Lysenkoism to its absolute triumph in 1948. Lobanov, for example, the Minister of State Farms who presided at the August Session in 1948, had graduated from the Timiriazev Agricultural Academy, the best in the country, at a time when it was entirely free of agrobiology. If we imagine him leaving the Academy with a genuine understanding of agricultural science and a genuine commitment to it, we must then imagine him suffering a conversion to agrobiology in the 1930s, and we must further imagine him undergoing still another conversion, back to faith in science in the 1960s. (Since 1965, as president of the Lenin Academy of Agricultural Sciences, he has been supervising the demolition of agrobiology.[121]) Better to disregard the problem of sincerity and genuine thinking, when dealing with such a man. He typifies the bureaucratic intellectual: his position influences his thought much more than his thought influences his position. One might even say that his position in the bureaucratic hierarchy determines his position in thought. A Soviet novelist — who shares some of this quality himself — called such people not *ortodoksy* but *vertodoksy*, not orthodox but weathercocks.[122]

It is startling to see such people refuse to obey their masters' voice, and indeed, there is no public evidence that Lobanov was so bold. He turned away from Lysenkoism publicly only after the Politburo did so. But other bureaucrats in charge of the agricultural establishment were not so timid. There was Benediktov, for example, who also graduated from the Timiriazev Academy in the 1920s, and also became one of the chief promoters of Lysenkoism in the late 1930s and 1940s, when he was Stalin's favorite Minister of Agriculture. When, in 1957, we find Khrushchev rebuking him for failure to support Lysenko's fertilizer scheme, we have no reason to assume that Benediktov had searched his professional soul and decided to return to the science he had been taught in the 1920s. What soul? What profession? He was a typical Stalinist bureaucrat. A changed situation was changing his thinking.[123]

Even Matskevich, the Minister who came closest to being a conscious reformer with extra-political principles, was not very different. He graduated from the Kharkov Zootechnical Institute before the rise of Lysenkoism, and returned as its director when a Lysenkoite curriculum was being imposed. He was sufficiently cooperative to become one of the agricultural chiefs of the Ukraine during the time of Lysenko's complete ascendance. In the 1950s, while he was Khrushchev's favorite Minister of Agriculture, he too was rebuked for obstructing Lysenko's fertilizer scheme, which may have been one of the reasons that Khrushchev finally demoted him at the end of 1960. Matskevich can hardly be considered a thoroughly independent thinker who sacrificed his job for the sake of science. He was one of the chief initiators of the "unrealistic scheme" to force corn on virtually every farm, with scant regard to climate and other local conditions. That corn campaign was a much more serious violation of sound agronomy than Lysenko's fertilizer scheme precisely because it was vigorously promoted by the bureaucracy over which Matskevich presided. In his case, as in Benediktov's, it is hard to imagine early training having decisive effect after years of neglect and abuse.[124] As for Volovchenko, the Minister that Lysenko attacked at the February 1964 meeting of the Central Committee, he was *trained* to Lysenkoism; and in 1963, when Khrushchev dragged him up from the chairmanship of a sugar beet farm to be Minister of Agriculture, he publicly endorsed Lysenko's fertilizer scheme.[125] But only a few months later, the Central Committee heard him denounce "unrealistic schemes," while Lysenko complained that he was obstructing widespread promotion of the notorious composts. One gets the impression of a geological transformation in the vast agricultural bureaucracy, making it impossible for the man in charge to launch any more agrobiological campaigns.[126]

In short, the heads of the agricultural establishment were hardly technicians resisting political authority because their professional conscience told them to. They were politicians rather than technicians, mediating between the highest politicians above and the specialists and farmers below. To some extent their abandonment of agrobiology was undoubtedly the result of pressure from specialists. The reader has already seen Khrushchev's outbursts against specialists who submitted a petition deploring the corn campaign, and against a few others who adopted a resolution condemning Lysenko's fertilizer scheme, with Matskevich taking Khrushchev's side in the matter of corn and the specialists' side in the matter of fertilizer. I doubt that the archives will disclose a large volume of similar petitions and resolutions coming from agricultural specialists, bracing ministers of agriculture to stand up more and more for genuine science until the Politburo finally gave in to the pressure.[127] I doubt it because the publications of the agricultural specialists indicate that only a very few of them were fighters. Most of the older ones, who knew the difference between genuine science and agrobiology, engaged in pliable defense of their disciplines, shading off, in many cases, into complete opportunism.[128] The younger specialists were increasingly incapable of distinguishing between genuine science and agrobiology. Yet agricultural science was saved from extinction, partly by the efforts of the dwindling minority who had the knowledge and courage to defend it, but more by the simple fact that the farmers were unwittingly on their side.

It was the farmers — peasants and farm officials became indistinguishable in this matter — who forced the abandonment of one Lysenkoite campaign after another, until, with the tree-planting crash in 1951–52, it was impossible to start any more. (The corn campaign was "scientific" rather than Lysenkoite in inspiration.) Even when superior political pressure brought the Ministry of Agriculture to issue a decree in favor of Lysenko's cows and bulls, enforcement was neglected.[129] Soviet farms were allowed to import far more milk cattle from Holland than they bought from Lysenko's farm,[130] for a very simple reason. Lysenko's bulls and cows did not endow a herd with the increased butterfat that he claimed they would.[131] For an equally simple reason most farms refused to compost mineral fertilizer with organic wastes and earth (in this case the Ministry took their side almost openly): The composts increased the farmer's burden without increasing his yield.[132] Chances are that most farms were not making conscious choices for scientific methods. They were not, in painful fact, doing a very effective job of fertilizing. Neither were they doing a very effective job of improving their dairy herds. In one case as in the other material shortages, ignorance, slovenly labor, poor rewards, and excessive centralization of authority were still mutually reinforcing brakes on the improvement of farm efficiency. But the poorest, most ignorant and slovenly farm management had nothing to gain from heeding the Lysenkoites, whether by loading earth into manure spreaders or paying premium prices for unreliable bulls and cows. That kind of misdirected zeal was mortally stricken in the campaigns of the 1930s, and died altogether with the "Great Stalin Plan for the Transformation of Nature" in the late 1940s.

Seen in that light, the ultimate puzzle is not why the agricultural officials followed the farmers and led the Politburo in turning away from agrobiology, but why they turned away so gingerly, why they were so grudging and slow in returning to the rule that scientific agriculture must be managed by autonomous specialists. It is not a sufficient explanation to invoke the bureaucrat's natural timidity, his habit of awaiting the truth from his superiors rather than seeking it in the advice of his subordinates. No doubt that is part of the reason, but Matskevich and Benediktov were bold enough to side with genuine specialists against Khrushchev on the fertilizer matter. Why did they rebuff the specialists who offered common sense criticism of the corn campaign? And why, long after they had ceased campaigning for any Lysenkoite recipes, did they continue to praise individual Lysenkoites for their practical aid to agriculture?[133] Not only individuals. Improbable as it may seem, Soviet agricultural authorities insisted that abandoned Lysenkoite nostrums had been practical successes, attributing their abandonment to changing conditions rather than to the worthlessness of the nostrums.[134] The obvious explanation is that these authorities shared some of the bullheaded irresponsibility of the Lysenkoites and of their superiors in the Politburo, qualities that were a product of the system.

The crucial feature of the Stalinist agricultural system, which threatened to transform dodged responsibility and arbitrary power from a tolerable vice into a fatal virtue, was the deliberate refusal to distinguish between political and technical authority. Stalin, it will be recalled, decreed the end of this distinction in 1929, at the same time that forcible collectivization began and Lysenko's first nostrum won the support of agricultural officials. During the next thirty-five years political and technical authority were never entirely fused — just as they will never be entirely separated — but they were sufficiently confused to create a generation of agricultural executives who were actually little more than table-thumping grain collectors, who knew just enough to shout down objections to their outrageous demands, who knew — let us be fair — just enough to be unaware of their essential ignorance. Khrushchev was the prime example of the type, but his ministers also belonged to it. They were creators and creatures of an agricultural system that extracted produce by force while preaching scientific management. What saved them and their system from permanent stagnation and possible disaster was their crude utilitarianism, their belief that scientific truth is measured by poods of grain. "I believe," said Khrushchev, "that theoretical and scientific disputes should be settled in the fields."[135] He had learned this primitive, wasteful, but ultimately effective criterion of truth from Stalin, and like his teacher believed that his authoritative glance could read the lessons of the fields. Using the same principle, with similar authoritarian impulses, Soviet managers of agriculture gradually and painfully learned the necessity of distinguishing between political and technical authority. The official who would achieve a steady climb in yields must know when to defer to the technical judgments of autonomous specialists. The powers that Stalin joined had to be separated. Khrushchev took steps in this direction but was spoiled by an immediate upturn in yields. A renewed stagnation, between 1958 and 1964, finally brought the lesson home to his colleagues.

At the end of 1964 the Stalinist pattern seemed to be entirely undone. The press was filled with anti-Lysenkoite articles and appeals for the restoration of fully autonomous, rigorously scientific methods in all fields of biology and agricultural science.[136] The trickle of Lysenkoites who had been deserting their cause became a flood. Lobanov, who had chaired the August Session in 1948 and defended Lysenkoism as president of the Lenin Academy from 1956 to 1961, resumed the presidency in 1965, this time to supervise the destruction of agrobiology. The most spectacular conversion was that of Stoletov, one of the original and most virulent agrobiologists. (In 1937 he had virtually called for the arrest of Tulaikov.[137]) He may have had a premonition of the turn against Lysenkoism, for he had shown signs of deserting it before the October meeting of the Central Committee in 1964.[138] One way or another, he kept his twin posts as Minister of Higher Education, RSFSR, and chairman of the Department of Genetics, Moscow University, closing the book of Lysenkoism and opening that of science with the dead pan that betokens the master bureaucrat.[139] Lysenko himself was deprived of his post as director of the Institute of Genetics at the Academy of Sciences and restricted to the experimental farm it had been running in the Lenin Hills, near Moscow. After some hesitation the Institute was simply dissolved, though the farm was left to Lysenko. Dubinin's project for a new institute of genetics, which had been crushed in the late 1940s and diverted to Novosibirsk in the 1950s, finally materialized in Moscow in 1966.

The sign that things had changed for good was the turnabout of the educational bureaucracy. It waited several months—perhaps for a decent show of thought, perhaps to make sure there would be no reversal at the top, or maybe just because a bureaucracy has to take time when turning around — and then came down solidly for a genuinely scientific biology curriculum in secondary schools. The only vestige of the dead past was reverence for Michurin, and a feeble suggestion that his "mentors" and "vegetative blending" might be scientifically justified.[140] Only three journals, the conservative literary monthly *October*, a seedman's journal, and Lysenko's own *Agrobiology*, tried to stand against the flood. Even these diehards felt obliged to make damaging concessions and pathetic appeals for compromise, trying to maintain some place for such favorite articles of faith as the stage theory (confessing that it was indeed incomplete) and the graft hybrids (granting that many claims had been exaggerated or mistaken).[141] They were reminded that science has its own kind of intolerance, harsher, in a sense, because it is spontaneously enforced by a community rather than imposed from above by an outside power. To a Lysenkoite's appeal for a synthesis of science and agrobiology the new editor of a biological journal replied with brutal simplicity: "What does he want science to be synthesized with — quackery, a bunch of absurdities?" [142] Without political protection against the criticism of professional scientists, agrobiology was doomed. The most it could hope for was the kind of marginal existence that "organic gardening" and other pseudoscientific sects maintain in countries where cranks are free to publish, even though they are excluded from the scientific community. Some people are struggling for that kind of freedom in the Soviet Union, but the Lysenkoite diehards show no signs of joining them. They seem content to publish an occasional item in the one or two journals that tolerate them.[143]

Lysenkoite diehards were not the only ones to feel the continuing power of the Stalinist system. "Nothing could be more harmful," warned a leading newspaper in the first days of the new freedom for biologists, "than the belief that the problems of science can be solved by a gang fight of two warring clans." [144] Scientists were therefore to refrain from uninhibited attacks on agrobiologists. There was an element of common sense in this order of restraint. Lysenkoites and ignoramuses probably constituted the bulk of biologists and especially of agricultural specialists; if they were subjected to massive ridicule, scolding, and dismissal, who would run the campaign for scientific agriculture and the thorough reconstruction of biological research and education? Better to try and salvage agrobiologists by persuasion and retraining.[145] But this rational element in the warning against "disorganizing the cadres" was combined as usual with anxiety for the principle of authority. Indignant exposés of agrobiological tyranny might easily get out of hand. Minister Stoletov and President Lobanov were obvious targets; behind them stood the men in the Politburo and *nachal'stvo,* even *vlast',** which remain sacred principles when their imperfect agents make mistakes. Indeed, Stoletov and Lobanov may have been kept in office for the same reason that Caligula sent his horse to preside over the Senate, to remind everybody what political power really means.

So the flurry of anti-Lysenkoite articles which followed the dismissal of Khrushchev was not allowed to turn into a steady storm.[146] The president of the Academy of Sciences declared that the Lysenkoites' immunity to criticism was ended, but at the same time he passed on to his subordinates a warning against destructive recrimination and "label-pasting." [147] The Academy resolved that there should be a conference to publicize the issues involved in the clash of science and agrobiology, obviously intending a counterblast to the August Session of 1948.[148] Such a meeting was not held. Instead, an expert commission was sent to investigate Lysenko's experimental farm in the Lenin Hills. The investigation was strictly limited to his butterfat and fertilizer schemes. The experts' devastating critique was presented to a small meeting of officials and specialists, which passed without public notice in September 1965. A few months later the critique and the record of the conference that endorsed it were published, shattering Lysenko's reputation beyond hope of repair.[149] He was allowed to keep the experimental farm at Lenin Hills, but his journal, *Agrobiology,* was terminated; and he ceased altogether to be heard along with a handful of diehards who would not — or were not permitted to — convert.[150] At the same time, efforts were made to repress further attacks on Lysenkoism. The biochemist, Zhores Medvedev, whose unpublished history of the Lysenko affair had won him threats of criminal libel in the summer of 1964, was rebuffed in his efforts to get it published, even after the manuscript had won the unanimous endorsement of a special commission at the Academy of Sciences. Finally, to head off foreign publication of an

* According to a recent Soviet dictionary (*Ozhegov*), *vlast'* is "the right and possibility of ordering someone or something, of subordinating him or it to one's will." *Vlast'* is also "the political regime . . . [for example] the Soviet *vlast'*." *Nachal'stvo* is the *vlast'* of a *nachal'nik* or official.

unauthorized version, Medvedev permitted Columbia University to bring out an English translation, whereupon he was fired from his job.[151]

In a sense it was quite appropriate to terminate the Lysenko affair with a small, very specialized conference, excluding the big issues of genetics, plant physiology, and soil science. All the noise that the Lysenkoites had made about those subjects was really beneath criticism. In 1965 as in 1929 they were cranks with agricultural nostrums, and the final conference limited itself to scientific appraisal of the last two nostrums. Lysenko tried, in a six-hour confrontation with the experts' commission, and then in an angry memorandum to the conference, to raise the issues above such pesky matters as the genealogy of his cattle and the rations he had given them, or the methods he had used to distinguish the effects of his composts from other factors that influence yields. He had never troubled himself to be precise and he insisted that precise data were not essential. Shrilly he called attention to "biological theory," to "the progressive biological theory" that had been developed "in unity with the practice of collective and state farms," and had "always enjoyed the support of the Party and the state." [152] In vain. The specialists stuck to their narrow task with devastating effect, for Lysenko and his colleagues were simply incapable of sustained reasoning with facts and figures. Their "favorite device," as the chairman of the investigating commission put it, was "to say a single thing and hush up all the rest, as long as everything looks good." [153] This was a perfectly obvious, and deadly accurate characterization, and would have been with respect to any of Lysenko's recipes at any time, beginning in 1929, when he first became prominent with his scheme for vernalizing grain. But that perfectly appropriate response to Lysenkoism was thirty-five years overdue. That is why it was scandalously inadequate. For once Lysenko was calling for discussion of a, or even the, most important issue. He *had* won "the support of the Party and the state" for the "biological law" that any agricultural scheme that looked good to political authority was in fact good. This confusion of political and technical authority was now declared to be a mistake; specialists were to resume their laborious efforts to distinguish really worthwhile methods from those that only seemed so. Period. The thirty-five year error was not to be examined or analyzed.

Those who revere Stalin, whether as redeemer or as satan, will say he taught his successors this style of self-correction as he taught them his ways of self-defeat. But it is unhistorical to attribute so much to a mere human being. All the Bolsheviks, Stalin included, were interacting with evolving situations when they moved close to a suicidal extreme of self-deceiving tyranny in agriculture and science, and when they began to flounder back toward self-correcting liberty. To note that they still have a long way to go is not as meaningful as to discover the route they traveled, for such a map may suggest their future path. The critical reader is probably dissatisfied with the map that has been drawn in the preceding pages. Almost no attention has been paid to the theoretical outpourings of the extended clash between science and pseudoscience, though they must surely indicate something about the evolving pattern of Bolshevik thought. On the practical side, which has been the main theme, the reader has probably come to feel a nagging sense of disbelief. Can it be true that Bolshevik officials, who prided themselves on their practicality, were simply deceived for thirty-five years about the value of agrobiological schemes? It may seem to the reader that a correction must be made. Either the Bolshevik leaders were swayed by theoretical considerations that have been improperly slighted so far, or agrobiological recipes did in fact generate some practical gains. The way to settle such doubts is to take a close look at the theoretical outpourings, and an even closer look at some representative agrobiological recipes.

7

Academic Issues: Science

PLANT PHYSIOLOGY

Outside the Soviet Union many people have sought the origin of Lysenkoism in the genuine difficulties and serious disputes of twentieth century biology. This refusal to face the fact that a complete crank won mastery over the Soviet community of biologists is sustained by the belief that Lysenko began as a scientist. He is supposed to have gained entrance to the community by doing good work in plant physiology; indeed, he is sometimes credited with opening a major new field of inquiry.[1] Then, in genetics, he is supposed to have gone wrong, by taking the Lamarckist side of the celebrated debate over environmental influences on heredity.[2] These stories have certain elements of empirical truth, as pictures of dragons have elements of real reptiles, but the composite results are myths. They express the anxieties of the artist rather than the facts of the empirical world, which are sometimes too bleak to be accepted unadorned. The painful truth is that Lysenko was never seriously involved in any genuine scientific problem, and therefore — not nevertheless — he won mastery over Soviet biologists. From first to last he and his devotees simply brushed aside or double-talked their way around serious scientific inquiry, which was nothing more to them than a scholastic impediment to quick solutions of agricultural problems.

An understanding of science is not essential to the perception of this basic fact. It is obvious in "agronomist Lysenko's" use of political force to effect entry into the communities of plant physiologists and geneticists, and to establish his dominion there.[3] Indeed, his strangeness in those communities was apparent in the name that was constantly applied to him until his mastery was beyond dispute — agronomist, not scientist. This should be a trivial matter; in the republic of learning titles should be insignificant, and no one should believe that pure science is superior to applied. Unfortunately this is not a trivial matter. The attitudes of academic snobs are exaggerated versions of values inherent in the scientific community, indeed, in society at large. Everywhere self-esteem rests on disdain for others. President Kalinin used the language of old-fashioned radicalism, but he was reacting to a fact that no one can avoid, when he told the students at the Timiriazev Academy that there was "a chasm between the people and the priests of science." He blamed it on the superiority that scientists feel toward the rest of mankind, and concluded with

> a wish to all our new scientists who are being created here, that they should make this the principle of their life, that they are not exceptional people, that they realize that they are carrying out a portion of work which is not superior in its significance to the work of an ordinary tailor or shoemaker.[4]

Kalinin won extended applause with his version of the ancient equalitarian faith, but the applauding students who went on to *higher* degrees in *pure* science — note how invidious comparisons are built into the language — almost certainly felt themselves superior to mere *praktiki.*[5]

A *praktik,* says a Soviet dictionary, trying to hide the hierarchy of values inherent in the word, is "a person who knows his business in a practical way, who has great experience in his specialty." [6] The examples of usage given by the dictionary — "he is both a learned theorist and a good *praktik;* he knows engineering as a *praktik*" — cannot hide the implicit sneer: a *praktik* lacks theoretical understanding of his skill, and is therefore inferior to the person who has such understanding. Soviet dictionaries are not likely to give an example of usage which was fairly common in the 1920s — "Marx and Plekhanov were theorists of socialist revolution, Lenin was the great *praktik*" — for this remark was officially denounced as a disparagement of Lenin.[7] And Stalin, who had the deserved reputation of being the chief *praktik* of politics in the 1920s, who insisted that practice (*praktika*) has priority over theory, sanctioned an inference that jolted Soviet philosophers at the end of the 1920s: Therefore Stalin is the chief theorist. Whatever the logical value of this syllogism, it performed the vital social function of elevating Stalin to a higher point than any mere *praktik* could occupy.[8] Paradoxes of that kind were made keener by the fact that the official ideology exalted the lowliest of *praktiki*, the workers and peasants, while boasting of the rapidity with which some of them were rising above their comrades, to the status of specialists and administrators. Nowadays this incongruity is hardly more disturbing than the traditional sentiment in Christian communities about the ultimate superiority of the humble and the meek. In the 1920s and 1930s, however, it was a strong element of the operative ideology, justifying such things as *vydvizhenchestvo* ("pushing up" workers and peasants into professional and administrative positions), concomitant efforts to drag down the *kastovye* and *zamknutye* (caste-ridden and ivory-tower) specialists, and Stalin's assurance to Stakhanovite workmen that they knew the capacity of machines better than engineers. In all these outbursts the persistent effort was not to realize Kalinin's dream, that is, to put everyone on an equal level, but to make selected tailors and shoemakers the superiors of learned specialists. In short, the pledge that the last shall be first was designed to reshuffle people within a traditional hierarchy of status, not to destroy the hierarchy itself.[9]

These paradoxes in the relationship between scientists and *praktiki* must be kept constantly in mind, if one wishes to understand Lysenko's interaction with scientists. He received his secondary education in a school of gardening, his higher education as an extramural student at the Kiev Agricultural Institute, working all the time at a rural experiment station.[10] He had no postgraduate training or higher degree, no formal claim to the title of scientist, yet he aspired to the theoretical heights from which, as he told a *Pravda* correspondent in 1927, practical problems could be solved by a few calculations "on a little old piece of paper." [11] Thus he awed the correspondent into granting him the rank of "barefoot professor," an "outdoor scientist," who "holds a plow with one hand, a flask with the other." The figures of speech were clumsy imitations of a pioneer in journalistic agriculture,[12] but the subject's picture of himself came through clearly enough. He and his press agents never stopped this overcompensating boasting of the difference between Lysenko and the usual scientist. "Science in the Hands of a Muzhik's Son" was the headline in 1936 over a public letter to Stalin from Lysenko's mother and father, thanking the revolution for making such a big man of their son, who would otherwise have been a gardener (*sadovnik*) all his life.[13] Three years later, a leading ideologist tried to effect a compromise between Lysenkoites and geneticists by imploring the Lysenkoites to overcome their anti-intellectual (*makhaevskie*) moods, the geneticists their seignorial (*barskoe*) attitude.[14] There is no evidence that they did. In 1948 an agrobiologist was furious that the Academy of Sciences still abounded with talk of "Michurin the gardener, Williams the ignoramus, Lysenko the illiterate."[15] As usual in cases of class consciousness, epithets were an exaggerated description of reality and simultaneously an influence upon it. The feeling and the fact of social distance interacted, each intensifying the other.

Assigned to a remote agricultural research station in the North Caucasus, and given the humble job of finding a good winter-habited crop for the locality, Agronomist Lysenko impatiently shunned the usual systematic testing of all the likely candidates. (Throughout his career he kept this antipathy to extended trials on little experimental plots.) He worked out a formula that would tell him "the amount of heat" which a plant needs to get through its stages of development from seed to the production of new seed. After failure with the heat formula[16] he made a discovery about cold, which he presented as a revolution in plant science: when he moistened and chilled the seed of winter-habited plants and then planted them in the spring, they completed their life cycle in a single season, as spring-habited plants do. He was mortified to discover that learned specialists in plant physiology were unimpressed. They were put off by Lysenko's shocking ignorance of previous work, which included many such transformations of winter into spring habit, by his sloppy reasoning, and by his poor Russian.[17] Even his famous coinage, *iarovizatsiia* (vernalization), seemed to one professor "not especially successful and grammatically not altogether correct."[18] At a major conference in January 1929 he was coolly brushed aside by N. A. Maksimov, the country's leading plant physiologist: "The results obtained by Comrade Lysenko do not represent anything new in principle; they are not a scientific discovery in the precise sense of the word."[19] In October 1929, after *Pravda* had carried an article by the Ukrainian Commissar of Agriculture praising "young agronomist and plant breeder Lysenko" for a discovery of enormous practical benefit (the Commissar thought it would put an end to winter-killing of grain),[20] the learned specialists showed some respect for the practical achievement, but persisted in brushing off the scientific. Lisitsyn, who had been Lenin's favorite plant scientist, made the same point as Maksimov: "From the point of view of plant physiology, Agronomist Lysenko has made no discovery." [21]

Many years later an admiring biographer reported Lysenko's reaction in appropriate Stalinist bombast:

> The superciliousness and sneers that rose in the way of the young researcher will not detain us. Those people [the sneering plant physiologists] learned that "the young man" would not repeat Gassner's experiment [which opened the field of inquiry that Lysenko claimed to be opening]; he would not add to the thousandth experiment his thousand and first. He would not in this way give them the satisfaction of proving him wrong. Of course it would have been reasonable to return to Gandzha [the agricultural research station in the Caucasus], and spend long years in stubborn labor, so that he might present them with a work resting on formidable columns of figures. They would graciously approve his success, in order to upset it by an experiment of verification. He knew what these experiments of verification sometimes amount to! An inexperienced fellow, without considering the conditions, the environment, the circumstances, and the time, pokes a grain in the soil. Without knowing the requirements of the plant, its needs, the fellow raises a little green monster, and announces that the theory has not been confirmed. No, he would not let them prove him wrong![22]

That might be called the essential law of Lysenko's methodology (and of agrobiologists in general): avoid verification, do not let them prove you wrong.

All this, so far, has to do with the emotional interaction of Lysenko and the scientists. It does not touch the substantive issues. The most distinguished scientists could have been wrong, for all their learned degrees and rigorous methods; the most willful agronomist writing the crudest Russian could have been right, if only by accident. The only way to judge is to examine the substantive issues, which are fortunately easy to understand. Winter-habited plants start their growth in the fall, stop during the winter, and start again the following spring, completing their life cycle with flowering and the production of seed (earing) in the summer. If they are deprived of the chilling period, their stage of vegetative growth will extend indefinitely; they will not ear, that is, they will not enter the stage of reproduction. This winter habit can be changed by moistening and chilling the seed or the seedling before planting; then, if planted in the spring, it will behave like a spring-habited plant, going through all the stages of its life cycle in a single season. This transformation of winter to spring habit was known as early as the midnineteenth century. The Ohio State Board of Agriculture reported experiments with it in 1857, and a Russian agricultural newspaper told of similar work in 1875.[23] These are historical curiosities with little scientific significance, because the experimenters' purpose was to discover whether the transformation of winter into spring habit could be of economic value; when that purpose was disappointed, interest evaporated. It was revived by plant physiologists in the early twentieth century, who wanted to explain the phenomenon rather than find an immediate profit in it. George Klebs saw it as an important test case for his mechanistic approach to the physiology of development, and Gassner in 1918 published the pioneering paper, that is, the paper that described the transformation of winter into spring habit with sufficient precision and theoretical relevance to start a number of people working on it.[24]

They were trying to discover what mechanism is activated by chilling winter-habited plants as they start to grow, and how this mechanism determines the onset of reproduction many weeks later. This was part of the general search for mechanisms, presumably biochemical, which regulate the progression of all plants through their diverse stages of development. The favorite technique in the search was not temperature manipulation but photoperiodic effects, that is, hastening or delaying the reproductive stage by manipulation of the plant's alternating periods in light and dark. It was established, for example, that winter wheat that has not been chilled can still be made to ear by subjection to continuous light.[25] Plant physiologists have preferred to work with photoperiodic rather than temperature aftereffects, because they are more diverse and involve many more plants. As a recent Soviet textbook points out, the scientists' preference is grounded in nature's: the changing length of day acts as a precise astronomical calendar, signaling forthcoming changes in the seasons much more reliably than randomly fluctuating changes in temperature.[26] In either case, whether working with light or temperature, students of plant development have seen "importance" or "significance" or "relevance" in research that has groped for the mechanisms of change from one stage to another. They have tended to assume that they are looking for biochemical substances, generated under various conditions of light and temperature, which stimulate or inhibit the various stages of development.[27]

That was the outlook and the state of knowledge that prevailed already in the late 1920s, when agronomist Lysenko entered the field with vernalization, *iarovizatsiia,* his coinage to describe what he believed to be a great discovery: the transformation of winter into spring habit by moistening and chilling the seed. He was deeply incensed at those who told him that he had made no discovery, not only because the transformation was long since known, but even more because his experiments did not help answer the question of how it worked. The most that learned critics could say for Lysenko was that he focused attention on two minor aspects of the problem that had not been given sufficient attention. First, the temperature aftereffect can be achieved by chilling germinating seeds as well as seedlings, which had previously been the favorite objects of experiment. And second, varieties of wheat differ in their degree of responsiveness to chilling. But even these modest tributes were spoiled by qualifications. Lysenko insisted that all varieties of wheat, spring as well as winter, will respond to chilling of the seed by hastening the onset of earing and therefore, he reasoned, their yield will be increased. His learned critics had data to show that the gradient of responsiveness ranges from the virtual requirement of extended chilling in varieties that are strongly winter-habited to no response or a negative response in varieties that are strongly spring-habited, with an intermediate range that includes "ambidextrous" plants (*dvuruchniki*), as Maksimov called the varieties that follow the spring or the winter habit depending on circumstances. The learned experts also had reason to doubt that hastening the onset of earing will always increase yields. In general they showed much more respect for the work of other students of temperature aftereffects, in particular, for Tolmachev, a professor at the institute that had trained Lysenko to be a provincial agronomist. Tolmachev published, about the same time as Lysenko, a report of very similar work with chilled seed instead of the usual seedlings. He won greater respect because he was more rigorous and far more knowledgeable in his discussion of the significance or relevance of his data.[28]

Lysenko responded to his supercilious critics by flouncing angrily away from any effort to play the professional scientists' game according to their rules. After 1929 virtually all his articles were published in mass newspapers, or in journals created for him by government decree, or in pamphlets and anthologies of which he was author and ultimate judge.[29] He would not submit to the usual procedure of scientific publishing, which is to send out manuscripts for confidential reviews by established specialists as a condition for publication. Neither would he submit to the usual rules of rational discourse within the institutions where he worked. First at the Odessa Institute, where he moved in 1929, then at the Lenin Academy and the Institute of Genetics in Moscow, where he moved in the late 1930s, he would tolerate no criticism of his views.[30] A dictatorial position in biological science was not thrust upon Lysenko in 1948; he sought it from an early moment in his career, and fought hard to keep it at the end.[31]

It is accordingly difficult, indeed it is impossible to stick to the substantive issues in analyzing his clash with the plant physiologists, for it did not follow the rules that enable a scientific community to stick to substantive issues. The usual efforts to maintain impersonal, rational discourse broke down, and the historian, like the participants, is obliged to try psychological analysis of the arguers along with substantive analysis of their arguments. That is what we are doing when we say—and we can hardly avoid saying it—that Lysenko called political power into play, forcing his scientific opponents to various types of evasion, hypocrisy, and dishonesty. From the start he would not comply with the first demand on a person offering knowledge to a scientific community: show the relevance to previous knowledge. He either ignored the demand or declared that he did not need to concern himself with trivial scholasticism, which had no relevance to the needs of Soviet agriculture. "It is better to know less," he told a group of critics in 1934, "but to know precisely what is necessary for practice, both for the present day and for the immediate future." [32]

It was Lysenko's self-appointed mission to create from scratch a new theory of plant development that would be of immediate benefit to farmers. He began by pasting the name of vernalization—that at least was indubitably his—on almost any kind of stimulation *and* almost any kind of resulting growth; in almost any kind of "seed" (tubers and cuttings as well as true seed); with raised temperatures as well as lowered; in light or dark; with moist or dry conditions. He drew the line, however, at the photoperiodic effect, which was the favorite experimental technique for studying plant development in the 1920s, and at the use of chemical growth substances or plant hormones, which came to be the favorite technique in the 1930s. It is impossible to understand these antipathies of Lysenko's without recourse to psychological analysis, for he never discussed photoperiodic and chemical effects at any length. He simply ignored them or brushed them aside, leading his disciples to declare that he had "disproved the theory, which used to exist among plant physiologists, of the existence of the so-called photoperiodic reaction." [33] Lysenkoite hostility to chemical stimulants and inhibitors was great enough to reveal the psychological cause:

> In essence the "hormonal theory" is a poorly masked attempt to distract our scientific research thought from the advanced and progressive theory of the stage development of plants, as worked out by Academician T. D. Lysenko.[34]

A purely logical Lysenko might have taken a more tolerant view of photoperiod and growth substances; his theory of stages was vague enough to encompass virtually anything. Indeed, in the time of declining power his learned followers began to stretch the stage theory to allow for phenomena that they had denied or brushed aside in the time of rising power. In the heady period from the 1930s to the mid-1950s Lysenko's vagueness was not an instrument of a bureaucratic chameleon or a political broker, who rise by offending nobody and changing nothing. In the agrobiology of that time, as in Stalinist thought generally, vagueness had a willful, activist function; it allowed the chief to tell his inferiors what new things must be believed and done at every turn in the hazardous struggle. Lysenko was annoyed by scientists who studied plant hormones as Stalin was annoyed by academic economists, because they undercut his claim that practical utility and theoretical truth were fused in every intuitive judgment that he made.

The reader may have the uncomfortable feeling that this is *ad hominem* nastiness rather than objective analysis of a man's beliefs. Unfortunately the analysis of ideological beliefs cannot avoid statements that are uncomfortably similar to accusations. Consider the following example. Lysenko pinned his concept of vernalization to the plain old practice of sprouting potatoes before planting them. Assume, as I do, that he was not consciously faking. His statements about the vernalization of potatoes must still be characterized as a verbal shuffle, whose function was to enhance his fame as a practical biologist. Can the reader discover any other meaning in the following passage?

> The speeded up development of such plants [that is, sprouted] we explain basically not by the fact that the eyes of the tubers are sprouted before planting, but by the fact that the sprouts (though

> they are very small) are subjected to the influence of certain external conditions, namely: the influence of light (of a long spring day) and of a temperature of 15–20° C. Under the influence of these external conditions (and that precisely is vernalization [*a eto i est' iarovizatsiia*]), in the potato tubers' eyes as they start to grow there occur those qualitative changes which, after the tubers are planted, will lead the plant to more rapid flowering, and hence to more rapid formation of young tubers.[35]

Aside from the factual error at the end — the implication of a causal connection between flowering and tuber formation[36] — this characteristic passage is inane. It conveys the feeling of an explanation without its substance. It has a striking similarity with the explanations offered by "peasant experimenters" for the supposed benefits of the seed stimulants that were tried in their "hut labs" during the 1920s. Their views were put in print by an organizer of the "hut lab" movement, who subsequently became one of Lysenko's chief disciples:

> Some peasants explain that, during the time of moistening, the seeds go through, as it were, "a part of their life's journey." Sown in the ground such seeds sprout more quickly than unmoistened ones, and in their further development outstrip them.[37]

Lysenko did not always write clear inanity about plant development. Sometimes he was obscure. But clear or cloudy, meaning must usually be sought in the social function of his statements, rather than their relevance to the life process of plants. The grand theory of plant development, which he promised in the early 1930s, never got past the first two stages, vernalization and "light." After a brief flurry of articles and pamphlets, none of them reporting anything like genuine experimentation or theorizing, he simply dumped the whole project on his disciples, while he went off to make a revolution in genetics. It is possible to read scientific meaning into these articles and pamphlets, which were quite careless of its requirements, but then inconsistency becomes the problem. One can, for example, compile a list of seven meanings that Lysenko gave, often inadvertently, to his key concept of vernalization:

1. the transformation of winter into spring habit by chilling moistened seed
2. the hastening of the reproductive stage, in spring- as well as winter-habited plants, by chilling seed
3. the hastening of the reproductive stage by warming seed

It was suggested to Lysenko that he make a meaningful generalization of these three, and of the behavior of perennials: Vernalization is simply another name for temperature aftereffect (or thermal induction), that is, the hastening or retarding of flowering by suitable alterations of temperature. But he angrily rejected this limitation of his concept, insisting that vernalization was something much broader:

4. the initial stage in the development of any plant or part of a plant, when certain conditions of air, moisture, and temperature are essential for the onset of the next stage and for ultimate flowering, but light is irrelevant

This amounted to a verifiable assertion, which could be proved wrong on many counts. There are seeds, for example, that need light to germinate, and — to take the example that especially enraged Lysenko — winter wheat will flower without chilling, if it is subjected to continuous light.[38] As if that were not trouble enough for the theory, Lysenko declared that potatoes are vernalized by warmth *and light,* thereby implying — if one wishes to be logical, as he did not — still more meanings for vernalization:

5. the stimulation of buds by warmth and light
6. the initial stage as in no. 4 but with light essential

And finally the logical reader must find a place on the list for such passages as the one quoted above, "Under the influence of these external circumstances (and that precisely is vernalization). . . ." That comes under number

7. null [39]

An equivalent list of meanings for the light stage would be less impressive, for Lysenko had far less to say about it. One example will suffice to show the difficulty he had in trying to discuss it without getting entangled in the distasteful concept of photoperiod:

> To get through the *light* stage some plants require *light,* temperature, moisture; other plants, of which millet is one, get through the light stage if *darkness,* moisture, air are available. . . . For millet to get through the *light* stage light is not a necessary factor. However, if when getting through the light stage, a millet plant is kept in the dark for 5 to 10 days, this will lead either to retardation of growth or to the death of the plant. Thus, light is necessary not for getting through the light stage, but for the process of nutrition.[40]

As Kuzma Prutkov might have said, one cannot avoid the unavoidable or embrace the unembraceable.

Within five years of starting to build his own stage theory Lysenko simply abandoned the effort. Some disciples and flatterers kept trying to complete it for him, but there is little point in reviewing their efforts. They were trying to breathe scientific life into a clot of inanity and inconsistencies, and they tended to quit as political bosses quit nodding their approval. From the mid-1950s to the end of 1964 the chief point of Lysenkoite talk about the stage theory was to prove that the data of hostile scientists could be accommodated within the master's doctrine.[41] Since his final fall in 1965 the few plant physiologists who have discussed it at all have been mainly concerned to keep a shred of self-respect as they abandon the doctrine in favor of standard science.[42] In short, Lysenko's stage theory has failed to survive his loss of political power, though some effort was made to achieve that miracle.[43] All that survives is the term "vernalization," now confined to sense one in the list above. Even that, in my opinion, is too much. Other terms, such as "thermal induction" or "temperature aftereffect," describe the phenomenon more sensibly and without unpleasant associations.

How then are we to account for the fact that many plant physiologists once credited Lysenko with important contributions to their discipline? At first glance the answer seems simple and uninteresting, depressingly void of significance for the history of science: Soviet plant physiologists were forced to praise Lysenko, and some of their foreign colleagues thoughtlessly took them at their word. Actually the matter is not so simple. Soviet geneticists were subject to equal or greater pressure to acclaim Lysenko's contributions to their discipline, yet very few did so, and hardly any of their foreign colleagues followed their example. Are we to conclude that plant physiologists are generally less perceptive or less dedicated to their calling than geneticists? Hardly. If we have the stomach to dig in the grubby record, to examine the actual reactions of these two communities as pseudoscience was forced upon them, we discover that the different patterns were shaped by essential differences in the two disciplines.

Consider the case of Maksimov, the most eminent plant physiologist at the end of the 1920s, when Lysenko first intruded in that community. Maksimov's initial response was to tell the simple truth: "The results obtained by Comrade Lysenko do not represent anything new in principle; they are not a scientific discovery in the precise sense of the word." [44] By 1933, responding to a British request for a report on vernalization, he was no longer straightforward, but he made a strenuous effort to be honest with his scientific colleagues abroad, almost none of whom could read Russian. He began with a display of great respect for Lysenko's work, stressing especially its practical value to Soviet agriculture, and then he spent the bulk of his article on the work of predecessors and contemporaries, bringing out in this indirect way the inanity and confusion of the contributions that were distinctively Lysenko's.[45] To this and other signs of Maksimov's continuing skepticism the Lysenkoites responded with a virulent attack,[46] culminating in a special meeting at his institute, in April 1936, at which he was subjected to a string of denunciatory speeches. At the end Maksimov rose to apologize for criticizing Lysenkoism, saying that he had changed his "point of view" since he wrote the British article.[47] Indeed he had. He never again criticized Lysenkoism in print. Worse yet, the few lines of tribute to vernalization that he had inserted in a 1931 edition of his famous textbook, grew to several pages of uncritical praise for the stage theory in subsequent editions.[48] Following the August Session of 1948 he made another abject public apology, no longer for criticizing Lysenkoism but simply for allowing non-Lysenkoite work within the Institute of Plant Physiology that he directed.[49] In 1949 he put his name to an utterly Lysenkoite effusion on the stage theory,[50] and died in 1952, too early to take advantage of the reviving possibility of criticizing this pseudoscience.

I do not set this down in order to perpetuate the humiliation that a fine scientist endured in his lifetime. My purpose is to analyze the pathological condition of a scientific community. There were far worse symptoms than Maksimov's insincere praise of Lysenkoism. In a sense his dishonesty was a service to his discipline. By humiliating himself he held on to administrative power, and so was able to protect men like Chailakhian, who did the research on plant hormones for which Maksimov made public apology in 1948. By inserting some pages of Lysenkoism in his textbook he managed to keep it in print, thus spreading many more pages of instruction in genuine plant physiology among students throughout the Soviet Union. In short, Maksimov was engaged in pliable defense of his science, letting the enemy in at points of greatest pressure to prevent the complete destruction of the whole enterprise. Of course this was a painful strategy and a dangerous one, shading off by insensible degrees into complete opportunism and total surrender to pseudoscience. A fine specimen of the complete opportunist is V. I. Razumov, who began his scientific career as a junior colleague of Maksimov's, earning Lysenkoite hostility for studies of photoperiodism that cast doubt on Lysenko's hasty generalizations.[51] In the mid-1930s, as soon as Lysenko's dominance was clearly established, Razumov suffered a complete conversion. He did not, to be sure, become a militant ignoramus like Lysenko; very few scientists did. He retained enough knowledge to provide the stage theory with sophistic arguments that the ignoramuses were incapable of. Thus he was especially valuable in the periods before 1948 and after 1952, when it was not politic simply to deny the existence of plant hormones or to banish the term "photoperiod." But crude or sophisticated, Razumov became so completely identified with Lysenkoism — by the end he was Lysenko's chief plant physiologist — that in 1965 he went into eclipse along with his master.[52]

Few plant physiologists committed themselves so wantonly to Lysenkoism. Vlasiuk may serve as our specimen of the canny opportunist, a far more common type. He turned against research in plant hormones not in the 1930s, as soon as Lysenko's hostility to them became apparent, but in 1948, as soon as it became official. Unlike Maksimov, who made a public show of turning against such research in 1948, Vlasiuk really did, which was a serious matter because he was a very important administrator of science.[53] He was, among other things, the president of the Ukrainian Academy of Agricultural Sciences, who shocked Khrushchev in 1961 by confessing that his endorsement of *travopol'e* had been insincere, the product of his loyalty to the Party.[54] It was extremely gauche to recognize that loyalty to political authority might be at odds with dedication to one's scientific calling, yet Vlasiuk managed to keep his career going, in part because he was one of the most energetic proponents of Lysenko's fertilizer scheme. In fact, he published an article on behalf of the famous composts *after* Khrushchev's dismissal had deprived them of political support.[55] But once Lysenko's fertilizers were formally investigated and officially condemned, Vlasiuk adapted himself once again to the new situation, and remained a major figure in the Soviet community of plant physiologists.[56]

Thus the forceful intrusion of a militant ignoramus into plant physiology reduced some of its adepts — such as Razumov and Vlasiuk — to opportunistic surrender, while provoking others — such as

Maksimov — to pliable defense of their scientific principles. To complete the typology we must add the intransigent specialist, the stiff-necked man or woman who would concede nothing to pseudoscience except silence, when political authority made honest speech or writing impossible. D. A. Sabinin, for example, professor of plant physiology at Moscow University, virtually refused to give printed tribute to Lysenko and made fun of him in the lecture hall, even after the university newspaper insinuated that Sabinin's loyalty was placed in question by this ridicule.[57] For a year in the late 1930s he was suspended from his post, and then, after the August Session of 1948, he was exiled from Moscow altogether, to wander without employment until 1951, when he shot himself.[58] He left behind a manuscript on the physiology of development, in which he firmly disposed of Lysenko's stage theory.[59] Another specialist of similar character was Chailakhian, who survived to see Sabinin's manuscript published in 1963. One of the country's leading authorities on plant hormones, Chailakhian made extremely few concessions to the Lysenkoites while defending his subject against them.[60] Indeed, he resisted the temptation to which Kholodnyi, his elder and rival in the study of plant hormones, succumbed. When they disagreed in their theoretical speculations about these obscure substances, Kholodnyi was not above an attempt to sick the Lysenkoites onto Chailakhian.[61]

Intransigent specialists did not control most centers of research and higher education in plant physiology, but neither did militant ignoramuses or complete opportunists. Pliable men of scientific principle seem to have held most of the important posts during the thirty-five years when Lysenkoism enjoyed the support of political authority.[62] They were obliged to skimp some of the biochemical research that was the cutting edge of their science, and they had to tolerate Lysenkoite nonsense in their journals and textbooks. Thus they dropped behind the world's leading centers of their discipline, and blushed to see foreign colleagues use "Russian" or "Soviet" as a synonym for Lysenkoite. But they managed to keep functioning as a scientific community. In this respect their fate was strikingly different from that of the geneticists, who were overwhelmingly intransigent in their reaction to Lysenkoism and were overwhelmed by complete disaster. Equally committed to two different disciplines, the two communities behaved in different ways, provoking different reactions from their common foe.

GENETICS

Plant physiology was — and still is — a sprawling, highly empirical discipline, relatively lacking in unifying theories. With his fuzzy concept of universal vernalization and his inchoate theory of stages Lysenko pitted willful vagueness against simple methodological injunctions (try to be consistent, look for biochemical stimulators and inhibitors, and so on), and against a great mass of factual information, not against a full-fledged theory of plant development. Such a theory is still music of the future.[63] Even so, the conflict between dogmatic vagueness and biochemical empiricism occurred in one field of the discipline, which could be separated from the others precisely because unifying theories were — and still are — lacking. A plant physiologist who wanted to keep out of trouble did not have to convert to pseudoscience. He could work in some other field than plant development. Even if he was bold enough to stay in that special preserve of the Lysenkoites, he could do empirical research and report the results as such, without waving theoretical red flags at the Lysenkoite bulls. He could, for example, write about "phases" rather than "stages," or about auxins and gibberellins rather than plant hormones or growth substances, and thus avoid provoking the Lysenkoites, except in the wildest years, 1948 to 1952.[64]

In sharp contrast genetics was — and is — a highly theoretical discipline, elegantly centered on a few basic concepts and theories. Some admirers have acclaimed its similarity with a formal or deductive science. Here for example, is the great mathematician Hilbert, reacting to the construction of genetic maps on the evidence of crossing over:

> The numbers [percentages of crossing over] are in accord with the linear Euclidean axioms of congruence and agree with the axioms concerning the geometrical concept "between." Thus the laws of heredity emerge as an application of the linear axiom of congruence, that is, of the elementary geometrical propositions concerning the displacement of line segments — so simply and precisely, and at the same time so wonderfully that no one could have imagined it in his boldest fantasy.[65]

Hilbert was probably exaggerating the axiomatic nature of genetics, as the Lysenkoites were, when they made "formal genetics" a favorite epithet. But exaggerated or exact, admired or despised, there *is* a pronounced formal quality in the science. When Lysenko attacked its axioms, the whole discipline was in mortal danger; there were no sidelines where geneticists could take refuge, as the plant physiologists did. Of all biological disciplines the geneticists came closest to the formal ideal that has attracted the scientific mentality since its origin, and they paid for this success. They had no empirical maze to hide in, when militant ignoramuses denounced rigorous thought as an obstacle to practical activity.

The simple principles of genetics seem so easy to grasp, to people who find them in textbooks, that the original difficulties of finding them in nature are usually overlooked. They are worth reviewing here, for they tend to recur whenever an untrained practical person has to deal with problems of heredity, as Lysenko did when he decided to breed an improved variety of wheat in three years. With the advantage of hindsight we can see that the basic difficulty was finding a way to distinguish between that part of an organism which is hereditarily determined and that part which is not, between the genotype and the phenotype, as Johannsen in 1909 named these inseparable aspects of an indivisible whole. The distinction must be made, if one is to think about heredity, yet there are no obvious morphological structures that stimulate the mind to visualize it, as there are, say, leaf and flower to prompt the distinction between vegetative and reproductive functions. Like the atoms of Democritus or Newton's particles of light, the genes are apprehended by abstract thought. Indeed this is a kind of abstract thought that suggests metaphysics, for the evidence of our senses seems flatly opposed to the argument that matter and light and living organisms are mosaics of particulate entities. Yet this method of thinking about heredity became increasingly necessary for practical breeders as well as theoretical biologists during the eighteenth and nineteenth centuries. The breeders were stumbling toward it by their intensified efforts to separate or combine, to emphasize or diminish, particular characters of domestic plants and animals; the biologists, by trying to arrange species in a natural system, defining their "essences" in such a way as to explain similarities, differences, and patterns of transition. The breeders, as we would say nowadays, were grappling with the problems of individual heredity and how it may be changed, the biologists with the problem of evolution, that is, changes in the pooled heredity of breeding populations. It seemed obvious that the second or larger problem could not be solved without making certain assumptions about the first or smaller problem. Indeed theoretical biologists rarely thought of them as separate problems.

Darwin and Mendel made the separation; that is how their great breakthroughs were effected. They were not quite conscious of what they were doing. Darwin worried a lot about the origin of individual variations, and repeatedly tried to formulate the theory of natural selection in which a way as to connect it with a theory of individual heredity. But logically the theory of natural selection was independent of his speculations about individual heredity. Mendel's thought process is much harder to analyze, for he left only a small record of it, but it seems fairly clear that he too worried about the connection between these two problems. In the introduction to his famous article he expressed the hope that some day a connection might be established between his laws of individual heredity and "the evolution of organic forms." [66] But that was only a fleeting glance at the big problem of his time; in the rest of the article he modestly confined himself to a problem so small and simple that it probably seemed trivial to his contemporaries, if they paid any attention to it at all: What is the most elementary statistical pattern of characters that persist through successive generations of hybrids?

Mendel was not only setting aside the grand problem of species transformations, which enthralled laymen as well as professional biologists in his time; he was not even facing the subsidiary problem of individual variations. He was simply trying to turn into precise ratios the banal observation that hybrids sometimes resemble one parent, sometimes the other, and sometimes an intermediate mixture. That seems to be the main reason why his work was ignored from its publication in 1866 until its simultaneous rediscovery by three different biologists in 1900. Biologists had to realize which extremely simple problems required solution before the big complex ones could be effectively dealt with. After a generation of inconclusive speculative debate over the sources of individual variations — and at the same time, a generation of intensive work at cytology — a number of biologists came all at once to appreciate the significance of Mendel's simple ratios and the method by which he had established them.[67] By counting individual characters, as they combine and segregate in successive generations of hybrids, Mendel had found the only way to make a precise distinction between what is hereditarily determined and what is not. Ultimately, said the enthusiasts, this method would explain not just the disappearance and reappearance of green peas in successive generations of hybridization with yellow; this was the way to solve all the big puzzles about individual variation and even about the transformation of species.

Pooh-pooh was the instant response of many biologists, including the celebrated Haeckel in Germany, and his analogue in Russia, K. A. Timiriazev, a very learned biologist and superb popularizer, whose influence among the Russian intelligentsia was enhanced by his political radicalism. Timiriazev respected Mendelism, especially for its answer to an antievolutionary argument that had disturbed Darwin: An individual variation from the species type, however advantageous, could not be the source of a new species because it would be swamped in breeding with the multitudinous individuals of the older type. Darwin had two answers to this objection, and Timiriazev refused to choose between them, as the extreme Mendelians insisted must be done, in order to achieve consistency. On the one hand, Darwin knew that some characters cannot be swamped because they do not blend in hybridization. As in the case of yellow and green peas, the progeny may have either one parent's distinctive trait or the other, and the supposedly lost trait will reappear in later generations, repeatedly available for natural selection. On the other hand, Darwin knew that many of the most important characters do not have this simple discontinuous nature. They blend in hybridization, which seems to substantiate the argument that a single variation, however advantageous, will get swamped. In that case the only defense of evolution seemed to be Lamarckian: A species is transformed as many individuals simultaneously change their heredity in the same direction, adaptively responding to a common environmental stimulus.*

Darwin, and Timiriazev after him, did not believe that this Lamarckian assumption was necessarily a teleological departure from the mechanistic theory of natural selection. More about the philosophical aspects of the problem later. Here the point is simply that the radical Mendelians of the early twentieth century shocked learned biologists with the extravagance of their mechanistic assertions. They denied the reality of blending heredity, which we can all see in our daily observation of parents and progeny. It is only an appearance, they insisted, the product of complex combinations of particulate hereditary characters. Indeed the most ex-

* It is unfair to Lamarck to perpetuate his name only in this meaning, but usage decrees the injustice whether we approve or not. For a fair appraisal of Lamarck's contribution to science, see the references in chap. 7, n. 69.

treme position was that hereditary characters are not only particulate but eternal; their combinations may be shuffled by selection, but the elements being shuffled are themselves changeless. To monistic radicalism of this sort Timiriazev replied with eclectic common sense (his radicalism was confined to human problems): Mendelian and Lamarckian assumptions are not mutually exclusive; they are complementary contributions to the great science of evolution, Darwinism.[68]

With the advantge of hindsight, we can see the mistake of Timiriazev and those who contined to think as he did through the 1920s. They were resisting the choice between studies of evolution and studies of individual heredity, which would have to proceed separately before they could be joined in a scientific rather than a speculative union. To keep them separate, to work out Mendelian ratios of simple discontinuous characters, while ignoring complex characters *and* the sources of individual variation *and* the transformation of species — this seemed extremely one-sided and simple-minded. It seemed to old-fashioned biologists that the Mendelians were making far too hasty and far too sweeping generalizations from their limited studies of one pattern of heredity. The expectation that other patterns would be discovered therefore continued into the 1920s. Relatively few biologists actually designed experiments to prove the Lamarckian case. To design experiments on heredity was almost inevitably to be trapped in the distinction between genotype and phenotype, which is death to Lamarckism.* But many biologists working in other fields than genetics, especially those in evolutionary systematics, continued to think that a synthesis of Lamarckism and Mendelism would be the ultimate result.[69]

And then, in the 1920s, geneticists discovered a way to explain complex characters *and* individual variations *and* the transformation of species, all on the basis of Mendelian concepts. They were very far from final solutions of these enormously complicated problems. They simply demonstrated, by some crucial case studies, that they *could* solve them; it began to appear that they knew how to go about it. That was enough. Lamarckism rapidly withered away, for

* The fatality is virtually imposed by definition alone. The genotype is defined as the hereditary aspects of an organism. Hereditary means self-replicating, which rules out innovation except as accidental aberrations of the self-replicating mechanism. If we think of it as capable of adaptive changes, we have ceased to think of it as a purely self-replicating mechanism; we are spoiling our effort to distinguish between that which is hereditary and that which is not.

it had lost its useful service. Speculation about the sources of variation and the mechanisms of evolution seemed no longer necessary, now that rigorous generalization was possible, generalization of experimental data that splendidly complemented and enriched the initial theoretical assumptions. The community of biologists unreservedly accepted genetics as a major discipline — many said it was *the* major discipline, which would revolutionize all the others. In any case the community began to hear less and less talk of Mendelism, Lamarckism, Darwinism, or any other eponymic doctrines, which are on everybody's tongue when a highly theoretical science is in the formative stage, as genetics was in the first three decades of the twentieth century.[70]

All this to give the patient reader some understanding of what Lysenko assaulted, *not* what he started with. The simple fact is that Lysenko's "genetics" originated entirely apart from intellectual processes in the community of biologists. His ideas about heredity were not derived from Lamarckism, or from any other trend in science, whether speculative or experimental, moribund or alive. Between 1933 and 1935 he created his own genetical concepts in a series of intuitive strokes, to suit the simple practical purpose of breeding an improved variety of wheat in two or three years, and to beat down the objections of learned critics. Any resemblance to genuinely scientific thought was purely accidental. Justice to the Lamarckists requires insistence on that fact, for many of them were serious biologists. They had nothing to do with the birth of Lysenkoism, and most of them kept their distance as it became a dominant force, if only because most of them were abandoning Lamarckism as the geneticists showed that they could solve the problems to which it was addressed. Komarov, to take the most notable example, was a botanist whose sympathetic interest in Lamarckism became increasingly historical during the 1930s, as it ceased to be a living trend in science. At the same time he became vice president (1930–1936) and then president (1936–1945) of the Academy of Sciences, where he protected genetics against the onslaught of Lysenkoism. And not only as an administrator. In successive editions of his popular books on plant evolution he stuck firmly to genuine science, showing how genetics was replacing Lamarckist speculation, and inserting only tiny, ritualistic nods to Lysenko.[71]

Furthermore, justice to *Lysenko* requires insistence on the fact that he did not derive his ideas from academic Lamarckism. He was very proud of the practical, agricultural origin of his genetical concepts. (Of course he would not tolerate the suggestion that he was merely juggling ancient superstitions and folklore of farmers. He insisted that he was transmuting the most advanced agricultural experience into revolutionary theoretical concepts.) For a time he denied any kinship with the Lamarckists. At the crucial meeting of December 1936, he asserted that "the geneticists, who have not really got to the bottom of Darwin's evolutionary doctrine, which we have mastered by action [*deistvenno*], . . . have no basis for calling our views Lamarckist." [72] He was wrong. There *was* an important similarity between his homemade "genetics" and Lamarckism; in the late 1940s he and his disciples finally admitted and even came to boast of it.[73] But the similarity was the unforeseen product of an essentially different approach to the problems of heredity.

Lysenko got involved in these problems without any intention of proving the inheritance of acquired characters, as the basic doctrine of Lamarckism is misleadingly named.* When he first reported on his work in plant breeding (January 1934), he praised Mendel, for showing that hereditary characters can be torn apart and put together again in many different patterns.[74] He criticized Mendel's latter-day disciples, the geneticists, for trying to inhibit practical activism. They had told him that he could not breed an improved variety of wheat in two or three years, because extended progeny tests are required to discover what will come of crossing various types. Lysenko was sure that he knew in advance which particular crosses would give him exactly what he wanted, and in any case he could select what he wanted from the first generation of hybrids, without worrying about segregation in further generations or about useful though recessive characters in the rejects. His theory of stages gave him this foreknowledge:

> By this means we have already succeeded in solving a problem that is quite unclear for formal genetics to this very day, i.e., the

* The name is misleading because it calls attention to irrelevancies, such as mutilations and other nonadaptive alterations. More serious is the false implication that environmentally induced change in heredity is affirmed by Lamarckism and utterly denied by genetics. Actually the geneticists have shown precisely how such change is effected. They differ from the Lamarckists in denying the "adequacy" or "specificity" of environmentally induced changes in the heredity of an individual organism. That is, they deny that an individual hereditary mechanism can make an adaptive response to an environmental influence, except by accident. On this basis they have shown how multitudes of breeding individuals, populations, can and do make finely adaptive responses to environmental influences.

problem of finding two parents such that early or late forms can be obtained by crossing them. The explanation of this, it seems to me, consists in the fact that we operate with the plant's development, i.e., with the interaction of the "internal" and a definite "external." But genetics, as it seems to me, operates all the time exclusively with characters, i.e., only with the results of development. It seems to me that genetics (formal genetics, of course) is essentially interested not in genes but in phenes.[75]

Scientific sense cannot be made of such effusions. This one joins factual inaccuracy — he did *not* have foreknowledge when hybridizing — with an emotional intuition of the Stalinist type: The geneticists must be wrong to tell him that he was wrong, for their academic prescriptions and prohibitions would slow down a practical breeder. Here one can see the beginning of his unintended approach to Lamarckism. His haste was a practical version, or perhaps we should say a gross anti-intellectual caricature, of the impatience that Lamarckist scientists had shown to solve big problems before the little subsidiary ones were settled. Theirs had been an academic impatience to know the connection between individual variation and species transformation before the simple rules of heredity were established. His was the impatience of a breeder — or a publicity hound — to get an improved variety without a lot of progeny testing. In both cases the result was a hasty leap over the distinction between what is hereditarily determined and what is not. When pressed on this point by geneticists, who showed precisely how the laborious distinction must be made, most Lamarckists quietly abandoned their speculations. Even if they were not completely convinced, they were professional scientists; they belonged to a problem-solving community and had to abide by its collective verdicts, if they could not dispute them rationally.[76] When Lysenko was pressed to distinguish between the hereditary and nonhereditary characters of his plants, he leaped to the conclusion that the distinction was an academic absurdity, and fought back irrationally, with power.

Thus in its practical consequences no less than its origin Lysenko's confusion of genotype and phenotype was drastically different from that of the Lamarckists. Far from submitting to the verdict of the scientific community, he used political power to create an opposing camp, an army of people who would unquestioningly recognize the truth of anything he said. (That is not an exaggeration; the moment a Lysenkoite began even limited criticism of the master's doctrine, he was denounced as a turncoat.)[77] The name of

Michurinism was fixed to this militant school, and genetics was totally rejected as Mendelism, or Weismanism (after August Weismann), or Morganism (after T. H. Morgan), or some hyphenated combination of those eponymic pejoratives. By 1937 Lysenko ceased to speak of genes or genotypes except to deny their existence, and chromosomes were mentioned only to deny that they had a special role in heredity. The basic concepts of genetics were replaced with notions so vague as to allow practical men complete freedom in altering living things to suit their needs (or publicity hounds complete freedom to create sensational impressions of such mastery). "In our conception," Lysenko wrote,

> the entire organism consists only of the ordinary body that everyone knows. There is in an organism no special substance [*veshchestvo*] apart from the ordinary body. But any little particle [*chastichka*], figuratively speaking, any granule [*krupinka*], any droplet [*kapel'ka*] of a living body, once it is alive, necessarily possesses the property of heredity, that is, the requirement of appropriate conditions for its life, growth, and development.[78]

If scientific content has to be read into such pronouncements, Lysenko can be interpreted as dissolving genetics into physiology, identifying the function of self-replication with all the other life functions. But such a paraphrase is misleading in its precision, for his understanding of the other life functions was almost as vague and evasive as his understanding of self-replication. The simple truth must be faced. Any part of biological science that Lysenko touched was turned into a vague, personal dogmatism.

To attempt a coherent outline of Lysenkoite "genetics" is thus a self-contradiction: it began and ended with opposition to clearcut thought and rational experimentation. What is more, Lysenko and especially his disciples shifted their stance as their political influence waxed and waned. By 1936 he overcame his initial reluctance to condemn genetics altogether, but he hesitated to come right out and say so, for the highest authorities decided at that time to subordinate genetics to agrobiology, not to abolish it. Lysenko and his zealots chafed at this limited victory even as they were winning it. "Why do we need this science?" one of them exclaimed at the crucial conference of December 1936; "it only hinders our work." [79] Within a few months Lysenko began an open campaign for complete abolition. He denounced "the whole logic of thinking with mechanical models, the very basis of the corpuscular theory in biology." [80] He went far beyond claims of foreknowledge in hybridization, as the term is ordinarily understood. He insisted that vegetative hybridization — alteration of heredity by grafting — is a fact that utterly subverts genetics (as indeed it would, if it were a fact, for stock and scion do not exchange germ cells).[81] His people reported that reruns of Mendel's famous experiments with peas did *not* yield the ratios Mendel claimed to find.[82] When the country's most eminent student of statistics showed that these people misinterpreted their own data,[83] Lysenko replied that mathematics has no relevance to biology. "That is why we biologists do not take the slightest interest in mathematical calculations that confirm the useless statistical formulas of the Mendelists." [84] He called genetics "real barefaced metaphysics" and rejected out of hand any attempt at compromise. "In my view it is time to eliminate Mendelism in all its varieties from all courses and textbooks." [85]

After another major conference, in 1939, reasserted the official effort at compromise, the Lysenkoites showed a little moderation, but dropped it as the great war with Germany ended. Now Lysenko endorsed the views of Olga Lepeshinskaia, an elderly physician who had been arguing since the early 1930s that she could grow cells from bits of egg yolk and other noncellular globs of matter.[86] She had also been demonstrating that there was an affinity between her views and Lysenko's,[87] but he had held back from agreement, for in the 1930s he conceded that his school ought to show how its notions of heredity accorded with the data of cytology.[88] In 1945 he finally summoned the courage to deny that unfulfilled obligation. The geneticists could have their scholastic delight at the correlation between Mendelian ratios and the intricate pattern of meiosis. By asserting that cells can grow from noncellular material, Lysenko and Lepeshinskaia made meiosis irrelevant, thereby opening the door to "cytological" justification of any kind of transmutation. In short, cytology was rejected along with genetics.[89] Even the theory of natural selection, which had seemed to be an unassailable part of the official theoretical ideology, was virtually rejected in the late 1940s, to make room for Lysenko's discovery that competition does not occur among organisms of the same species.[90] Though the August Session of 1948 was not supposed to give official sanction to these extreme views,[91] the Lysenkoites behaved afterward as if it had, as if they had won unlimited political support for their wildest speculations. A new law of species formation by direct transformation was announced. Wheat, for example, was transformed directly into rye when raised in "appropriate" conditions.[92] That was equivalent to saying that dogs give birth to foxes when raised in the woods, and very soon Lysenko's followers were making even more extravagant claims. One man turned viruses into bacteria, while another changed plant tissue into animal tissue, and a third drew chicken from a rabbit.[93] In effect living matter was becoming structureless goo, ready to be shaped at will into anything the Soviet farmer (or the publicity hound) might wish.

Lysenko never explicitly repudiated or amended the extremes that were reached at the apogee of his power in the late 1940s and early 1950s; he merely refrained from active propaganda for some of them as his political influence ebbed. He also tacitly approved the efforts of his more learned disciples — the ones who did not desert him — to revive something like the compromise of the late 1930s. They dropped Lepeshinskaia and conceded the legitimacy of cytology, of mathematical calculation, and even of biochemical research in nucleic acids. They tried to maintain the dominance of Lysenkoism by arguing that none of those supposedly new trends proved the existence of genes or justified the resurrection of genetics.[94] When political support evaporated, Lysenkoism collapsed altogether. The leader and a few of his disciples lapsed into enforced silence, while most of the school quietly abandoned their former notions in order to hold places in or around the scientific community. It is a distasteful and depressing sight; even the man who once turned rabbit into chicken continues as an editor of an important biological journal.[95] The admirer of persistent cranks can take heart at the occasional appearance in an agricultural journal of a Michurinist article defending vegetative hybrids.[96]

Through all the politically motivated shifts a characteristic style of thought and stubbornly unchanging doctrines persisted in Lysenkoite "genetics." They resist coherent presentation — to the end an authoritative textbook was never achieved[97] — but they do exhibit regularities, as, say, American advertising does. Of course, the Lysenkoite style of thought was never as cynically realistic and cunning as that of American advertising. Sometimes it resembled the ordinary stream of inarticulate consciousness, sometimes the savage thought process that Lévi-Strauss has named *bricolage*,[98] sometimes the kind of emotional thinking that is called feminine (as if women alone have strong opinions which they refuse to defend by reasonable argument). But none of these types of plainly "prerational" thought is as close to the characteristic style of Lysenkoism as the one that might be called "surrational" (or masculine, if we wish to keep an even score in the battle of the sexes): a show of rational discourse camouflaging a basic refusal to meet the tests of genuine reason.

The Lysenkoites endlessly expatiated on a few basic ideas, piling up reasons and facts (and "reasons" and "facts") to confirm them, but steadily evading or resisting the necessary first step to confirmation: an effort to define the ideas precisely. The beginning and the end of their thinking about heredity was the simple insistence that it is a property of the whole organism, with the supposed corollary that the organism contains no special substance or matter of heredity. Biologists had passed so far beyond this kind of simple-mindedness that they were inclined to respond with impatient gibes: Does the fuzz on a leaf play the same role in heredity as the pollen or the ovule in the flower? Does the cell wall have the same function in self-replication as the cell nucleus? Leaving aside sarcasm, as Soviet biologists were usually obliged to do, and considering the basic Lysenkoite idea dispassionately, as they were usually unable to do, one can find in it vestiges of scientific thought. Darwin, for example, pondering heredity when views on it were "utterly formless, diffuse, indefinite,"[99] assumed that the whole organism must contribute elements to the germ cell. Otherwise, how can we explain the replication by the germ cell of the entire organism with all its different parts? Darwin answered with the hypothesis of pangenesis: microscopic particles or gemmules flow from all the parts of an organism to the germ cell, which becomes in effect a package of seedlets, each one ready to recreate the part that formed it. That kind of speculation was rampant in the late nineteenth century, but it could not be tested, and the further development of cytology made it utterly unrealistic. However, when Darwin made it, it was a scientific speculation. It did not flout known facts. It was offered tentatively, to be accepted as true only if it could be proved true. And it was essentially mechanistic: to understand heredity as a property of the whole organism Darwin tried to reduce it to the interaction of constituent elements.

On all counts Lysenko's basic idea was not scientific speculation but an atavistic simulacrum of it, recalling Albertus Magnus[100] rather than Charles Darwin. Even when he seemed on the verge of reducing the hereditary property of the whole organism to the action of special particles within it, in his famous explanation of species transformation, Lysenko kept from falling into the mechanistic mode of thought:

> We conceive the matter as follows: In the body of a wheat plant organism, under the influence of appropriate conditions of life, granules [*krupinki*] of a rye body are born. But this birth does not arise by means of a transformation of the old into the new, in this case of wheat cells into rye cells, but by means of the rise in the depths [*nedrakh*] of the body of the organism of a given species, out of substance [*veshchestvo*] that does not have cellular structure, of granules of the body of another species.[101]

Lysenko's granules, unlike Darwin's gemmules, dissolved into an intangible substance even as he talked about them. It was pointless to ask how one might test for them, not only because the facts of cytology ruled out their existence, but even more basically because the mind had nothing to grasp in thinking about them. Anyhow Lysenko was far from requesting or tolerating tests. He ignored even the modest request for clear photographs of rye seeds growing on wheat plants; a careful search by a reliable man had turned up rye seeds in harvested wheat, and that was proof enough for Lysenko.[102] We will ignore his defense of simple frauds, who doctored photographs to show hornbeam trees producing hazelnut limbs and pine spruce,[103] as we will ignore his tacit approval of complex frauds, who made belated, grotesquely illogical efforts to prove that the master's doctrine of heredity was compatible with data concerning nucleic acids.[104] We are trying to understand Lysenkoite thought at its best, in its purest formlessness.

Just as the Lysenkoites resisted the attribution of the hereditary function to particular material forms, so too they resisted the reduction of species formation to the mechanistic processes of natural and artificial selection. At first their resistance seemed little more than a verbal shuffle: they insisted that selection does more than favor certain variations from the type, it "creates" them in some unspecified way. Then they specified: variation is directed by grafting and by alteration of the environment. Then they came to the sudden creation, "under suitable conditions," of new species in the depths of the body of the old. At the same time Lysenko's cluster method of tree planting also led on to his grand summation, "the law of life of a biological species." He saw that individual saplings do not die because they are deprived of light and moisture by surrounding trees; the scrawny ones die so that the healthy may live and the species flourish.[105] Even the roots of the self-sacrificing saplings know their duty; they graft themselves onto the roots of healthy trees to help the species live though the individual dies.

> If we proceed from the law of life of a species, then we begin to understand not only the causes of roots growing together, in other words, not only the transfer of roots from a tree that is internally ready to die to those that are not as yet ready to die, but also [we understand] why the growing together goes on not at the time when the tree is drying up and dying, but at the proper time, in some cases many years ahead of that. The roots grow together not because they touch one another. On the contrary, not infrequently they touch in order to grow together.[106]

The farsighted roots act in accordance with the species' law of life:

> The entire life of an organism and of any part of its body is directed in one way or another to the multiplication, to the increase of the mass of that biological species, one of whose forms of existence is the given organism, the given living body. . . . In any being (organism), in any particle of a living body, everything is directed to one and the same thing, to one "purpose" — to the increase of the living mass, the numbers of the biological species; but not any species, only that one which the given organism or the given particle of the living body is, the one to which it belongs.[107]

One is tempted to say that the Lysenkoites were simply vitalists, though prevented from admitting it by the Marxist philosophical tradition.[108] Yet their mode of thought was essentially anti-intellectual; they were incapable of creating a coherent, fully developed doctrine, whether mechanistic or vitalistic. They simply would not be pinned down, which is the function of any articulated body of thought. Consider, for example, their talk of "marriage for love" among plants and animals. The phrase was intentionally playful, designed to offend the mechanistic sensibilities of most biologists, but the underlying thought was quite serious. Fertilization, in the Lysenkoite view, was not a random process; sperms seek and eggs select their most suitable mates. This anthropomorphic language was not merely a handy way of pointing to the subtle mechanisms by which, say, pistil and pollen of the same species recognize their compatibility and mate. Biologists frequently use such anthropomorphic shorthand without ceasing to be scientists, for most of them are constantly searching for the mechanisms that accomplish the goal-directed feats of mindless protoplasm. Even those few, the vitalists, who insist that mechanistic explanations are not enough, begin their arguments for something more (life force, entelechy, or whatever) at the point where the mechanists allegedly fall short. They try, in short, to speak to the scientific community in its own language. Not so the Lysenkoites. Very belatedly they made some clumsy efforts in that direction, during the last decade of their politically sustained existence as a school. As their power to repress or ignore biochemical studies of life processes dwindled, they conceded that these studies were useful. Some Lysenkoites even looked through them, hunting proof of their inadequacy, insisting on the need for purely biological laws, such as "selective fertilization," the dignified name for "marriage for love." Lysenko himself never bothered with such donkeywork, and the assistants to whom it fell were so tendentious and unconvincing as to make one wonder whose side they were really on.[109]

The plain fact is that "marriage for love," like most of the pet notions of Lysenkoism, originated and ended as an intuitive rationalization of an agricultural practice, in contemptuous indifference both to the scientific study of fertilization and to the vitalists' intellectual dissatisfaction with that study. In the 1930s Lysenko decided that pure varieties of rye and other cross-pollinating plants do not have to be planted as far from each other as the seed laws required — until 1938, when he got them changed.[110] There was no reason, he said, to fear that proximity would pollute varietal purity, for the pistils of each variety could be counted on to choose the pollen most suitable to them:

> It has been proved by Darwinism, and also confirmed by special experiments, that in all those cases when the economic demands presented to a given plant coincide with its biological demands, in all those cases it will never be useful, and in some cases it will even be harmful, to limit the freedom of cross-pollination.[111]

Passages like this suggest not the scientific but the bucolic type of vitalism, such as D. H. Lawrence's hymns to the natural union of men with their living environment. And in fact the Lysenkoites did get a reputation for opposing artificial insemination of livestock on the ground, as a critic sneered, that it does not satisfy "the 'demands' of the cow, the ewe, or the mare."[112]

But once again we are dealing with a vestigial or abortive suggestion of coherent thought, in this case poetic, in all cases the product of an incoherent eagerness for sensationally quick agricultural progress (or for quick sensations of progress). Lysenkoite talk of "marriage for love" was an escape, by adman fantasy, from the painfully slow struggle to achieve varietal purity in peasant agriculture. In this case as in much else Lysenkoite vitalism was neither scientific nor bucolic. It was an agitprop (Soviet adman) pseudo-activist version of the famous *nichego!* (no matter!), with which Russian peasants responded to criticism of their slovenly fields. The traditional peasants may not have romanced about "marriage for love" in the plant world, but they did tell stories of cultivated plants changing into weeds long before Lysenko came along with his semi-literate version of the ancient fable. They saw it happen all the time.

THE AUTONOMY OF SCIENTISTS

This then was Lysenko's "genetics." Small wonder that Serebrovskii shouted *"Mrakobes!"* ("Obscurantist!" or, literally, "Demon of darkness!")[113] in his first debate with Lysenko, or that M. M. Zavadovskii issued a sober declaration of war in 1936:

> Proceeding from the need to be honest with myself and in the interests of the proper search for truth, I am obliged to say that Lysenko proposes to us the replacement of Mendel's conception by a miserable, wretched, primitive conception, unworthy of his remarkable powers.
>
> Compare, Comrade Lysenko, your crude homemade fabrication, which allows the most extreme teleological and vitalist explanations . . . with the coherent, finished, splendidly unambiguous conception of contemporary science. . . .[114]

The solemnity of this declaration, with its inconsistent tribute to Lysenko's unspecified powers,[115] suggests what must be obvious to anyone who has read the previous chapters of this book. Great courage was required to speak these plain truths in 1936,[116] and even more in the years following, until 1948, when criticism of Lysenko's doctrine was totally prohibited and the courageous specialists were those who fell into refractory silence.

It is hardly surprising that only a handful of geneticists and allied specialists (M. M. Zavadovskii was an embryologist) consistently displayed such courage, not only in the years of accumulating disaster, but also in the years of painfully slow recovery, when terror no longer threatened the scientific critic of Lysenkoism, but jobs and other prerogatives could still be forfeited by radical truthtelling.

Some people will automatically assume that the handful of bold radicals spoke for all their scientific comrades in genetics and the allied disciplines. This romantic expectation will be doubted by those who regard scientists as a species of white collar worker — *sluzhashchie,* serving people, is the Soviet term for the class — dedicated above all to climbing the ladder of reputation and position. Cynics will even expect to find mass conversion from science to Lysenkoism between 1935 and 1952, and the other way round during the next thirteen years, as political support moved to and from Lysenkoism. Fortunately we need not limit ourselves to idle guessing, for scientists make a public record of their behavior when they are faced with a conflict between professional and political considerations. When we examine this record, we find elements of truth in all three expectations. There were intransigent professionals, there were complete opportunists, and there were a lot of people who tried to avoid the choice between those extremes.

When Lysenko first invaded breeding and genetics, a number of important people in those fields tried to reach a compromise with him, as they had with Michurin. The most important would-be comprimiser was Vavilov, director of the Institute of Genetics at the Academy of Sciences and simultaneously of VIR (All-Union Institute of Plant Industry), which were the country's two most important centers of pure and applied genetics respectively. In striking contrast to bold alarmists like Serebrovskii, Vavilov tried to defend science by appeasement of its enemies. He praised Lysenko's theory of stages as

> a major achievement in plant science. It discloses broad horizons. We have not yet fully utilized this radical new approach to a plant, which, in addition to its significance as an agricultural technique, has supplied a new theory of breeding, and allows us to utilize more fully the world's resources of varieties.[117]

Vavilov chose his words carefully, so as to allow a little gentle criticism along with the strong praise, trying to win Lysenko away from his most flagrant attacks on science. But abuse was all he got for his pains. At the decisive conference of December 1936, a protégé who was defecting to Lysenkoism ridiculed Vavilov for his "heterozygous condition, or rather, the condition of a vegetative chimera, whose individual parts are incompatible. He is both . . . a Michurinist and an anti-Michurinist; he is both a Lysenkoite and an anti-Lysenkoite."[118] That may have been the unbearable limit of degrading diplomacy for Vavilov; or maybe, personal feelings aside, the conference as a whole proved to him the impossibility of compromise. Whatever the cause, he changed. His concluding remarks were quite different from the bland soothing syrup with which he had opened the conference. Firmly and neatly he took the side of the embattled geneticists, and stayed there until his arrest in August 1940.[119]

Vavilov was a very unusual individual, but the pattern of his response to Lysenkoism was symptomatic. Many other scientists, especially those in administrative positions, tried to protect their institutes by flattering the Lysenkoites. In plant physiology such pliable men of principle held most of the most important posts through the whole period from the rise of Lysenkoism in the 1930s to its collapse in the 1960s. In genetics a different pattern became apparent very early. Not only did this discipline produce utterly intransigent critics of Lysenkoism — the severest plant physiologists were not nearly as hard-hitting — but would-be compromisers like Vavilov were quickly driven to choose between utter intransigence and utter capitulation. Zhebrak, to take another example, was a peasant's son and Party member from the age of seventeen, had experienced the condescension of aristocratic scientists,[120] and was capable of seeing reactionary connotations in Vavilov's law of homologous series.[121] Yet he felt compelled to follow Vavilov into the camp of intransigence. Even as Zhebrak wrote a major appeal for reconciliation, in the spring of 1937, he was attacked for "denigrating" the achievements of Lysenkoites.[122] That was part of their campaign to drive genetics from higher education. Zhebrak, who had been "pushed up" at age thirty-four to be chairman of the Department of Cytogenetics at the Timiriazev Academy, the country's leading institution of higher education in agriculture, became one of the most unyielding, durable opponents of Lysenkoism.[123] His was one of the last eight centers of research and teaching in genetics that still remained in 1948, when the August Session precipitated their closing, and Zhebrak felt obliged to publish a wishy-washy self-criticism in *Pravda*.[124] The editors demanded total surrender, and he responded, when he got a professorship of botany at the Moscow Pharmaceutical Institute, by creating a little center of genetics there *sub rosa*.

If the terror had not snatched Meister in 1937, he too might have joined the militant defenders of genetics, though he seemed at the time of his disappearance to be the conciliator most favored by the ideological establishment. A plant breeder with a strong interest in theoretical genetics, Meister was codirector of the All-Union Institute of Grain Culture in Saratov, the celebrated creator of several famous varieties of wheat. His effort to recast genetics on the basis of dialectical materialism[125] — he had joined the Party in 1930 at age fifty-seven — was treated respectfully by the ideological establishment, and his summation of the Conference of December 1936 was officially billed as "the platform that can unite all the participants in the discussion." [126] But a careful examination of Meister's arguments makes it clear that he could not have maintained much longer his appearance of evenhanded criticism and praise for geneticists and Lysenkoites alike. On all essential issues he stood with the geneticists. Yet the chances were small that he would have said so openly and strongly. He was after all a plant breeder, with an option that was unavailable to most of the pure geneticists: he could occupy himself with wheat improvement and keep his mouth shut about the principles of genetics. That was the dominant trend among eminent Soviet plant breeders. At the Conference of December 1936, fourteen of the twenty-four people who criticized Lysenkoism were plant breeders; by the August Session of 1948 the ratio dropped to one of nine. And already in December 1936, almost half of the seventy-seven plant breeders who were prevailed upon to speak evaded the disputed issues, limiting themselves to a description of their creations.[127]

Of course, silence can be a form of protest. At the August Session, for example, though only one plant breeder criticized Lysenkoism, only one with any serious claim to distinction spoke for it, which went far to sustain the geneticists' argument that the best breeders were not Lysenkoites.[128] Two of the anti-Lysenkoite speakers went further; they explicitly claimed for genetics the celebrated varieties produced by Konstantinov and Shekhurdin, famous plant breeders who had publicly criticized Lysenko in the 1930s and now, at the August Session, sat quietly in the crowd of seven hundred. When Lysenko and his lieutenants called on them to repudiate any connection with reactionary Mendelism, and still they sat quietly, silence became a distinctly eloquent form of protest.[129] It tells us, to be sure, even less than polemical speeches about the roles that were actually played by genetics and by Lysenkoism in the day-to-day work of breeding improved varieties. That practical question will be considered later on. Here we are examining the academic issues that Lysenko raised, as he gathered political power to crush the science of genetics. Plant breeders could rise to the defense of the science, as a dwindling minority did; or they could avoid the conflict, as most of them did, by falling silent on contested issues or by making occasional Lysenkoite noises. They could even enlist in the Lysenkoite army, as a minority did, without necessarily and completely betraying their professional commitment to breed improved varieties.[130] That range of choice was denied to pure geneticists. Whether they engaged in militant defense of their discipline, or made compromising efforts to appease its enemies, or quietly concentrated on their special research and teaching, sooner or later the stark choice was forced upon them: to honor their professional commitment, or to betray it. If they honored it, they would almost certainly be fired from their jobs, and might follow Vavilov to prison or Agol to execution. If they betrayed it, there was far less chance that the terror would snatch them, and there was virtual certainty that they would get or keep high position.

Only the barest handful of people who were trained geneticists in the mid-1930s, when the trial began, chose to betray their professional commitment. Of the thirty-five people who were staff members at the Academy of Science's Institute of Genetics in 1937, only four turned Lysenkoite and kept their jobs past 1940, when Vavilov was arrested and Lysenko took his place as director.[131] Those four* are almost the complete list of Soviet geneticists who became Lysenkoites. Some others mouthed the phrases of conversion, especially after the August Session of 1948 made this a clear test of political loyalty.[132] Usually honest hypocrisy was evident in the confessions of these minor Galileos, as in Zhebrak's. And even in the rare cases when a Communist geneticist made a strenuous effort to be simultaneously loyal both to his scientific and to his political disciplines, the scientific won out over the political.

Alikhanian is a fine specimen of this rare but significant type, whose behavior illuminates the clash of the two disciplines. During the turmoil of the late 1930s he was a shrewd young geneticist and Party leader at Moscow University, taking a major part in "the liquidation of wrecking," and enthusiastically endorsing a chimerical textbook that had Lysenkoite passages interspersed in the usual ex-

* They deserve to be known: R. L. Dozortseva, K. V. Kosikov, Kh. F. Kushner, and N. I. Nuzhdin.

position of genetics.[133] To answer the charge that genetics was practically useless, he began to breed an improved variety of chicken, and won official praise at the conference of October 1939, where he described his chicken and criticized the Lysenkoites for trying to abolish genetics.[134] Subsequently he dropped chickenbreeding and proved his loyalty by sacrificing a leg in the war against German aggression. At the time of the August Session in 1948, Serebrovskii had just died and Alikhanian had taken his place as chairman of Moscow University's department of genetics, though he was only a *dotsent* in academic rank. Thus he was a prime, vulnerable target for Lysenkoite hostility toward the university that had been a major center of anti-Lysenkoite agitation. (As late as February 1948, a large conference had gathered there to defend natural selection.) Great pressure obliged Alikhanian to mount the tribune and tell the August Session that genetics was a true and useful science, fully compatible with Michurinism.[135] Lysenko shook him up with heckling, but he rallied and stuck to his views until the very end of the conference, when Lysenko announced that the Central Committee had read and approved his report. The audience responded with a stormy standing ovation, and Alikhanian joined two other anti-Lysenkoite speakers (out of a total of nine) in instant recantation.[136] His seemed the most complete, utterly lacking the quibbles and cavils of the other two, entirely rejecting his science in favor of "the morality of the state, the morality of the people [*narod*]," as another repentant biologist at another meeting described the force that brought him to his knees.[137]

It seems likely that Alikhanian was entirely sincere as he delivered his recantation:

> It is important to understand that we must be on this side of the scientific barricades, with our Party, with our Soviet science. . . . I, as a Communist, cannot and must not, in the heat of polemics, directly oppose my own personal views and concepts to the whole forward movement of the development of biological science. . . . As of tomorrow I will not only begin to emancipate all my own scientific activity from the old reactionary Weismanist-Morganist views, but I will also begin to remake, to transform all my students and colleagues.
>
> It is impossible to conceal the fact that this will be an extremely difficult and agonizing process. . . . I categorically declare to my colleagues that in the future I will fight with those who formerly thought as I did, unless they understand and go along with the Michurinist trend. . . . We will transform Moscow State Univer-

> sity into a center of propaganda of the Michurinist doctrine, into a center of work on Michurinist biology.[138]

He was a dutiful Communist. But he was also a professional geneticist, and as soon as he got the chance he resumed research in his science, within the shelter of the Institute of Atomic Energy. He was almost certainly insincere when he included in his description of the results patent nonsense about vegetative hybridization — of *Penicillium*.[139] As the Lysenkoites lost the support of Party leaders Alikhanian dropped such talk, and eighteen years after the August Session, when he was restored to his position at Moscow University, he published the study of Michurin he had outlined in 1948, a tendentious argument that Michurin's views were fully compatible with genetics.[140] That has been the official view since Lysenko's fall. Once again Alikhanian has every reason to believe that there is no conflict between his political and scientific disciplines. After all, he is a geneticist, not an historian of ideas.

The stubbornness of the people who were already trained geneticists when Lysenkoism invaded their field is clear evidence that their professional commitment was stronger than their political, even in the rare case of Alikhanian, who made a great effort to be loyal to both. Nevertheless the strength of the professional commitment should not be exaggerated. It provoked only a small number of geneticists to challenge the authorities directly, by publishing criticism of Lysenko when that was frowned upon, and by falling into eloquent silence when demands for recantation were addressed expressly to them. The majority of geneticists did their work quietly until they were pushed away from it, and returned to it just as quietly — those that survived — when the political bosses allowed them to return. Even the minority who publicly criticized Lysenkoism were usually very circumspect, carefully avoiding an open and explicit attack on the political authorities who stood behind it. With very few exceptions they shied away from the basic issues of agricultural recipes and of academic autonomy, confining themselves almost entirely to criticism of Lysenkoite views in pure science. When the authorities reimposed a ban even on scientific polemics, between 1959 and 1964, almost no defiant articles were published by geneticists.[141] That kind of daring was left to a small group of Varangians, as an admiring journalist called the chemists, physicists, and mathematicians who ventured out of their special fields in defense of genetics.[142] The Varangian phenomenon has, naturally enough, occurred most among the scientists who have been least pressed to defend academic autonomy within their own disciplines. Among the hard-pressed biologists, as among social scientists, the rule has been don't-trouble-trouble, while an uncomfortably long list of such specialists gave endorsements to Lysenkoism when it was dominant, and stopped giving endorsements as political support ebbed away from it. If the political authorities allowed biologists a bare minimum of autonomy necessary for performance of their special services, most of them were obedient *sluzhashchie,* white-collar workers.

Of course the commitment to scientific inquiry did keep most biologists from turning into complete opportunists, it did turn nearly all geneticists into intransigent though largely silent resisters, and it did turn a few scientists into knights errant of the free spirit. But anyone who is inclined to draw romantic generalizations from these facts should read the stenographic record of a gathering at the Academy of Sciences in 1948, the aftermath of the more famous August Session.[143] The real issues had been made crystal clear and utterly unarguable by Lysenko's announcement, on the last day of the August Session, that the Central Committee had read and approved his report. It was impossible in Stalin's Russia to engage in debate with the highest political authority. Now the heroic free spirits were those who stayed away from the meeting, in particular four men whom the Lysenkoites repeatedly denounced as virtual enemies of the people for their refusal to make the required self-criticism.* Those who came and spoke were of three types. Some were Lysenkoites, celebrating their total victory with uninhibited frenzy.[144] Some were honest hypocrites like the plant physiologist Maksimov or the ecologist Sukachev, who humiliated themselves in order to save their institutes. And some were complete careerists, intent on securing their positions regardless of other individuals or principles. A prime specimen of the last type was the histologist Khrushchov, director of the Institute of Cytology, Histology, and Embryology, which had continued to be one of the main centers of resistance to Lysenkoism even after it had been taken from its founder and director, Kol'tsov. Director Khrushchov labored to

* They deserve to be known: the evolutionary theorist Shmal'gauzen (or Schmalhausen), and the geneticists Dubinin, Rapoport, and Zhebrak. I. V. Pan'shin and Timofeev-Ressovskii, who could not attend because they were in prison, were denounced as actual enemies of the people. The principal denouncers were Glushchenko and Kushner.

prove that not he but this, that, and the other subordinate, working behind his back and over his head, had fostered Mendelism-Morganism in the Institute.[145] Similar speeches were delivered by a neurophysiologist and two leading philosophers.[146] Each philosopher tried to convict the other of responsibility for the evasions and compromises that had been the characteristic reaction of the philosophical establishment to Lysenkoism. Only an obscene metaphor can do justice to this meeting. The Lysenkoites had forced political salts into the bowels of Soviet scientists, and some began to void themselves in public, the honorable ones fouling themselves alone, the dishonorable trying to rub it off on others.

In judging the strength of the scientist's professional commitment when it conflicted with political loyalty, one is tempted to ignore the army of Lysenkoites. The bulk of them were probably ignorant opportunists, that is, people who understood only political discipline, who could not betray a scientific commitment for the simple reason that they did not understand it. Lysenkoite publications are the main evidence for this judgment; they tended to be as primitive caricatures of scientific reasoning as the model set by Lysenko himself. That is hardly surprising; as late as 1948 the Minister of Higher Education characterized the Lysenkoites as mostly young, mostly in agricultural lines of work, and mostly in the provinces.[147] Increasingly they were the products of an educational system that was reorganized in the late 1930s and 1940s to produce Lysenkoite pseudoscientists. But some of these ignorant opportunists, if they worked hard at beating down the criticisms of scientists, became learned opportunists — in plain English, conscious frauds. They learned the scientific standards for distinguishing right and wrong merely to use them sophistically, while actually adhering to the political standard (truth is the opinion of influential people). This type, which understands the scientific commitment but places no value on it, is worth considering. What prevented them from destroying the scientific community, from turning it into a mere instrument of the political bosses?

Consider the eminent example of Glushchenko. Trained as an agronomist, he was one of the first young men to do postgraduate work under Lysenko at Odessa in the 1930s. There he was given the job of proving that the master's intuitions were right. Simply ignoring the elementary rules of genetical experimentation, he "proved" that self-pollinating plants such as wheat will degenerate unless forced to crossbreed. In the same crude way he "proved" that cross-pollinating plants such as rye may freely breed according to their nature, without fear of varietal degeneration, for they practice "marriage for love." [148] In 1939, after the Council of People's Commissars had ordered Vavilov to start some Lysenkoite work in the Institute of Genetics, Glushchenko moved into that main center of research in pure genetics.[149] When Vavilov was arrested and Lysenko became director, Glushchenko became head of the laboratory of plant genetics, the country's leading authority on vegetative hybridization. That was the equivalent of being the country's chief authority on squaring the circle or building perpetual-motion machines, and it is a measure of Glushchenko's native intelligence and industry that he got to the point where he did not sound like a semiliterate crank. He got to be almost as skillful as Kushner and Nuzhdin, who had been properly trained in genetics before they became Lysenkoites.

All three were masters of sophistry designed to prove that genetics was breaking down because of a growing contradiction between its empirical data and its metaphysical concepts.[150] Glushchenko became the Soviet Union's leading "geneticist" in foreign lands, as the absolute isolation of late Stalinism gave way to restricted intercourse.[151] (Real geneticists were almost always judged unsafe for travel abroad.) Thus he reinforced the belief of a few foreign biologists that the Lysenkoites constituted "the Soviet school of genetics," somewhat akin in their views to Hinshelwood or Waddington, who doubt that natural selection acting on random mutations and genic recombination can fully explain such things as the great persistent patterns of evolution or the adaptations of microorganisms.[152] Of course, belief in that kinship does not survive a single hour of comparative reading of the alleged cousins. But the fact that Lysenkoite opportunists tried to foster such a belief during their final decade — at the height of their power they spat on the very thought of world science — betrayed their fatal weakness. As they lost the power to say "we are the boss," Lysenkoite opportunists had nothing to say except "we agree with the boss." When Lysenko and the hard core of militant ignoramuses lost all power, Glushchenko and the few other learned opportunists simply dropped their Lysenkoite sophistry and began, or resumed, the speech of ordinary geneticists.[153] In an important sense it is meaningless to say that they were frauds. Like the ideal type of politician they were simply indifferent to any rule except the necessity of pleasing influential people.

Thus we can see why learned opportunists were unable to destroy scientific communities, whether of geneticists or plant physiologists, or any other where they provided militant ignoramuses with apologetics. The growing need for apologetics was itself a sign that their days were numbered. Political bosses, increasingly aware that pseudoscience was not passing the tests of "practice," were sullenly preparing to restore autonomy to the scientific communities, as the only way to get real help in the improvement of agriculture. It was very hard for them to admit that they had made a losing bet on the militant ignoramuses, and they had some worries about the political consequences of restoring full autonomy to more and more communities of scientists. But Stalin's successors had no more interest than he in the theoretical issues of natural science. They backed ignoramuses in a number of fields for a long time — in some of the social sciences they are still doing so — as protection against the discouraging truth that came from autonomous communities of science. The sophistic chatter of learned opportunists merely served to fill the empty days of worry, while the bosses tried to resist the painful admission that they were dependent on yet another autonomous community of truth-tellers. The moment the fateful admission was muttered,[154] the Lysenkoite army of opportunists dissolved, the learned ones showing the way for the ignorant to follow. Scientific communities were there to receive them. The stubborn adherence of white-collar workers to the rules of their trade, the heroic agitation of a minority, and the fraternal help of an even smaller minority of Varangians had kept those communities alive through a thirty-year period when Stalinist bosses believed them useless. That is one of the most remarkable achievements of Soviet scientists. It is also evidence of some residual rationality in the outlook of Stalinist bosses.

8

Academic Issues: Marxism

In the West, Marxist theory has usually been considered the chief source of Lysenkoism, even though the most well-known Lysenkoite writings lay overwhelming stress on agricultural practice as the chief source. In these writings arguments from Marxist theory are a minor, adventitious theme, which has been greatly exaggerated and misunderstood by Western readers.[1] After all, it hardly seems possible that a ridiculous travesty of science could be of practical benefit to agriculture, while Marxism — especially in its Leninist version — is as readily invoked to explain stupidity as, in the Communist world, it is invoked to explain wisdom.

There is another, more basic obstacle to recognition of the minor, adventitious nature of Marxist theorizing in Lysenkoite writings. For most anti-Communists, as for most of their adversaries, there is no difference between theoretical Marxism and the mentality of Bolshevik leaders. If political chiefs claim warrant for their beliefs in the Marxist theoretical tradition, then the warrant is assumed to be there. It was one of Stalin's greatest accomplishments to establish that political standard of truth not only in the minds of his contemporaries, but also in the succeeding generation. Indeed, it is still a question whether theoretical Marxism can be fully revived after its long stultifying subjection to the will of political bosses. In fields of inquiry where it has genuine and important relevance, as in economics, the liberation of the intellect has been proceeding by detaching particular problems from Marxist theory, which remains for the most part what Stalin made it, a treasury of incantations suitable to bless any statement that political bosses may wish to make. To discover why the bosses have wished to make certain statements, and to change them in time, one must examine the actual thought process of these anti-intellectual bosses, rather than the ritualistic incantations that have adorned it. Readers who have no inclination to doubt this basic rule of Soviet intellectual history should skip this chapter, unless some morbid sympathy moves them to retrace the political mortification of Marxist thought. The sympathy will be rewarded with a minor epiphany: unsuspected similarities between Soviet and Western trends of thought will appear.

PHILOSOPHY

Before 1917 Marxists showed only a desultory interest in biology; social theory and politics were their consuming passion. Even the philosophical problems that are entangled with biological issues provoked only occasional comment, mostly from Engels and mostly in his notebooks and private correspondence, with Marx modestly deferring: in matters of natural science, "I always follow in your [Engels] footsteps."[2] The most persistent philosophical issue was reduction. Is it the chief method of natural science, and if so, does it mean that biological events are not finally explained until they are reduced to chemical and physical events? Engels was inclined to accept the mechanist's yes to both questions, but he recoiled from the inferences that many contemporaries piled on this yes. He was most concerned to deny that social events are reducible to biological events, as the social Darwinists claimed, not to speak of the thermodynamic processes to which some extremists wished to reduce human history. Of course biology does have sociological implications, which will be examined later, and it does raise other problems for the *Weltanschauung* of modern thinkers. If the biologists succeed, as Engels assumed that they would, in reducing thinking to the physiology of the nervous system, will thought be deprived of its reality? Engels exhumed the Hegelian concept of qualitative levels to prove that higher levels have some new and distinctive reality though they rise out of lower levels. Likewise with the belief that life is deprived of reality as it is reduced to the biochemistry of proteins, which seemed in Engels' time to be *the* stuff of life. And likewise for Darwin's theory of natural selection, which raised an analogous problem by reducing the *Zweckmässigkeit,* the goal-directed structure and functioning of living things, to random patterns of chance phenomena. Marx and Engels rejoiced at Darwin's "mortal blow to teleology," but they were quite unwilling to accept the death of human purposes, the bleak fatalism, which some people derived from Darwinism and from associated theories of cosmic evolution.[3] These strivings to accept the mechanistic outlook yet get beyond it, to agree that science is the only way to truth and at the same time to reject the usual *Weltanschauungen* that clung to scientism, were summarized in the odd name, dialectical materialism, which became the standard label for the orthodox Marxist *Weltanschauung* toward the end of the nineteenth century.[4]

Until the Bolshevik Revolution changed the standpoint from which Marxists viewed biology, very few of them went beyond the desultory interest that Engels had shown. Karl Kautsky was exceptional, for he began his intellectual development as a social Darwinist, and maintained to the end a genuine concern for the sociological implications of biological theories.[5] Plekhanov, "the father of Russian Marxism," was closer to the type. He defended the theory of natural selection against Chernyshevsky, the patron saint of the Russian populists, who felt that Darwin's stress on competition undermined the naturalness of mutual aid. But Plekhanov began his defense with the declaration that biological questions need not be considered. He accomplished that feat by strict adherence to the rule that Marxism begins where Darwinism leaves off. One is the science of social, the other of biological evolution. Plekhanov was concerned only with "the philosophy of biology," which he interpreted so vaguely as to avoid entanglement with any biological issues.[6] Lenin criticized Plekhanov for ignoring the acute epistemological questions raised by modern physics, but he was himself indifferent to the problems raised by biology. Like most Russian radicals he admired Timiriazev, and was much impressed by his translation of an American book on the agricultural marvels that modern biology promised.[7] But he had no reaction whatever to the rise of genetics, as reflected in Timiriazev's writings or anyone else's.

Neither did the other leaders of orthodox Marxist thought in the generation between Engels' death and the Bolshevik Revolution. Kautsky and Plekhanov were pleased to note that De Vries attributed evolution to mutations, which went to show that nature does proceed by leaps, reinforcing the orthodox Marxist insistence on revolution as the natural way to improve society.[8] But these were only passing comments, which evoked no response from prerevolutionary

Marxists. The clash of Mendelism and Lamarckism did not win even passing notice from the leaders. In the remote Caucasus Stalin, an obscure Georgian revolutionary, casually nodded approval of neo-Lamarckism, in the course of an article that was not published in a major language until 1946. Stalin had a confused notion that neo-Lamarckism stood for development by leaps against neo-Darwinian insistence on utterly gradual evolution. Perhaps he had misread Kautsky's comment on De Vries, or maybe he had heard of Korzhinsky, a vitalist with a theory of directed mutations. In any case Stalin showed no interest in the clash of Lamarckism with Mendelian genetics.[9] Neither did the other prerevolutionary Marxists. Kautsky, pleading with the Russian comrades not to make a party issue of modern physics and philosophy, cited Weismann's theory of heredity as a good example of an interesting issue to which no one attached political relevance.[10] In short, the historical facts simply do not support, they refute the widespread notion that Marxist thinkers have always clutched at the inheritance of acquired characters as an essential element of their *Weltanschauung.* Before the revolution they could not have cared less about it or any other technical problem in biology.

1917 undermined that indifference. It did not suddenly open Marxist eyes to the real philosophical issues involved in the Mendelian rejection of Lamarckism. It changed the social relations of Marxism and natural science. Marxism became the official creed of a state, whose scientific specialists were almost entirely "bourgeois" in outlook, that is, hostile or indifferent to Marxism. The response of the Bolshevik state was to launch campaigns for the creation of "red specialists," in the first place by peaceful persuasion, by efforts to prove to scientists that the Marxist *Weltanschauung* was perfectly compatible with their own. The initial result, during the first fifteen years of Soviet rule, was not only a great intensification of Marxist interest in natural science, but also something quite new, the appearance of natural scientists with an interest in Marxism.[11] Naturally enough they paid serious attention to problems that Engels or Plekhanov had brushed aside as merely technical.

The intellectual results were hardly earthshaking, whether for Marxism or for natural science. In the case of biology, between fifty and seventy-five specialists were drawn into the discussion groups organized by the Communist Academy and the Timiriazev Institute for the Study and Propaganda of Natural Science from the Viewpoint of Dialectical Materialism. (In ordinary parlance the title was mercifully reduced to Timiriazev Institute or even Timir-In.) The specialists tended to equate Marxist philosophy with the simple mechanistic materialism that most natural scientists take for granted. They spent most of their time discussing issues then agitating biologists qua biologists, with repeated glances at their possible significance for social theory and political action. At first most of these Marxist biologists were inclined to support the rearguard action that Lamarckists were then fighting against genetics, but by the end of the 1920s the vigorous polemics of Serebrovskii, Agol, and a few other Communist geneticists won them over, as the large community of non-Marxist biologists was being won over, to enthusiastic support of the new science. As for Soviet Marxist philosophers, they were willing to put their seal of approval on the argument of Agol and Serebrovskii that genetics was the realization of dialectical materialistic principles in biology. Whether the philosophers were mechanists or neo-Hegelians, most of them were sufficiently broadminded — or vague — to approve of any scientific theory, as long as the scientists professing it gave some sign that they also professed the Marxist *Weltanschauung*.[12]

Nevertheless, rudimentary though it was, the philosophical argument that genetics and Lamarckism provoked in the 1920s turned on a real issue, which continues to disturb thinkers even now — outside of biology. Indeed, it is possible to show that Empedocles, Aristotle, and Epicurus debated the philosophical problem that underlay the clash of genetics and Lamarckism.[13] It can be stated very simply. (The philosophical complexities that arise on close examination[14] can be ignored, as they were by Soviet Marxists and biologists.) If we are to avoid teleology, the attribution of human purpose to inhuman nature, we must reduce *Zweckmässigkeit,* the goal-directed structure and functioning of living things, to the purposeless interaction of material units. This was the view of Greek atomists as it is of modern geneticists. For a long time it was debated on a purely speculative level, until Darwin's theory of natural selection showed how to turn it into scientific propositions that could be tested empirically. That is one reason why the controversy over Darwinism was so intense and so widespread, while that over Mendelian genetics was comparatively mild, restricted largely to the professional community of biologists. The Darwinists broke down the larger society's resistance to the mechanistic view of living beings; the Mendelians, profiting from this victory, had little more to answer than technical objections within their professional community.

A couple of examples will serve to illustrate this historical pattern. In 1885 the conservative publicist Danilevskii published a full-length refutation of Darwinism. "Selection," he exclaimed in summation,

> that is the seal of senselessness and absurdity stamped on the brow of the world system [*mirosozdanie*], for that is the replacement of reason by contingency [*sluchainost'*]. No form of the crudest materialism has descended to such a low world view [*mirosozertsanie*], at least, none has been consistent enough for this.[15]

By 1927, when Chetverikov wrote his pioneering article on population genetics, showing how to explain natural selection on Mendelian principles, publicists with Danilevskii's outlook had learned to leave biology alone. (The Scopes trial was a shocking anachronism.) They did not cease to feel revulsion against mechanistic explanations of living beings. A twentieth-century American analogue of Danilevskii, looking in his daughter's ear, saw God's design and the devilish folly of Communism.[16] But he did not undertake a polemic against biologists who seek a mechanistic explanation of the ear's design; he confined his polemics to *Weltanschauung* and politics.

For their part the founders of population genetics did not find it necessary to make an extended defense of their basic philosophical assumption. Chetverikov was aware of the objection to it.

> After all, nature is not an urn containing marbles with which we carry out our experiments in probability theory, nor does life flow along a bed of mathematical formulas. And do we have the logical right to base a *systematic* process of evolution on the *chance* appearance of mutations? [17]

His response was mainly the characteristic one of the scientist: I can show how this assumption works. But he was also a learned man, aware of the intellectual tradition that underlay his assumption, and he briefly invoked such precedents as the kinetic theory of gases and the statistical regularities of suicide. That is as close as he came to a philosophical defense of his basic assumption, and the other founders of population genetics — Sewall Wright and Ronald Fisher — came no closer.[18] It was unnecessary. In the twentieth century those who are offended by the biologists' mechanistic picture of living things take their complaints elsewhere. There are a few who quarrel with the biologists, but they are generally ignored.[19]

In these basic respects Soviet Marxist biologists and philosophers of the 1920s were very similar to their non-Marxist contemporaries. The biologists tended to take their mechanistic assumptions for granted, and to concentrate on the scientific issues, which were being rapidly settled in favor of genetics and against Lamarckism, in part because Lamarckism could not be purged of teleology.[20] The Marxist philosophers, when they commented at all, were usually vague and syncretic, prepared to give their Marxist blessing to whatever views seemed likely to bring the most important part of the scientific community to profess Marxism. Thus, in 1930, Agol, Serebrovskii, Levit, and Prezent — three Marxist geneticists and one philosopher — appeared at a major conference of zoologists to argue that Lamarckist teleology conflicted with the Marxist outlook, while dialectical materialism was implicit in Mendelian genetics.[21] Their opponents were a few old-fashioned non-Marxist biologists; within the disciplined community of Soviet Marxists those who still favored the Lamarckist outlook were keeping it to themselves.[22]

Yet at this very time, in 1930, a major upheaval was shaking the Soviet Marxist community, which soon put an end to the happy preaching of a natural concordance between the Marxist outlook and the assumptions of geneticists. Stalin began the upheaval at the end of 1929 by denouncing the lagging of "theory" behind "practice," a cry that was soon taken up by young militants on the philosophical front. They insisted that philosophy and all other parts of "theory" must be refashioned in order to be of immediate service to the chiefs of revolutionary "practice." Henceforth the central principle of philosophy was to be *partiinost'*, partyness, which was now interpreted with an authoritarian anti-intellectual narrowness that made the older Marxists gasp and splutter. They had understood *partiinost'* to include a sociological proposition (philosophical views are determined by class interests) and a moral injunction (the philosopher should discover and support views that correspond to the interests of the lower classes). Stalin's young men now added two corollaries: the Central Committee of the Soviet Party is the unanswerable judge of class interests and of philosophical views; and the philosopher's chief duty is to work for the Central Committee. After a fierce, brief controversy Stalin let it be known that he endorsed those corollaries, and the Central Committee legalized them with an appropriate decree. The autonomy of philosophers was virtually destroyed and remained so until 1950, when Stalin himself expressed dissatisfaction with the consequent paralysis of thought.[23]

For biology the triumph of the new *partiinost'* and enthronement of "practice" over "theory" seemed, at first, to have merely verbal consequences. Of course, biological science was also a sector of the theoretical front that was disappointing the chiefs of practice. Attempts to win over "bourgeois" specialists by persuasion were declared inadequate. As part of the new effort to force their conversion, Marxist philosophers were no longer to give their blessing to "bourgeois" science. They were to reconstruct science on the basis of dialectical materialism; and, in so doing, they were to show how it could be of immediate use to agricultural and medical practice. The staff of the Timiriazev Institute gathered in 1931 to hear its new director, the embryologist Tokin, lecture on the necessity of

> reconstructing bourgeois science. We can no longer tolerate the division of Marxists into schools according to their adherence to this or that little group of professors. We need to create a single Marxist-Leninist school in biology. This Marxist-Leninist school, mastering the method of dialectical materialism, must assume leadership of the mass of scientific workers, must really reconstruct the material that is at hand in accordance with our tasks, and must no longer tail along behind bourgeois science.[24]

In accordance with these obstreperous sentiments both Lamarckism and Morganism were condemned, but no one had the faintest notion of what should take their place. The majority of Marxist biologists continued in fact to hold views that were indistinguishable from those of eminent geneticists like Morgan or his student, H. J. Muller, who came to work in the Soviet Union in 1933.

Thus, in the early 1930s, talk of revolutionizing biology seemed to be nothing more than a pledge of loyalty to the Marxist "classics," as the writings of Marx, Engels, and Lenin were called, and to the living political leadership. At the time, neither the dead nor the living authority had anything meaningful to tell biologists, though the Timiriazev Institute strove to prove the opposite. It published, in 1933, a thin anthology, *Marx, Engels, and Lenin on Biology,* which was supposed to be a guide to the revolutionary reconstruction of the science. But the only thing in it that might have been relevant to the actual work of biologists was a hint of Lamarckism in a fragment of Engels', and this, the editors were at pains to point out, did not mean what it seemed to mean: "Engels . . . did not pose and did not solve the particular concrete problem of the inheritance of characters; he was not concerned with the problems of the biological 'mechanism' of the inheritance of characters."[25] Only revisionists, the editors insisted, tried to interpret Engels' remark as Lamarckism.

Neither did the current chiefs of Marxism-Leninism have any concrete advice on the transformation of biology. Bukharin, who was still a major figure on the theoretical front in spite of his political disgrace as a leader of the defunct right wing, celebrated Darwinism in 1932 with a tribute to genetics as its magnificent continuation. Momentarily he expressed disapproval of those who exalt the role of contingency and ignore "the search for *necessary* laws."[26] (It is not conservatives alone who are disturbed by the dissolution of natural laws into statistical regularities.) But he made no effort to explain his philosophical worry in terms that might be meaningful to geneticists and other students of evolution, and he concluded with a hymn to the great agricultural benefits that would soon flow from the united efforts of socialist farmers and practical geneticists like Vavilov.[27] At the time, that was standard Communist ideology, propagated even in children's books.[28]

Thus dogmatic insistence upon the "classical masters" and the living bosses as the proper source for the necessary reconstruction of biology, and strenuous demands for immediately practical research were compatible in the early 1930s with continued devotion to standard biology. Inane editing of "classical" texts was a small part of the work at the Timiriazev Institute, or the Biological Institute of the Communist Academy, as it came to be called in the 1930s. Its main work was sound research and writing by genuine scientists. M. S. Navashin was studying the influence of aging on the rate of mutations, hoping to discover something of immediate use to breeders and of fundamental significance for genetical theory. Krenke was correlating the appearance and the age of plants in a precise, quantitative manner, trying to apply dialectical notions such as the unity of opposites and development through contradiction. The Institute's Division of General Biology, under the Hungarian émigré, E. S. Bauer, and the German émigré, Max Levin, both of whom soon perished as "enemies," had the delicate job of preparing encyclopedia articles, textbooks of general biology, and methodological treatises for those who would revolutionize the science. But there is no point rehearsing the complete roster.[29] The only one of the Institute's divisions that was directly preparing the way for Lysenkoism was Olga Lepeshinskaia's. She claimed to be disproving the basic rules of cytology, and she was vigorously criticized by Director Tokin in a public wrangle.[30] As Lysenko came to dominate Soviet biology in 1936 the Timiriazev Institute was dissolved. Except for Lepeshinskaia and perhaps two other people, the center for the dialectical materialist reconstruction of biology yielded no supporters for Lysenkoism, only opponents.[31]

Nevertheless, in three important ways the Marxist biologists and philosophers of the early 1930s were unwittingly cooperating in the drift toward Lysenkoism. They were endorsing the principle of a uniquely Marxist biology, drastically different from that of "bourgeois" scholars. They were endorsing, in principle, the end of intellectual autonomy for scientists. And they were agreeing to the principle that the prime duty of biologists is to serve the country's immediate practical needs, as determined by its political bosses. At the time these principles seemed to have no important consequences, if only because the political chiefs had no interest in biological theories. Hindsight is required to see the insidious significance of a few obscure articles that elaborated these themes in unusual ways. For example, a minor writer in the journal, *For Marxist-Leninist Natural Science,* in 1932 gave a routine compliment to geneticists for their great services to socialist construction, but spent most of his space criticizing plant physiologists for resistance to Lysenko, who was showing how to master plants in the interests of socialism.[32] Another minor author, reviewing a genetics textbook in the chief philosophical journal, *Under the Banner of Marxism,* also paid the usual tribute to the great services of geneticists in agricultural improvement. But then he held them partly responsible for the sorry state of many farms: the geneticists' concentration on the genotype and indifference to the phenotype undermined the importance of good care for plants and animals.[33] In the first case the philosopher was echoing, in the second he was anticipating, a favorite Lysenkoite theme. In both cases the ominous feature of these early, haphazard contacts between Marxist philosophers and Lysenkoism was their implicit appeal to Stalinist practice as the criterion of scientific truth. On all other grounds, as the record of the Timiriazev Institute makes clear, there was no basis for an alliance of Marxism and Lysenkoism. But the Stalinist version of Marxist philosophy was dissolving all other grounds, releasing a great urge to unite Lysenkoite biology with the almighty theory of practice.

The first philosopher to perceive this clearly was Prezent, the same man who appeared before a conference of zoologists in 1930 to prove the concord of Marxism and genetics. For two years after that he continued to adhere to the official line on biology, rapidly changing his position to do so. He denounced the spontaneous concord of Marxism and genetics, insisting instead on the need for a reconstruction of the science to make it harmonize with dialectical materialism. At the same time he continued to denounce Lamarckism, and edited a non-Marxist textbook of genetics, explaining in the preface that such works would have to be used until the Marxist reconstruction was accomplished.[34] He had no more meaningful idea than any other philosopher of what that reconstruction would be, except for his sudden realization that it would be based on the new Stalinist criterion of practice.

Over thirty years after the event Prezent told me that the insight came to him in October 1931, at the All-Union Conference on Drought Control, where the Commissar of Agriculture said that the mass trials of Lysenko's vernalization had revolutionary significance for science as well as agriculture.[35] Soon afterwards, in an otherwise humdrum pamphlet concerning *Class Struggle on the Front of Natural Science,* Prezent gave a superb definition of the Stalinist criterion of practice. Probably without realizing it, he was anticipating the argument that would, after much struggle, win for Lysenkoism the official blessing of Marxist-Leninist philosophy:

> Only productive practice is the criterion of truth and the essence of concrete cognition. And this same practice, which produces an object in accordance with a postulated law, makes possible the production of this object on the *mass* scale necessary for society. And only such socioeconomic practical mastery [*obshchestvenno-khoziaistvennoe prakticheskoe ovladenie*] is the true meaning of cognition.[36]

One point must be added to make this a perfectly accurate characterization of Stalinist epistemology: The boss knows best. If the agricultural and political bosses certified that Lysenko was achieving "socioeconomic practical mastery" of agricultural plants and animals, then it followed that Lysenko was automatically achieving cognition of the truth about them. By 1934 Prezent was Lysenko's chief theorist, helping him spell out the scientific meaning of his practical triumphs.[37]

Many observers then and since have blamed Prezent for Lysenko's assault on genetics. Lysenko himself testified that Prezent taught him to read Darwin, which, in context, came to this: Prezent helped Lysenko see how his intuitive ideas about plants could be expressed in pseudoscientific language and given the title of Darwinism.[38] The relationship can be described another way: In the 1930s Soviet philosophy fell victim to the bosses' imperious contempt for the usual rules of rational discourse, and Prezent was admirably qualified to bring this style of thought to Lysenko. But Prezent was hardly an indispensable agent. Before the famous collaboration Lysenko had already shown great native ability in pseudoscientific, authoritarian presentations of his intuitive ideas, and there is little reason to doubt that he could have continued his progress even without Prezent's assistance.

Other agrobiologists were learning the new style of dialectical materialist argument without Prezent. Williams, to take the most notable example after Lysenko, in a routine report of his work in 1938, wrote that he viewed soil science "from the positions of dialectical materialism, i.e., from the viewpoint of the significance of soil in all branches of agricultural production. . . ." He further described his school as "biological," in contrast to the opposing "chemical" school of plant nutrition, which, "guided by the conclusions of formal-deductive logic, fertilizes the soil and not the plant." [39] All the essentials of Lysenkoite or Stalinist epistemology are implicit in this offhand argument. Williams' science is dialectical materialist simply because it accords with the dominant concerns of the Soviet bosses, in this case their concern with boosting agricultural production. Williams' school is "biological" simply because it relies on the intuition of a practical agriculturist, and is contemptuous of the rigorous methods of academic people, such as chemical analysis and "formal-deductive logic." And no serious attention is given to the arguments of critics; innuendo and demogogy, which accumulate in passages of Williams not quoted here, are combined with free distortion for the perfectly unashamed purpose of beating "them" and boosting "us." In short, the Stalinist political methodology, which saturated the Soviet atmosphere in the 1930s, was carried into biological science by a number of agents. Prezent can hardly be considered an obligate vector.

Yet it is a striking and significant fact that Lysenko had great difficulty in winning the unreserved support of the Soviet ideological establishment. Prezent was for a long time a loner, offensive to his fellow philosophers no less than academic biologists.[40] When the editors of *Under the Banner of Marxism* called a meeting, in October 1939, to appraise the war between Lysenkoism and genetics — they had been shilly-shallying for the preceding four years — their chief, Mitin, heartily agreed with those who blamed the war on Prezent. He accused Prezent of "conceit that passes all bounds," in trying to fasten his "scholasticism" and "bombast" [*slovobludie*] onto the work of Comrade Lysenko.[41] This acrimonious comment won applause from the audience and an objection from Lysenko: "In order to knock [our] work, must you knock Prezent?" Mitin tried to reassure him: "We are not criticizing you, Comrade Lysenko. We are criticizing a tendency of some of your colleagues toward scholastic imposition of philosophical categories on concrete material." [42] Once again the audience applauded, approving Mitin's effort to withhold from Lysenkoism the irrefutable sanction of dialectical materialism, to avoid a Marxist-Leninist-Stalinist anathema on genetics.[43]

Mitin could offer no sensible alternative, for he agreed that Lysenkoite agricultural successes, indisputable because they were certified by the political bosses, were conclusive evidence that Lysenko had discovered theoretical truth. Lysenko was therefore right in his criticism of genetics, and geneticists were wrong to criticize him. Yet Mitin escaped from the logical conclusion by refusing to believe that Lysenko was calling for total anathema on genetics. That was Prezent's "scholasticism." Lysenko tried to make the facts clear: "The cause of the fight with genetics is not such 'bad' Michurinists as Prezent . . . , but the falsehood of the fundamentals of Mendelism." [44] Still the philosophers would not take him at his word, not even when he collected texts from Engels in an effort to prove that the "classics" were on his side.[45] Mitin had already forestalled that argument, saying that there were obsolete ideas "even in the holy of holies of our theory, in Marxism." [46] He did not specify what they were; he was waiting for the living word of the truly unanswerable source of truth, the current Party bosses.

The leading philosophers continued these efforts to straddle until 1948. As late as December 1947, when Mitin was once again obliged to make public comment on the biological controversy, he tried once again to praise Lysenko and condemn his critics without condemning the science of genetics.[47] Since it was the explicit, proudly proclaimed duty of Mitin's corps of philosophers to be agents of the political bosses, we are entitled to draw an important inference about the bosses' state of mind. They were sure that Lysenko was a great aid to agriculture, but they were not quite sure that he was absolutely right in his biological theories. Their confidence in their own ability to discover truth by the act of bossing — that is the essence of the Stalinist epistemology — was virtually absolute in agriculture as in other fields of political economy; it was very great in the social sciences and the arts; but it was comparatively feeble in the natural sciences. For almost twenty years the bosses and their philosophical agents praised Lysenko for his practical achievements, but inconsistently withheld the holy seal of dialectical materialism, which would have certified that his theoretical views were irrefutable. They were holding back from thorough application of their own criterion of practice. Or maybe the criterion, if carefully examined, would be found to make special exceptions for abstruse matters like heredity. No one could tell, for the Stalinist criterion of practice excluded careful rational examination; it made the intuitive judgment of the boss at any given moment absolute and unquestionable.

In the summer of 1948 the chiefs made their absolute and unquestionable decision for Lysenkoism in biological theory as well as agricultural practice. At that point Mitin and the other high priests of Marxist-Leninist philosophy dutifully endorsed the revelation, solemnly raising the condemnation of genetics to an official principle of the theoretical ideology.[48] Stalinist etiquette forbade Prezent even to hint that he had long ago called for such an elevation — one must never anticipate the chiefs' decision — but he could not resist a little mocking of his dilatory colleagues: "Many philosophers have constantly oscillated on these questions, but oscillations, as we know, must have a certain limit. You cannot be pendulums on questions of science!"[49] Now the philosophical congregation applauded *him*, while Mitin and another leading philosopher fell into a subtle wrangle, trying to blame each other for the previous failure of their institutions and journals to give all-out support to Lysenkoism.[50]

It is too easy to dismiss Stalinist philosophers as "pendulums," mechanically measuring the impulses imparted to them by the political bosses. They were that, but they were also men with minds, which they were prepared to twist so far from self-respecting consistency that one feels at times an involuntary shudder of sympathy for them, as for Ignatius Loyola. Consider, for example, G. F. Aleksandrov, the leading philosopher who tried in 1948 to outdo Mitin in a demonstration of eager support for the new line. "The biggest battles are still ahead," he declared with a warrior's joy, forecasting Lysenko-type assaults on physics and chemistry.[51] He threatened all the caste priests of ivory-tower bourgeois pseudoscience, who merely interpret the world, instead of changing it for the benefit of the toiling masses; they would soon learn the lesson that had long since been taught to social scientists and was now being taught to biologists:

> Just as the revolutionary practice of the proletarian masses overturned the fatalistic views of bourgeois sociologists in practice, while materialism destroyed them in theory, so the many-sided rich practice of socialist agriculture has utterly dissipated the views of formal geneticists, while Michurin and the Michurinists have destroyed them also in theory.[52]

There is an awesomely fantastical quality in men who could stand in Stalin's Russia in 1948 and declare its agricultural system in advance of all others, so far in advance as to have broken down all barriers between labor and learning, revealing to Soviet biologists truths invisible to their bourgeois ex-colleagues in the crisis-ridden countries of the decadent West. To assert this sort of thing from afar, as foreign Communists did, was to indulge in daydreams, religious daydreams of paradise as a distant place that one may someday go to by airplane or hearse. To assert it inside the Soviet Union in the face of overwhelming contrary evidence — meat and vegetables were scarcer at the end of Stalin's reign than they had been in the best year of Nicholas II, and there was even reason to doubt that the grain problem was solved — required a breathtaking audacity, like that of believers who find the kingdom of God within the grubby mortal world we all know at firsthand. Or maybe the Stalinist philosophers were nothing more than cynical admen, who will say anything necessary to get their bosses' praise and pay. In that case it was Stalin and the other bosses who had the astonishing faith that overcomes the evidence of the senses.

Voluntarism, the belief that willing makes things so, was a vital element of the Stalinist theory of practice. "If there is a passionate desire to do so, every goal can be reached, every obstacle overcome."[53] Thus Stalin in 1931, when the fantastic goals to which he was driving the country made it impossible to doubt the sincerity of his voluntarism. Speaking to the Central Committee in 1961, Lysenko may have been no less sincere in claiming that his school could change living things "according to desire, according to practice."[54] But it is highly probable that many of the bosses listening to him no longer shared his simple old-fashioned faith. A few years later they dumped him and told their philosophers to preach the end of "subjectivism," that is, the end of the Stalinist criterion of practice in biological and agricultural science. Obviously, authoritarian voluntarism, which would have been a suicidal mania in undiluted form, was constantly tempered by varying amounts of realistic self-doubt. We can see this not only in the bosses' final abandonment of Lysenkoism and restoration of "relative autonomy"[55] to biological science; we can see it also in their long hesitation before Lysenkoism was granted total support, in the brevity of the grant (1948 to 1952), and then again in the long period of renewed hesitation before the bosses' final decision to exempt biologists from the Stalinist criterion of practice.

No Soviet philosopher or sociologist has given us a substantial analysis of voluntaristic self-assurance and realistic self-doubt as they mingle and conflict in the Stalinist criterion of practice. None has gone beyond such general descriptions of the criterion as Prezent gave in 1932, or L. F. Il'ichev, a boss of theoretical ideology, gave in 1963. He objected to the belief that intellectual stagnation is evidenced by the chronic failure of Soviet philosophers and social scientists to write outstanding books — no philosopher, for example, has ever won a Lenin Prize.[56] Il'ichev insisted that the Soviet accomplishment in these fields is not to be judged by the writings of scholars, for philosophy and social science are *"the business of the entire Soviet people, of the entire Party, of its theoretical staff, of the Central Committee of the CPSU."* Knowledge must be judged by the practical results achieved on the basis of it.

> Practice and only practice is the sole criterion of the truth and the social value of theories, both in the natural and in the social sciences. [Note how the natural sciences disappear from his further explanation of this allegedly sole criterion.] The index of the progress of social science is above all the theoretical documents of the Party, the new Party Program, the level of consciousness, of education and culture, among our people, the degree of influence of Marxist-Leninist socioscientific thought on the processes of world development. One must judge the level and role of socioscientific thought in the first place by those sociospiritual transformations that are the result of the leadership of social development, the direction of social processes.[57]

That is as close as any Soviet writer has come to a clear statement of the basic proposition in Stalinist epistemology: Truth is apprehended by the act of bossing, which is the highest form of practice.

There seems to be matter for serious epistemological and sociological investigation in this proposition, which has an intriguing resemblance to American pragmatism or to Bogdanov's version of Marxism. But Soviet thinkers who have written on the criterion of practice have done little more than embroider it with fashionable illustrations.[58] Usually they do not even come as close as Il'ichev to a clear statement of its basic proposition. They must pretend not to see the element of bossing in the relations of leaders and masses, unless the leaders are wickedly or stupidly trying to prove untruths. And they must take it for granted that the truthful judgments of wise and good leaders do not need critical examination by philosophers or sociologists. Indeed, it has been one of the chief social functions of Stalin's stunted theory of practice to keep intellectuals from judging the intuitive judgments of bosses.

What then have Soviet philosophers talked about since 1952, when they were once again charged with the task of judging the contest between biology and Lysenkoism, but were as ever unable to discuss the chief philosophical issue in it? The answer must be given in two parts, for these have also been the years of tentative little steps toward the restoration of autonomy to philosophers. The old-fashioned philosophical establishment has been constantly in evidence, trying at all times to maintain an official line. At the same time individual thinkers have managed to express their views, if only on less dangerous topics than the criterion of practice. In both cases the years between 1948 and 1952 were shown to be an inconclusive interlude; the attitudes and arguments of the period before the August Session reappeared after 1952, with only minor variations. Once again the official chiefs of philosophy tended to oscillate, though now they had very little room for it, since they were bound by the verdict of the August Session. And once again the favorite philosophical topic was reduction, for or against, though now the positions taken on both sides were firmer. The great progress of population genetics and the breakthrough to chemical understanding of the gene — Watson and Crick published their famous article in 1953 — pushed both sides to extremes, the one of confidence, the other of desperation. The defenders of genetics became much more assertive in their mechanistic reductionism than they had been in the 1930s and 1940s, when Kol'tsov's chemical picture of the gene had been merely materialistic speculation.[59] The political bosses were sufficiently impressed by the worldwide excitement to legitimize research in nucleic acids, though still hindering the revival of genetics; and the Lysenkoites tried desperately to turn the implied distinction between biochemistry and pure biology into a permanent wall protecting them from scientists.

There is little to be gained from a detailed review of the subtle twists and turns of the philosophical leadership. On the one hand the decision of the August Session was a binding commitment; to cast doubt on it was to cast doubt on the authority of an ex cathedra pronouncement signed by themselves and sealed by the Central Committee. On the other hand the highest bosses showed signs of possible second thoughts, and it was the duty of the philosophical leadership to serve the political bosses. So more often than not the leadership favored Lysenkoism, but time and again by little indirect signs, such as silence at a critical moment or publishing a defense of genetics in the leading philosophical journal,[60] they showed their potential readiness to turn. They were obviously waiting to see on which side the bosses might finally come down. At the same time the handful of philosophers who had, after the August Session, joined Prezent in strenuous arguments for Lysenkoism stayed with him in the period of uncertainty following 1952. But there was little more substance in their extended arguments than in the brief comments of the philosophical leadership. An anti-intellectual near vitalism was the steady burden of Lysenkoite philosophizing, with its tone changing from aggressive to plaintive as the end approached.

"Biological thinking," a Lysenkoite educationist lamented in 1962,

> has essentially "bogged down" in physicochemical terms and concepts. The living thing has disappeared and is represented to us merely as "a complex system of physicochemical events," subordinated to the laws of physics and chemistry, in spite of the most subtle reservations and kowtowing to "dialectical transitions," "leaps," "new qualities," etc.[61]

Another sore spot was the growing importance of biometry, or statistical analysis. The Lysenkoite philosophers insisted that statistical regularities could never take the place of truly causal, purely biological laws. But since they could not define such laws, they were equally incapable of explaining their philosophical significance. A personal dogmatism, as vague in its propositions as it was positive in asserting them — indeed, all the more assertive because of its vagueness — was the essence of Lysenkoite philosophizing as it was of Lysenkoite pseudoscience.[62]

For different reasons the opponents of Lysenkoism did not get much further in the philosophical defense of science. They almost always preferred scientific arguments, if only because nearly all of them were scientists, inclined to dismiss philosophical attacks with a smile. A novelist caught their mood very neatly:

> LYSENKOITE: Does your doctrine of plant hormones accord with materialist philosophy?
>
> PLANT PHYSIOLOGIST: Yes. The hormones are material, and so is the increase in crop yield that they bring.
>
> Loud laughter in the auditorium.[63]

The philosophers' shabby record in the biological controversy was one of the reasons why their discipline was contemptuously ignored by the usual defender of genetics. But this indifference to philosophy is probably a special instance of a worldwide phenomenon. Very likely Linus Pauling expressed the opinion of most natural scientists, when he told Mitin and some other Soviet philosophers in 1962: "I have been studying science forty years already, basically in America; I've lived also in England, and I cannot recall a single instance when philosophy exercised any influence on the work of any scientist." Of course, when his Soviet hosts pressed to see if he was himself entirely free of philosophical assumptions, it transpired that he was not. He excluded vitalism, for example, from the viewpoints a scientist may take, and he also bluntly excluded dialectical materialism. "From my point of view dialectical materialism is a dogmatic theory that imposes certain restrictions on a person. . . . Besides, this is a very vague philosophy for all its dogmatism." [64] Nobody can say how many Soviet scientists inwardly agree with this characterization of the philosophy they are required to profess. Many may consider dialectical materialism an obviously true philosophy, uncorrupted by the stupidities of its worst advocates. One frequently hears such comments from Soviet scientists. In any case, whatever name they may prefer for their philosophy of science, the majority do show clear signs of sharing the elementary mechanistic assumptions that came from Pauling when he was pressed to say something about the region where science shades into philosophy.[65]

These trends of thought were most freely expressed at a major conference in the philosophy of natural science in October 1958, when no one knew that the Central Committee would soon fire anti-Lysenkoite editors. Indeed, the blatant use of power on behalf of Lysenkoism may have been precipitated by the alarming evidence of its weakness at the philosophical conference. Mitin and most of the other philosophers who spoke there simply ignored the clash between Lysenkoism and genetics.[66] If a visiting Bulgarian and G. V. Platonov, a latter-day Prezent, showed that some philosophers were still willing to speak for Lysenkoism, a visiting Rumanian showed that it was possible to make an effective philosophical defense of genetics. Kedrov, a member of the philosophical leadership, attempted a diplomatic compromise, evenhandedly rebuking both the Lysenkoite attack on natural selection and the scientists' defense of "the mythical gene." [67] Aside from philosophers, many scientists spoke at the conference, only twelve of whom discussed the biological controversy. Seven defended genetics, and two attempted a compromise; only three were unabashed Lysenkoites. The seven scientific defenders of genetics confined themselves to science, ignoring Lysenkoite arguments from philosophy, such as this mystifying criticism of the Watson-Crick model of DNA: "It deals with the doubling [*udvoenie*], but not the division (*razdvoenie*) of a single thing into opposites, that is, with repetition, with increase, but not with development." [68]

There was a near exception to this pattern in the paper of two mathematicians specializing in cybernetics. They approached philosophy with the argument that the inheritance of acquired characters is an antiscientific, teleological doctrine, because it assigns causative force to events that come *after* the events that are to be explained: future progeny are supposed to inform present parents of future needs, which the parents are supposed to provide for by providential alterations of their germ plasm.[69] Neither the two mathematicians nor any other speaker, whether pro- or anti-Lysenko or trying to stand in between, gave this familiar argument the kind of careful philosophical analysis that it deserves.[70] The country's most well-known writer on the philosophy of practice, who had formerly hailed Lysenkoism as a great triumph of that supreme criterion of truth, confined himself to a mild request for clarification: Did the mathematicians really mean to say that cybernetics rules out the inheritance of acquired characters? [71] The best philosophical comment on biology was given by the visiting Rumanian, who defended genetics with the obvious principles of mechanistic reduction, "the recognition that any property must be correlated with a definite structure . . . , a higher form of motion must be broken up into its components," and so on.[72]

The final fall of Lysenkoism has not greatly improved the philosophical analysis of biology, though a few instructive articles have appeared.[73] In some respects this little field of philosophy has become murkier than ever. Prezent fell or was pushed into a dignified silence, while Platonov made an effort to continue the defense of Lysenkoism as a recognizable doctrine.[74] But the philosophical leadership has tried to preserve its self-respect by obfuscation. Briefly conceding that "the August Session of the Lenin Academy (1948) felt the negative influence of the cult of the personality," they spell out the negative influence entirely in the field of science, and hasten to claim that "even in those conditions" there was creditable work on "the methodological problems of biology."[75] To sustain this assertion they rely on convenient forgetfulness, but even worse, on double talk. "Soviet Marxists," we are told,

> have proved the inadmissibility of separating biology from physics and chemistry, and along with that they have shown the unjustifiability both of underrating the study of the physics and chemistry of living things, and also of a complete reduction of the laws of biology to the laws of physics and chemistry, and have emphasized the necessity, when studying the physics and chemistry of living things, of taking into account the specificity of the phenomena and the laws of life and accordingly — of biology.[76]

The reader who fears that a poor translation obstructs the authors' meaning should consult the original Russian, which is, if anything, somewhat more opaque than this English version. *Kommunist*, the theoretical organ of the Central Committee, hired an ex-Lysenkoite to pour equally viscid oil over the Marxist-Leninist interpretation of biology.[77] The only function of such talk is to give the impression of continuity in Marxist thought from the Lysenkoite past to the non-Lysenkoite present. Nobody seems to care if the thought is meaningful in any other sense.[78]

At first glance, the wheel seems to be coming full circle. Soviet Marxist philosophers seem to be returning to the kind of interest in natural science that characterized their prerevolutionary ancestors. That was the desultory interest of outsiders, who wonder occasionally what significance natural science has for their *Weltanschauung* but hardly imagine that their *Weltanschauung* will alter natural science. The insistence that Marxist philosophy must transform not only natural scientists as citizens but also their scientific disciplines recedes into the past, a weird excrescence of an especially intense phase in the conflict between political bosses and the intelligentsia. Insistence on distinctively Marxist natural sciences brought repeated turmoil to the philosophical outskirts of physics, and would have done no worse to biology, except for the famous criterion of practice. That was the ideological lever which brought the political machine through the outskirts of biology into the science itself, and it was agrobiological cranks who worked that practical lever, not philosophers. Now the political machine has been moved back out of biology, as the political bosses have changed their intuitive understanding of the criterion of practice.

Everyone would benefit, if the philosophers and sociologists would analyze this evolving thought pattern of the political bosses, but Soviet scholars are unlikely to judge their bosses in the near future. So the usual situation is taking shape. The biologists do their biology, and the political leaders do their politics, leaving it to the philosophers to worry about the implications of biology for the *Weltanschauung* that they all nominally profess. In this respect the political leaders are different from their prerevolutionary ancestors and similar to the non-Marxist rulers of other countries: they make no attempt to interpret the philosophy of nature, leaving that job to academic professionals. Bukharin was the last Bolshevik leader to try his hand at prophesying from natural portents. Now the Soviet bosses are ordinary rulers who like their prophets dead, and keep a corps of priests or professors to guard the embalmed prophetic spirit against the corruption of live thought.

If Soviet philosophers of science ever win a large measure of academic autonomy, they too will probably forget their prophetic tradition and divide into schools very similar to those in the West. The ground for this expectation is not only their recent striving in this direction, since they have won a small measure of autonomy,[79] but also their failure in the past, when they were most strenuously asserting their absolute uniqueness, to produce any unique ideas. The Stalinist concept of practice is the only exception, and it was the product of the bosses' thought, not the philosophers'.

Waddington is not overgenerous, he is simply wrong in his tribute to Communist philosophizing about nature:

> In the late twenties and early thirties the basic thinking was done which led to the view that saw life as a natural and perhaps even

> inevitable development from the non-living physical world. Future students of the history of ideas are likely to take note that this new view, which amounts to nothing less than a great revolution in man's philosophical outlook on his own position in the natural world, was first developed by Communists.[80]

He mentions Oparin's early publications on the origin of life, but he cannot have read them, for they show that Oparin was neither a Marxist nor a Communist. They show his frankly acknowledged derivation from a number of nineteenth-century mechanistic writers on an ancient theme.[81] In the 1930s, when confession of Marxism was imposed on the Soviet intelligentsia, Oparin became one of the most active confessors. He began by claiming that Engels was one of the originators of his approach to the origin of life.[82] But that was petty and fairly harmless hypocrisy. Oparin also joined the Lysenkoite movement, the only really eminent biologist to do so. From 1948 to 1955 he was in charge of Lysenkoite firing and hiring within the Academy of Sciences.[83] He altered his speculations on the origin of life to suit the Lysenkoite creed, suppressing consideration of the origin of genetic systems. He even began to repeat the Lysenkoite stress on the gulf between the chemical laws that govern nonliving matter and the purely biological laws that are supposed to govern living things, which was quite inconsistent with his own biochemical approach to the origin of life.[84]

But let us draw the veil, as nineteenth-century writers would have said, on the degradation of an excellent scientist. We had to examine it because it does tell us something about the history of ideas: Contemporary chemical hypotheses concerning the origin of life are not the product of Marxist philosophy. They are the increasingly experimental continuation of age-old mechanistic speculation concerning "spontaneous generation." In this as in many other fields philosophy has been transformed into experimental science, and Marxist philosophers have helped the transformation as little as any other school, because they have not been very interested in the scientific aspects of the problem. With the exception of Lysenkoite philosophers — a small, ephemeral group — they have simply approved the biochemists' progress toward the creation of living matter in the laboratory. Engels,[85] and a few other pre-Stalinist Marxists, worried a little about the implications that reduction might have for their humanistic *Weltanschauung.* Do we lose our sense of human dignity and purpose, if it can be demonstrated that atoms spinning in the void by chance gave rise to self-replicating jelly, which by chance gave rise to us, who will by inexorable chance return to the void with atoms no longer spinning?

Engels consoled himself with the assurance that human life, which must ultimately fail on our planet, must also rise anew elsewhere. He would not accept the cosmology that uses the principle of entropy to predict the ultimate extinction of life everywhere. Other naturalists in search of reassurance have preferred the exhilaration that comes from consciousness of an extremely improbable victory over chaos, all the more exhilarating because of its temporality. Teilhard de Chardin has used biological science to revive the faith that human beings are striving toward a distant goal other than mere disappearance. But these different ways of finding reassurance in nature have dwindled to the vanishing point in the Soviet Marxist community of intellectuals as elsewhere. Biochemists are interested in matter-of-fact questions such as: How can the transformation of nonliving into living matter be demonstrated experimentally? Philosophers have mostly given up the effort to read human meaning in the nature of things. If they try at all, they generally come quickly to the same feelings as the nineteenth-century poet, obscure in her own time, who said

> . . . 'twas like Midnight, some —
>
> When everything that ticked — has stopped —
> And Space stares all around —
> Or Grisly frosts — first Autumn morns,
> Repeal the Beating Ground —
>
> But, most, like Chaos — Stopless — cool —
> Without a Chance, or Spar —
> Or even a Report of Land—
> To justify — Despair.[86]

The gain for science is obvious. Biogenesis, as the chemical creation of life is now aseptically named, no longer needs to be defended against the churchman's complaint that it is a blasphemous concept. And twentieth-century radicals would not dream of aping Pisarev, the nineteenth-century materialist who controverted Pasteur's germ theory because Pasteur thought he had finally disproved spontaneous generation and established the divine origin of life.[87] The Lysenkoites tried to revive that sort of reasoning — Lepeshinskaia was even simpleminded enough to cite Pisarev's argument in support of her atavistic version of spontaneous generation[88] — but they failed. In the Soviet Union, as in other advanced industrial powers, natural science has become a mighty institution like industrial production or the military enterprise, quite immune to metaphysical criticism. Efficiency is the only standard of judgment applied to these institutions; even the morality of the scientist, as of the manager and the soldier, is usually seen as an aspect of efficiency. Indeed it is a sign of the times that the Lysenkoites, who were virtually the last people in history to controvert the findings of natural science, invoked metaphysical principles belatedly and halfheartedly. Their original and favorite appeal was to efficiency, or practice, as they called their weird understanding of efficiency.

Bukharin, the last Bolshevik leader genuinely concerned to find metaphysical reassurance in nature, seems now as antique as Engels, Voltaire, or Lucretius. Their tradition of humanism founded on cosmological vision appears nowhere more thoroughly dead than in Soviet philosophical missals, which continue the tradition as a leaden catechism.[89] Soviet philosophers with real minds actually at work are as far from this tradition as the great bulk of philosophers in other lands. Trying, for example, to reduce mind as well as living body to cybernetic mechanisms, they are overwhelmingly indifferent to the tremorous reactions of poetic spirits.[90] Occasional expressions of a shallow Promethean joy — see our computers, how mighty they are — are hardly a continuation of the metaphysical exaltation that the antique materialists shared with their mystical opponents.

> The Atoms of Democritus
> And Newton's Particles of light
> Are sands upon the Red sea shore,
> Where Israel's tents do shine so bright.[91]

The biblical figure of speech would have annoyed Engels or Bukharin, not the substance of the thought. They were struggling to beat down the characteristic vision of the twentieth-century poet.

> Space ails us moderns: we are sick with space.
> Its contemplation makes us out as small
> As a brief epidemic of microbes
> That in a good glass may be seen to crawl
> The patina of this the least of globes.
>
> . . .
>
> We're either nothing or a God's regret.[92]

The analytical philosopher may object that we are *not* turned into nothing by the biologist's reduction of life to the chemical reactions of mindless molecules.[93] Logically he is right, but his logic hardly touches the gloomy spirit of a faithless age. And the superficial sociologists, who rejoice to see the great *Weltanschuungen* peter out into jejune shibboleths of practical politicians, fail to see the inarticulate ideologies that drive these politicians along their bloody ways.

THE HUMAN ANIMAL

Whatever else they may be, human beings are animals, and the Communists are widely supposed to base their calculations on that fact. In anti-Communist novels and films drugs, reflex conditioning, even test-tube babies are used to exemplify the Communist method of social control. The technique of "brainwashing" has been attributed to Pavlov's school of physiological psychology, and Lysenkoism is endlessly ascribed to "the thesis that the environment can produce physiological and mental changes in man which can be passed on to later generations . . . , a happy device for hastening the achievement of a benevolent Communist society." [94] Some day a student of anti-Communist thought may explain how this fable got started and why the faithful accept it as literal truth against a constant flow of contrary evidence. Surely they must be struck by the paucity of arguments from human biology in Communist writings, including the Lysenkoite species. Maybe Frankenstein is a clue. He is a major figure in the popular culture of the West, personifying a hope of power and fear of domination through mastery of human biology, a romantic individualistic failure to understand that the social relations of human beings are the best instrumentalities of power, not their bodies. At any rate that is how the human animal is usually appraised by social theorists and by practical politicians, Communists included. The matter of their materialist interpretation of history is not germ plasm but social relationships, with Marx focusing on productive, the Bolsheviks on political relationships. The search for utopia through the biological transformation of human beings is a creation of their opponents' fantasy, not the Communists'.

What follows, then, is for those who find it hard to accept simple negation of a popular fable, who want to see for themselves how Soviet Marxist thought has dealt with the biological determinants of human history. Also for those who are disturbed by that final phrase. Anyone who is inclined to insist that there are no biological determinants of human history will be embarrassed, perhaps instructed, to see how Soviet Marxists have tied themselves in knots trying to hold to this simple rule.

Their standard formula derives from Marx and Engels: Natural selection explains the development of the organic world up to the human level; human development is explained by social regularities or laws (*Gesetzmässigkeiten, zakonomernosti*). This is the obvious meaning of Marx's famous remark to Engels in 1860: Darwin's *Origin* "contains the basis in natural history for our view" of human history, or, as he put it a month later to Lassalle, "the basis in natural history for my understanding of the historical struggle of classes." [95] In the next breath Marx explained that he had in mind Darwin's "mortal blow to 'teleology' in natural science," but some people, ignoring the explanation, have read social Darwinism into Marx. In their view, he tried to justify class conflict as the continuation of the biological "struggle for existence." In fact Marx expressly scoffed at such talk, considering it meaningless.[96]

Anyone who is inclined to discount what Marx said he believed, in favor of what he really believed, should consider Marx's scheme of human history. Human societies do not emerge from the subhuman level already involved in class conflict. They begin as classless groups; they divide into classes only as they achieve a little surplus beyond bare survival; and the culmination of their history will be another classless society, sharing an enormous surplus rather than the primitive minimum of bare survival. However one judges this famous scheme of history, it can hardly be described as an effort to justify class conflict on the analogy of Darwinian competition. Of course there was an important principle common to Darwin's method in biology and Marx's in social science. They both concentrated on the development or evolution of great numbers over long periods of time, and both sought explanation of the evolutionary process not in the intentions of the organisms under study, and certainly not in some supernatural intention, but in the unforeseen patterns of complex interactions. Engels claimed that Marx, as well as Darwin, not only sought but found. In his graveside eulogy he called Marx's historical materialism and his law of surplus value the equivalent in social science of Darwin's contributions to biology.[97] That became the conventional wisdom of the Marxist movement, and millions of Soviet school children are still learning to repeat a simple form of it: Darwinism is the science of biological evolution, Marxism of social.

Like many simple formulas, this one gets very complicated when a person thinks about it. To begin with, there is the problem of the transition from ape to man, essential for paleontologists and anthropologists, beguiling to Engels, perhaps even puzzling to intelligent Soviet school children. It is very hard to keep biology and sociology separate while thinking about this transition. We can hardly imagine that biological evolution ceased at some divine instant, when beastly flesh was suddenly endowed with human spirit, free at last from the patterned chaos of natural selection. To avoid such a leap we must recognize the intermingling in reality of the biological and social patterns of evolution which we are trying to separate in thought. We must ask by what mechanisms and through which stages human social relations emerged from the purely biological relationships of our subhuman ancestors. The usual answer has been that reproductive groups, feeding groups, and defense groups were the biologically determined beginnings of human society. It has been extremely difficult to turn this speculative, general comment into specific, verifiable propositions about human origins, to show rather than guess how something like baboon or gibbon groups became something like human society. Speculation has always been easy; indeed, it was fairly common even before Darwin. Engels read enough of it to pick up favorite themes and pass them on to his followers.[98] One theme began with the definition of man as a tool-using animal, and went on to ascribe his emergence to the interacting evolution of hand, brain, and social labor. Another favorite was the arrangement of reproductive groups in an evolutionary sequence. Both subjects were and are extremely speculative, which was one of the reasons why Engels and succeeding Marxist leaders wrote only a few tentative comments on them. But the main reason was the lack of practical urgency; once we accept the fact of a great transition from ape to man in the distant past, how and why become an academic problem of no great moment to the Marxist leader.

But that is not the end of the difficulties. Given the existence of an unmistakably human society, and assuming that its development is to be explained primarily by its own laws or regularities, which are not reducible to biological laws, the social scientist is still not quit with biology. He can hardly ignore the population problem: How does a human society avoid self-destruction by breeding? Or, if one wishes to state the problem with due attention to human consciousness: How does a human society try to avoid the worst and achieve the best size and quality of population? It is a mistake to think that this is a new problem just because it has only recently been stated in rational terms. The problem has existed during all the ages when it has been dealt with by unreasoning custom and governmental fiat, ranging from the mating rules and infanticide of aborigines to the birth-control movement and mechanized killing campaigns of contemporary civilized nations. If the social scientist wishes to sort out truth and error and social function in the diverse beliefs that underlie human patterns of breeding and killing "our own" people and "the others," he can hardly begin with the flat decree that biology is irrelevant. The beliefs under investigation usually contain biological assumptions — racial, for example — and the actions, the selective breeding and killing, have biological effects, whether or not we are aware of them.

The birth-control movement is a good example of the way that modern social theorists and political leaders get caught willy nilly in biological considerations. As the campaign for contraception emerged from the underground in the late nineteenth century, Marxist leaders tended to deride it as an insulting bourgeois palliative, an effort to blame the lower classes for the misery that the capitalist system imposed on them. The birth-control zealots were telling workmen to breed less and work more, so they might improve their lot without changing the social system. Kautsky, the leading theorist of orthodox Marxism in the prewar generation, responded with the usual indignation, but he also warned his followers not to forget the underlying reality of the population problem. In any social system, he insisted, the size of the population must be controlled. Under socialism it would also be necessary to give serious thought to the quality of the population, for a uniformly high level of health and education would heighten the importance of hereditary differences, making it necessary to plan for the reduction of poor and the increase of good hereditary types. It would even be possible to breed an especially beautiful race, like the ancient Greeks or the original Germans, though without their warlike instincts.[99] Thus Kautsky indicated his ethnocentrism, and his belief that eugenics was compatible with a leftist position in politics. Similarly with the Russian radical biologist, Timiriazev, who showed great respect for the anticlerical and prointelligentsia ideology that he discerned in the eugenics movement of the prewar period. He also admired the scientific study of human heredity which it stimulated.[100]

It must not be forgotten that Kautsky, who began his intellectual development as a social Darwinist, was unusual among orthodox Marxists in his biological interests.[101] Though he accepted the view that Darwinism leaves off where Marxism begins, he insisted that a wall must not be built between them. He was most concerned to correct the Marxist tendency to brush aside the population problem, but he also called attention to more academic questions. In sketching a Marxist theory of ethics, for example, he argued that moral instincts have been determined by biological as well as social evolution.[102] He did not get very far with this line of thought. Ethical theory has been a persistent weakness in Marxist thought, which tends to dissolve it in sociology, and the possibility of grounding a naturalistic ethics in biology has been either ignored or impatiently brushed aside by most Marxist authors. As for the interaction of biological and social factors in the transition from ape to man, Kautsky offered some speculations in his final theoretical work, published in 1927.[103] But by that time the First World War and the Bolshevik Revolution had opened an unbridgeable gulf between him and the Bolsheviks, whose thought is our concern here.

They managed to forget that they had once considered Kautsky "the most outstanding theorist of Marxism after Engels,"[104] and were left with little more theoretical heritage than Engels' posthumous speculations on "The Role of Labor in the Transition from Ape to Man."[105] Lenin left them nothing more than a repetition of Marx's contempt for the use of Darwinian phrases in social analysis, and a sweeping inference from it:

> *In general* the transfer of biological concepts into the field of the social sciences is a meaningless *phrase*. Whether such a transfer is attempted with a "good" aim or with the aim of reinforcing false sociological conclusions, a meaningless phrase does not cease to be a meaningless phrase.[106]

Yet it is noteworthy that Lenin never objected to Kautsky's efforts to apply biological concepts in social analysis. Neither did Plekhanov, who was in any case less categorical than Lenin in drawing a line between biology and social science. Plekhanov retailed Darwin's thoughts on the evolution of moral and esthetic instincts; he rebutted Chernyshevsky's ethical disapproval of natural selection and Masaryk's argument that Darwinism is incompatible with Marxism; and he ridiculed racist theories as another foolish effort to explain historical change by reference to changeless states of human nature.[107] In all these brushes with biological issues, Plekhanov managed to avoid serious involvement by repeating the simple old formula: "Logically Marx's inquiry begins precisely where Darwin's inquiry ends."[108] He was willing to grant that man's biological evolution is still continuing, but it had become negligible by comparison with man's social evolution: "his zoological development is finished, his historical life's journey has begun."[109]

The Bolshevik Revolution put an important strain on that facile principle, as it did on many other aspects of the new rulers' *Weltanschauung*. Trying to win over a predominantly hostile intelligentsia, the Bolsheviks felt obliged to engage in discussion of issues they had previously brushed aside as technicalities. To be more precise, the rulers delegated the conversion of the intelligentsia to a corps of Marxist intellectuals organized in the Communist Academy and kindred institutions. For the first twelve years of the Soviet regime these Marxist thinkers enjoyed considerable autonomy, as did the larger communities of non-Marxist intellectuals to whom they addressed their arguments for syncretic Marxism. Lively discussions resulted, in the borderland between biology and sociology, as elsewhere. Everybody knew from the start that the overwhelming majority of the intelligentsia longed, in Lenin's scornful words, for "an orderly bourgeois republic,"[110] but many believed it possible to ignore political differences and find agreement on problems where the outlook of Bolshevik and anti-Bolshevik intellectuals harmonized.

Public health seemed to be such a field. Commissar of Health Semashko outlined an ambitious program that must have won widespread approval among his fellow physicians, however far from him they were in political outlook. There may have been an avant-garde that disapproved of the Victorian ideology that he brought to the problem of sexual hygiene. His first principle was the purely procreative function of sex: "it must not be relished from the viewpoint of the depraved, perverted philistine."[111] His first proposal, taking precedence even over the struggle against venereal disease, was a campaign against masturbation, "a harmful and hideous vice."[112] If there was a dissenting minority on the sexual issue, where medical science is clearly inseparable from ideology, it was either too small or too unsure of itself to put its views on record. Similarly in the case of eugenics, which Semashko pictured as the culmination of a socialist program of public health:

> Only at the communist stage will it become possible to bring health to society as a whole. Hygiene, the study of the health of man and society, will be converted into eugenics, the science of making the human race healthy. In matters of health eugenics will place the interests of the whole society, of the collective, first, above the interests of individual persons.[113]

No one challenged this forecast, perhaps because it was too vague to criticize, but the physicians and biologists who tried to define it fell into discord with Bolshevik ideology.

The discord seems to have been an embarrassing surprise to all parties. Kol'tsov, who had Semashko's support in founding the Russian Eugenics Society and a eugenics division within his Institute of Experimental Biology, thought to preclude conflict by a deliberate separation of science and ideology. The ideals of eugenics, he declared, were given by ideology, the methods by science. For socialists the ultimate ideal was "such adaptation to the social structure as ants and termites have accomplished."[114] For individualists the ideal was the universal man. As a scientist Kol'tsov professed to be above that clash of ideals, ready, like the cattlebreeder, to develop whatever type of organism the customer might desire. He seems to have been naively unaware that he was violating ideological neutrality in the very act of proclaiming it, by attributing a termite ideal to socialists while granting their opponents exclusive rights to the dream of universal man. But the ultimate ideal never became a major issue. For one thing it was as fantastically remote from the actual level of scientific knowledge about human heredity as it was from the Bolshevik dream of social improvement, which included the ideal of universal man.[115] But more important as a cause for immediate discord was Kol'tsov's research program. He and Filipchenko, the other chief organizer of eugenics research, started with the assumption that the most precious element of the population was the intelligentsia. They took it for granted that the task of eugenics in the immediate future was to study the intelligentsia — its biological sources, the threats to its continuation, the best ways to foster its size and quality.[116]

Repeatedly the "bourgeois" eugenicists, as these self-conceited members of the intelligentsia were inappropriately called, tried to reassure the Marxist worshipers of the lower classes, but they only offended more. They tried, for example, to show respect for Russia's modern revolutionaries by studying the genealogy of the first ones, the Decembrists — and came out with a solid record of noble, even royal descent.[117] Ridiculous genealogy aside, on the very serious issue of birth control, Kol'tsov used an argument that set Bolshevik teeth on edge. He opposed a campaign for contraception on these grounds: The most active elements of the population, the intelligentsia and upward-striving individuals in the lower classes, who were already reproducing at an inadequate rate, would be most responsive to a campaign for contraception, while the great mass of ignorant and sluggish peasants and workers would go on outbreeding their betters at an even greater rate.[118] As Kol'tsov perceived that such an attitude toward the lower classes provoked official hostility to eugenics, he tried to recover good will by advocating genealogical studies of Party militants and *vydvizhentsy,* workers and peasants who won advancement because they displayed "the gene of activism." After all, Kol'tsov reasoned, that precious gene was distributed in the lower classes too, for men like Pushkin and Tolstoy had fathered illegitimate children on peasant girls.[119]

In short, the leading Russian eugenicists, like many of their colleagues in other lands, felt that low-class genes were concentrated in the lower classes, a feeling aggravated in Russia by the extremely great cultural difference between the lower classes and the upper. Many of the original Bolshevik leaders were themselves *intelligenty,* who shared the culture of the upper classes, but they built their political plans on the advancement of sons of the toiling masses, which was indeed taking place, moving toward the bloody climax of the 1930s, when the revolutionary *intelligenty* would be decimated. Already in the late 1920s the initial sympathy that the Bolsheviks had displayed for eugenics was turning to suspicious coldness. In 1927 a Marxist eugenicist gave the main reason very plainly:

> The eugenics movement that exists in our country has the castelike character of the intelligentsia, which expresses itself, among other ways, in celebrating the eugenic value of the intelligentsia by comparison with all other social groups. This is of course its *decisive* weakness. Precisely for this reason we see in our eugenicists a complete inability to grasp the *social* levers by which they might interest the masses and the governing power in eugenic measures.[120]

He called for a socialist eugenics to take the place of the discredited "bourgeois" movement. The Communist Academy issued the same call.[121] The Institute of Social Hygiene and other centers of medical research were supposed to create the new socialist eugenics, while the groups organized by Kol'tsov and Filipchenko either dissolved or abandoned the study of human heredity in favor of wheat, chickens, mice, and the inevitable fruit fly.[122]

The socialist or Marxist trend in eugenics never got beyond a few programmatic exhortations. Actual research in human heredity was confined to such specialized fields as medical genetics and twin studies, while talk of population improvement became simply talk of public health, with little or no attention paid to its hereditary aspects.[123] In part this was the common sense of physicians, who did not see the feasibility or urgency of efforts to prevent hereditary defects, much less to encourage hereditary virtues. In part the withering away of socialist eugenics was the result of a growing feeling that it was a contradiction in terms. Eugenics seemed to cast doubt on the value of any effort at social improvement other than genetical: restraint on the reproduction of undesirables and encouragement to the breeding of desirables. No one could be sure which was which, as the clash over the relative merits of the intelligentsia and the lower classes revealed. Genuine knowledge of human genetics was slight, limited in the main to rare hereditary diseases. The result was that eugenics, the effort to improve human society by improving its biological heredity, was either grossly ideological, as in the preference for certain classes or races, or simply pessimistic, postponing hopes of basic and permanent improvement of the human condition until the distant time when geneticists might know as much about breeding humans as they already did about corn.

This discouraging note could occasionally be heard even in the defense of eugenics by Marxists. "Honestly," one of them protested to the Communist Academy in 1926,

> the fact that . . . the germ plasm is, as it were, the base, while the organism, its realization, is merely, so to speak, the superstructure, ought not disturb us in the least. Why, . . . the whole mass of cultural values is the property and achievement of phenotypes alone. However the genotypes have been formed, it is phenotypes that live, suffer, love, fight for world revolution, and build socialism. Let us not fear the truth, but let us try to penetrate the secret of the genes, let us master them, and then the field of eugenics will become as important [*aktual'noi*], as deserving of holding the center of our attention, as euthenics is now, which directs all its forces to the improvement of the phenotypic mass.[124]

Reverberating through this declaration was the uneasy feeling that the "bourgeois" eugenicists might be right; "phenotypic improvement," including the building of socialism, seemed insecure and imperfect in the absence of eugenic improvement, which was postponed into the distant future.

The Bolshevik leaders of the 1920s and 1930s were not sufficiently anti-intellectual to ignore or laugh off the shadow of academic doubt that eugenics cast over their great program of social transformation. On the other hand, they were not sufficiently intellectual — or sufficiently sure of the intellectuals' sympathy — to encourage uninhibited discussion, which might have dispelled eugenicist gloom by force of logic. The growing Bolshevik tendency was neither to ignore nor to encourage the discussion of eugenics, but to suppress it, giving reasons that were really little more than gestures of peremptory irritation. That is, of course, a very common response to academic questions that seem to cast doubt on the value of one's moral commitments. Commissar Semashko, repeating his endorsement of eugenics in 1927, showed the trend very clearly. He specified this time that he did not endorse programs for "castrating sick people . . . , encouraging reproduction of the comfortable classes, and reducing the birth rate of the proletariat." The improvement of the social system by revolution, and postrevolutionary improvement of maternal health and child care, he stressed, "are the most important and effective eugenic measures. Not castrating sick people and criminals, but improving the social conditions of life of the population — that is the real path of eugenics." [125] To some extent he must have known that he was distorting. Almost all Soviet eugenicists, "bourgeois" as well as Marxist, disapproved of enforced sterilization,[126] which, Dr. Semashko must have known, is not the same thing as castration. Yet he did not utterly condemn the movement that he was beginning to malign, and his fellow Bolsheviks continued for several years his strangely confused condemnation of any existing eugenicist movement, while approving socialist eugenics, which was either indistinguishable from public health measures or a vague project for the distant future.[127]

American intellectuals would have turned this abortive discussion into their much loved theme of nature versus nurture, linking grim political conservatism with the nature side of the argument, optimistic radicalism with the nurture side. Soviet Marxists did not usually talk in these terms, in part because they inherited from Marx the conviction that they had transcended the famous antinomy between the world as it is and the world as we want it to be. When a partisan of Lamarckism asked whether we are slaves of the past or creators of the future, he was lectured on the Marxist understanding of freedom as the recognition of necessity, which includes biological as well as socioeconomic necessity.[128] There were only a few Lamarckist eugenicists in the Communist Academy during the 1920s, who argued that the environmental improvement resulting from the revolution would be transformed into hereditary improvement of the race.[129] The most extreme presentation of this view was given in an obscure article by a minor physician, who got most of his ideas about the creation of supermen from H. G. Wells.[130]

Far more important as symptoms of the Soviet mentality were a comic novel by the well-known author, Ilya Ehrenburg, which ridiculed capitalist efforts to breed a race of slaves, and the minor physician's enthusiasm for a kind of psychosomatic medicine.[131] This was the trend that A. D. Speransky started. Embryonic in the 1920s, it received strong official support in the mid-1930s, and remained a major embarrassment to Soviet physicians until the mid-1950s.[132] Stalinists responded eagerly to the notion that the nerves decide whether or not we become diseased. They were indifferent or hostile to the notion that germ cells decide man's fate, whether in the Lamarckian or the Mendelian manner. Lamarckian eugenics aroused only a brief flicker of interest among a few Marxist intellectuals in the 1920s. Their critics did not respond with disquisitions on the effects that improved medicine and differential reproductive rates might have on the frequencies of desirable and undesirable genes. Genuine knowledge in that area was — and is — extremely deficient. They responded with the usual criticism of Lamarckism as a biological doctrine. On the sociological level they simply argued that Lamarckism has no logical connection with the left, nor Mendelism with the right.[133] And they were right.

Just think. If the Lamarckian doctrine is true, if environmental influences evoke adequate or specific responses in the hereditary mechanism, then generations of oppression, malnutrition, ignorance, and disease must have made the lower classes and the subject races genetically inferior to their superiors, who must, by the same logic, be genetically as well as socially superior. This has indeed been a common attitude among upper-class people since ancient times, and many subject people have been led by the same logic to take a low view of their own capacities, falling thus into a self-fulfilling prophecy. Feeling inescapably inferior, they prove that they are, while their social superiors go through the opposite experiment in social psychology. Modern genetics has brought the good news that the hereditary stuff of the lower classes and oppressed races may persist undamaged through generations of maltreatment, ever ready to produce superb human beings whenever environmental conditions become favorable. This would seem to justify a correlation of Mendelian genetics with the political left and of Lamarckism with the right, and some left-wing geneticists have so argued — H. J. Muller and J. B. S. Haldane, for example.[134] But such a correlation has little more genuine logic than its converse. Lamarckian doctrine may push one toward a low view of lower-class heredity in the past and present, but it can also be used to paint a bright picture of future possibilities. And Mendelian doctrine can be used to bolster a right-wing outlook, if one assumes that desirable genes are concentrated in the upper classes and the dominant races.

The known facts and genuine logic of the matter can be summarized in two sentences. The Lamarckian doctrine gives no logical support to the political right or left, because it is factually wrong. Genetical science supports nothing more than a vague equalitarianism, because genuine knowledge of human heredity is inadequate for anything more precise. Many people — outside the Soviet Union, for the most part — have tried to evade this simple acknowledgment of ignorance. They have tried to lay biological foundations under a variety of political beliefs, ranging from aristocratic snobbism and racism to the blessed immortality of the great middle class. In sober fact, as Dobzhansky has shown, biology does not support the zealots of any class, nation, race, or party.[135] At the same time he and other sensible biologists have perceived some very general political implications in their science. Most important is the new support it gives to an old observation, that individual differences are more important than group differences. A biologist would put it this way: Hereditary differences in human capabilities among individuals within a breeding group are known to be far greater than average hereditary differences between breeding groups may prove to be.[136]

That is a dash of cold water on nationalists, racists, and aristocrats, who assume that anyone in their favored group must be genetically superior to anyone in despised lower groups. But beyond that, when we ask the geneticist exactly what, if any, are the average genetic differences in important capabilities, he can tell us very little. The greatest triumph in the study of human gene frequencies has been the correlation of sickle-cell hemoglobin with populations that originate in malarial areas and show unusual resistance to malaria. Scientifically that is very interesting, but it hardly touches the old argument that "equality," a similar social environment for all breeding groups, is the best way to discover what average hereditary inequalities they may have in important capabilities. If and when social equality is achieved, we shall discover if different breeding groups produce different fractions of superior and inferior individuals — assuming that we will ever agree on an objective definition of those loaded words. This argument for equality of opportunity long antedates the science of genetics and has received only a little reinforcement from it, if only because equality has become a fetish, ritualistically endorsed by nearly everybody, including racists, nationalists, and upper-class snobs. It has social functions analogous to those of Christian love in older societies, relieving the angry and guilty feelings that are engendered by inequality. Moreover genuine believers in various types of meaningful equalitarianism can still use purely spiritual or ethical arguments, as Jesus did, or as an eminent geneticist of our century does: "The declaration that 'all men are created equal' was a fine one and remains so, even though and in the best sense *because* it is untrue in the biological sphere."[137]

We are dealing not just with the nature-nurture controversy, which is usually a futile exercise in emotional illogic. We are touching on the much larger and more important problem of combining ethical and factual judgments. True understanding of the problem has been very limited, but human minds have shown great diversity, intense passion, and comparatively little logic in combining beliefs about moral behavior with assumptions about the facts of life. So far science has had very little influence on these diverse combinations. Whether taking a stand on incest, miscegenation, or intelligence testing, we make genetical assumptions, but they are still overwhelmingly ideological rather than scientific.

The reader must not imagine that Soviet Marxists have ever given this issue as much emphasis as it has been given here — for the benefit of the many American intellectuals who are obsessed with the nature-nurture controversy and imagine that all the world has the same preoccupation. Human heredity played a relatively small role in Soviet Marxist discussions of biology, both in the 1920s, when they were fairly free, and especially since that time. In the 1930s two accidents produced a scandalous extreme of the refusal to think about human heredity in any form, Lamarckian or Mendelian, ethical, political, or even medical. First there was a brief effort to turn the desultory talk of socialist eugenics into something real. Then, and more important, there was the upsurge of Nazi racism.

Serebrovskii, the eminent Communist geneticist, was the center of the brief storm over socialist eugenics. An enthusiastic partisan of scientific aid to the first Five Year Plan, he was not content to plan an ambitious program of livestock improvement. In 1929 he published a bold appeal for the biological improvement of human beings. To begin with, he urged a crash program of research in human heredity. Once good and bad human genes had been mapped, he forecast the separation of sexual pleasure from procreation. The pleasure would be achieved as usual by self-selected loving partners; the procreation would be carried out according to social plan, using methods of artificial insemination "that are widely applied just now only in horse and sheep breeding. . . . From one outstanding and valuable producer it will be possible to obtain up to 1,000 or even 10,000 children." [138] When this radical proposal appeared, Commissar of Health Semashko was being pushed out of his post, along with most of the other prerevolutionary Bolshevik leaders. Their crude young Stalinist replacements were much freer in the use of political force, but they were even more straitlaced than Semashko in their public views on sex and procreation. Serebrovskii was sharply rebuked for his "insult to Soviet womanhood," and especially for his presumption in lecturing the Party leaders on a supposed method of social improvement that they had overlooked.[139] But the leaders still refrained from a clear and explicit ban on talk of eugenics. Serebrovskii's program was denounced as "bourgeois," and a few bold spirits still felt free to advocate a socialist eugenics until 1936.[140] Then total prohibition fell on any advocacy of eugenics, indeed on virtually any talk about human heredity.

Many people imagine that Nazi racism was one of the main reasons for the Bolshevik turn to Lysenkoism. Actually it was an indirect cause of the turn; reason was conspicuously absent. The Soviet ideological authorities woke up to the seriousness of the Nazi threat in the mid-1930s, when Lysenkoism was on the rise, but the Lysenkoites had no part in the awakening. The campaign against Nazi racism was begun by geneticists, especially those who were engaged in research on human heredity.[141] And then, very suddenly, in November 1936, just as the campaign was gaining momentum, the ideological bosses in the Moscow district blew it up. They denounced Levit and his Institute of Medical Genetics for fostering racism rather than opposing it.[142] This frightfully absurd inversion of the truth had one clear implication: any study of human heredity would henceforth be associated with racism. And not only racism. The Central Committee had recently condemned intelligence testing for casting fatalistic doubt on the learning ability of many children, especially those of worker and peasant origin. Now the ideologists held Levit responsible for fostering the wicked notion that heredity has something to do with learning ability.[143] In this case too, no *reason* was given, only blindly emotional guilt by association. When the administrators of terror arrested Levit and declared him an enemy of the people, any study of any aspect of human heredity fell under suspicion of connection with enemy ideology. At that point the Lysenkoites picked up the insinuation, and made it a minor theme in their campaign against the science of genetics.[144] They did not invent it, and they added nothing to it except scurrility. They dug up the eugenicist writings of the 1920s and linked them with Nazism. When Kol'tsov refused, at a public meeting, to retract or apologize for anything he had written, Prezent flailed him in *Pravda,* declaring that "It doesn't matter who has taught whom, the fascists Kol'tsov or Kol'tsov the fascists. The fascists, following Kol'tsov's program, are physically destroying thousands and thousands. But they are merely 'thousands of phenotypes'!" [145]

The Lysenkoites never went beyond the official, rigorously anti-intellectual position on human heredity. *Any* study of it was indiscriminately lumped with racism and fascism. The Lysenkoites were quite content with the simple old formula, taken now to its wildest extreme: Biology is utterly irrelevant to an understanding of human society, which is governed by purely social *Gesetzmässigkeiten.* Even medical genetics was made to seem very suspect. The intransigent Dr. Davidenkov was virtually the only person who pursued research in the subject past the mid-1930s, for which he and his sponsor were savagely attacked in 1948.[146] The Lysenkoites made no effort to replace the scientific study of human heredity with their own doctrine, which they limited to subhuman animals and plants. In 1963 for the first time one of them proposed "Michurinist" study of human heredity. Medical genetics was then being revived — a foreign textbook had been translated in 1958, and a fight was under way over publication of an original Soviet book[147] — and this one Lysenkoite, a physician, pleaded for something more constructive than simple opposition. He granted that the study of human heredity "has proceeded up to now almost exclusively from the positions of Weismanism-Morganism and formal genetics," but he believed that the monopoly could be broken.[148] Nothing came of his appeal, and not just because Khrushchev's fall in 1964 put an end to Lysenkoism even in the agricultural field, which was its original home and final citadel. The simple fact is that study of human heredity had no more place in Lysenkoism than in theology. In the one brief denunciation of Nazi racism that Lysenko ever issued, he endorsed the official theology with blessed simplemindedness: "Man, thanks to his mind, ceased long ago to be an animal."[149] Biological science, whether genuine or pseudo, has nothing to say about such a creature.

What then happened to the tradition of desultory Marxist speculation about the transition from ape to man, about possible biological determinants of moral and esthetic instincts, about the population problem — in short, about the complex interweaving of biological and social *Gesetzmässigkeiten* in the evolution of human beings? The answer is depressingly simple: It was virtually interrupted for a generation, from the mid-1930s to the mid-1950s.[150] The transition from ape to man is the closest thing to an exception. Discussion of it was never absolutely stopped, but it was driven to ridiculous scholasticism by the requirement of a complete break between biological *Gesetzmässigkeiten* on the subhuman level and social on the human.[151] A handful of authors, including former participants in the eugenics movement, were able to keep Soviet anthropologists informed of progress in studies of human gene frequencies — until 1948, when they were forced to stop and apologize.[152] At that time the leading journal of Soviet anthropology even felt obliged to promise serious study of Engels' speculation about the effect of meat-eating on the evolution of the central nervous system.[153] In fact the anthropologists did little of that. They marked time until Stalin, in 1950, denounced the Arakcheev regime in science, whereupon they quickly revived the scientific discussion of human genesis.[154] They were also aided by Stalin's observation that the concept of a revolutionary leap, such as the one from ape to man, need not be construed literally. A leap from one qualitative state to a new one could also take place gradually. That made things hard for the theoretical ideologist who might have wanted to preserve some meaning in the concept of a leap, but the situation of the anthropologist was considerably eased.

Nevertheless the musty odor of scholasticism still clings to Soviet discussions of human genesis. A recent review, for example, begins with a ritual survey of "the classics of Marxism," and summarizes the controversy among Soviet anthropologists with obfuscating finality:

> The theory of two leaps [from subhuman creatures to fossil hominids and then to modern humans] won a victory in the course of the discussion and was generally accepted. . . . Basically the discussion turned on the question whether there exists a special period of the rise [*stanovlenie*] of human society, distinct from a period of development of society in formation [*sformirovavshegosia obshchestva*]. . . .[155]

The trouble is not that Engels left texts that restrict freedom of thought. His and other "classic" texts are fragmentary and vague, and did not preclude a fairly free discussion in the 1920s and early 1930s. Even after a premium was placed on bullheadedness and the "classics" became the sacred writ of the ancestral bulls, it was still possible for an official ideologist to declare that there were obsolete ideas in "the holy of holies of our theory, in Marxism." [156] It is also possible for Soviet anthropologists to publish large treatises on human genesis that display very little scholasticism.[157] But detailed monographs of this kind do not try, any more than their counterparts in the West, to find our ultimate purpose and justification in the awesome story of our ultimate origins. The anthropologist's story is fragmentary, confusing, and nearly pointless for those in search of humanity's ultimate purpose. The resulting situation is one that Soviet ideologists are quick to decry in other societies, but refuse to see in their own. The scientific study of human origins and the official, reassuring picture of them are at worst in conflict, at best irrelevant each to the other. Anti-intellectual rulers keep a church or other ideological bureaucracy to maintain a petrified faith they do not wish to think about.

It is a constant surprise, for an outsider, to find intellectuals who take such faiths seriously and get themselves into various difficulties trying to think about them. Alfred Russell Wallace, for example, the codiscoverer of natural selection, tried to reconcile his science and his religion by conceding the biological evolution of the human body while insisting on the divine origin of the human spirit.[158] The Soviet anthropologist who derides Wallace's inconsistency only a few pages later falls into a similar one of his own. His picture of human evolution is capped by the declamation that "man is the transformer of nature and has finally emerged from a purely animal condition." [159] In another work the same scholar asserts that "the search for the qualitative difference" between man and animal is the distinguishing feature of Soviet anthropologists, whose "sacred duty" it is "to consider hominids as people actively forming themselves rather than as animals stubbornly resisting their transformation into human beings." [160] This sort of rhetoric could be poetic comment on human genesis; in this case it is an objection to the scientific effort to explain humanity as a product of natural selection. The main social function of such objections, in the Soviet Union as in Victorian England or contemporary Arkansas, is not to nourish the poetic spirit. It is to preserve anti-intellectual rulers and churchmen against doubts about their unique power to explain the human mystery. Shallow anti-Communists will find comic satisfaction in this situation, as shallow Communists do in the history of the Christian churches' response to biological evolution. In fact there is comedy, but little cause for satisfaction in these aspects of the universal failure to discover reassurance in the nature of things.

9

The Criterion of Practice

The most difficult job for the historian is to develop a double vision, seeing his subjects' choices both as they saw them and as he, the retrospective outsider, sees them, free of the pressures that made them gasp and rage. He does not increase wisdom by laughter at their folly, by indignation at their tyranny, or by sentimental substitutes for ridicule and anger. The historian's scorn, rage, or facile charity are really self-congratulation at bottom. He is surreptitiously asking why the Bolsheviks were not as intelligent or humane as himself. It is very difficult to convert this self-conceited rhetoric into a genuine question: What were in fact the choices that the Bolsheviks faced, and why in fact did they choose as they did? It is very difficult to make this conversion, but it is worth trying.

POTATOES

Virus diseases have plagued the potato in hot dry summers ever since it moved from the Andean highlands to become a staple of industrializing Europe. Of course, eighteenth- and nineteenth-century agriculturists knew nothing of viruses, but experience taught them control measures and explanations to suit. The symptoms — foliar curl, wrinkling, speckling, and an attendant drop in the harvest of tubers — are cumulative when the potato is grown in the usual way, by planting tubers; but they are cut short when the potato is propagated sexually, with true seed. The natural inference from this fact was that "regeneration" cures because extended vegetative reproduction causes a clone to "degenerate," that is, to grow feeble with age, and the diseases were lumped together under the term "degeneration." But this cure had serious economic disadvantages, and facts inconsistent with the explanation were readily apparent. For example, in cool moist areas potato plants suffer from little degeneration, and their tubers, when shipped to hot dry areas as "seed potatoes," produce fairly healthy plants for a year or two. The natural inference from *these* facts was that "degeneration" is actually a physiological disorder caused by an unfavorable ecology and curable by a favorable ecology. This cure was economically advantageous in industrializing countries, where transportation costs were declining and the urban market for potatoes was rising. But the rising premium on healthy seed potatoes called attention ever more insistently to facts that the theory of ecological depression could not explain. For example, the fact that a favorable climate mitigates but does not eliminate the "degenerative" diseases was increasingly annoying to the potato grower's desire for higher yields as well as the plant scientist's desire for complete and consistent explanations.[1]

The escape from this puzzle into the richer puzzles of virology came at the beginning of the twentieth century, when several countries sharply increased government supported research and government controlled regulations in such agricultural enterprises as the production of seed potatoes. Historians of biology have paid little attention to these developments, which is a pity, since many salient facts suggest that there is rich material here for comparative studies in the symbiotic development of biology and agriculture. Consider, for example, the contrast between Russia, one of the major potato growing nations but one of the most backward, and Holland, long a world center of intensive, scientifically oriented agriculture, potato growing included. A Russian discovered plant viruses at the turn of the century, while studying a disease very similar to the degenerative diseases of the potato.[2] He lost interest in the subject, which is apt to happen to an individual scientist in any nation, but in Russia no one picked up where he left off. Plant virology, both as a theoretical study and as applied to agriculture, simply lapsed in Russia until 1930, when the first virological laboratory was established, as part of the characteristic effort to catch up with the West.[3] In Holland the significance of the new subject was grasped early in the century; there plant virology began a continuous development both as theoretical and as applied science. Indeed government officials and commercial producers of seed potatoes did not wait for many essential questions to be answered. As soon as it had been demonstrated simply that some types of degeneration are infectious, they began to apply prophylactic measures, such as the isolation of seedbeds, the systematic discovery and elimination of diseased tubers, and the development of rules for official certification of healthy seed.

A thorough history of these trends in Great Britain and the United States as well as Holland would probably show that scientists and potato growers have become increasingly specialized units of an increasingly interdependent complex, with government becoming ever more the planner and director both of science and of agriculture. This has been an unpalatable truth for Republican potato growers and proud Doctors of Philosophy, but a palpable truth nevertheless, causing awkward shifts in the ideologies of all concerned. It seems just as likely that social support for virology and its applications to potato culture has depended not so much on a rational understanding of the science and its complex relations with economic developments, as on the simple assumption that promoting science boosts yields. This is a great oversimplification, and a potentially dangerous one, as Russian experience shows.

Nearly all the Soviet Union is more northerly than Maine, but its southern area, the steppe, has the hot dry summers that are bad for potatoes. Nature being fairly constant, economic and scientific developments have shaped the southern peasants' changing response. Until collectivization they tended to leave potatoes alone. Only a minority, whose farms were close to the slowly growing rail lines and urban markets, found it profitable to import northern seed potatoes. Fewer still, who had bottom land that was flooded in the spring, got tolerable yields by planting in midsummer, a practice also known in parts of southern France and Oklahoma.[4] The forced industrialization and collectivization that began at the end of the 1920s drastically changed that situation. To supply the urban markets, now growing very rapidly, the government pushed the peasants into large-scale collective farms and demanded a large volume of produce from them; but, to get funds for industrialization, the government held agricultural procurement prices at a very low level. The peasants shirked, concentrating their efforts on the little household plots that had been left them. In 1935, for example, the peasants of Saratov district planted less than half the acreage of potatoes planned for their collective fields, while overfulfilling the plan for their household plots by 63 percent.[5] At the same time an acute transportation crisis made Saratov and other cities in the steppe highly dependent on local farms for potatoes, while the local farms found it increasingly difficult to import the northern seed potatoes that were essential for tolerable crops. Yields per acre fell, and the government forced a great increase in acreage, enlarging its problems rather than solving them.[6]

A continuous shouting filled the agricultural press. The inherent efficiency of large-scale farms and the government's great investment in modern equipment, education, and research were providing everything necessary for a steep rise in yields; peasants and local farm officials were bunglers, slackers, or even criminal "wreckers," if yields perversely fell.[7] Not just peasants and local farm officials, but scientists too. From 1930 the press reflected a gradual shift in official attitudes toward Vavilov's Academy of Agricultural Sciences and the large network of research stations and institutes that it headed, a shift from benign trust that they were doing great things for Soviet agriculture to a short-tempered suspicion that their costly puttering was irrelevant to the crushing problems of socialist agriculture. In August 1931 a decree "On Plant Breeding and Seed Production" testily ordered plant scientists to prove the obvious truth that socialist organization made possible a rate of agricultural improvement undreamed of in capitalist countries. With the parts of the decree that prepared the ground for Lysenko's repudiation of genetics we are not here concerned. (Scientists were ordered to breed improved varieties within 4–5 years rather than the 10–12 till then regarded as minimal.) For the potato, the decree laid it down that each district, including those in the south, was to become self-sufficient in the production of its own potatoes, whether for food, industry, or for planting. Scientists were to discover how to raise healthy seed potatoes in the southern districts, and they were to do so within 4 years, the period set for the complete conversion of Soviet potato plantings to the use of seed tubers certified for varietal purity and health. At present, the decree complained, less than 1 percent of the potato crop was planted with certified seed.[8]

Less than 1 percent of the potato crop was planted with certified seed tubers of existing varieties, but the Soviet Union was then winning a reputation as a, or even *the,* leading center for the scientific breeding of improved varieties.[9] Evidently there was truth in the official cliché, endlessly repeated from 1930 on, that plant scientists were "divorced from agricultural practice." But they were not "lagging behind socialist agriculture," as the cliché charged; they had run far ahead of it. They found themselves berated for their obvious failure to help agriculture — who could deny it? were not yields declining? — just when they were winning worldwide fame for advanced research. And the irony bites deeper than coincidence. Vavilov, whose legendary enthusiasm sparked the rapid progress of Soviet plant science, was inspired in part by the conviction that socialism makes possible a rate of coordinated scientific and agricultural progress undreamed of in capitalist countries — the very conviction that now inspired distraught officials to bark impossible demands at him.

Soviet potato specialists were not entirely nonplussed by the demand for the production of healthy seed potatoes in the south, where degenerative diseases cannot be held to a tolerable minimum by the usual prophylactic measures. For such districts they suggested the use of true seed, pointing out the advantages (tubers are infected by any virus in the parental plant, but true seed is not) and the disadvantages (frequent male sterility in the potato flower, a tendency not to breed true, and the extra labor of raising sets in greenhouses and transferring them to the fields). Perhaps this was the crucial mistake of the potato specialists, that they presented their solution in a tentative way, calling attention to its disadvantages and urging careful trials of various techniques before some of them might be firmly recommended to southern farmers.[10] Lysenko made no such mistake when he presented his solution of the problem early in 1935: within two years, he declared, the south could become entirely self-sufficient in the production of healthy seed tubers, by adopting the very simple method of planting in midsummer.[11]

Potato specialists must have gaped in astonishment at this announcement, which was first presented to them not in a technical journal but in the official newspaper of the Commissariat of Agriculture, and without the tag, "For discussion," that would have made it legitimate to disagree. Lysenko presented summer planting as his discovery, failing to mention, much less evaluate, the experience of farmers who had long practiced various forms of it in scattered spots of the United States, France, and his native Ukraine. He declared degeneration to be not the result of infectious diseases but an "aging" or "enfeeblement" of the plant caused by excessive heat at the time of tuber formation; summer planting kept potatoes from degenerating because the tubers formed in the cool of autumn.

Briefly and vaguely he reported the following experiment in support of his explanation: healthy tubers stored at 30–35°C. produced very unhealthy plants.[12]

In later recollections the Lysenkoites pictured the experimental tubers as cut in halves; the control halves were retrospectively stored in a cellar and produced healthier plants than those stored at 30–35°C.[13] As Lysenko warmed to his subject in further newspaper articles, speeches, and pamphlets, he added the famous "stage theory" to this explanation. He claimed to have revolutionized plant physiology with a new understanding of the stages of development in a plant's life, stages that are nowhere defined without ambiguity or inconsistency. Vague enough to begin with, the "stage theory" became little more than a phrase when Lysenko tacked it on his explanation of degeneration. Heat, he now said, caused tubers and the plants they produced to become "stage-aged or feeble." Briefly and vaguely he recalled the following experiment in support of this explanation: cuttings taken from the top of a potato plant flowered sooner and produced less tubers than cuttings taken from the lower part of the stem, which proves that the growing tip of a plant is "stage-older" than the lower parts, and therefore that the eyes of the tubers formed in hot weather are "stage-aged." The reader who fears that this is a willful travesty of Lysenko's method of experimenting and reasoning should read the original.[14] When the bombast is cleared away, one finds nothing more than a combination of the obsolescent theory of ecological depression and the obsolete theory of aging.

Of course, the officials in the Commissariat of Agriculture had little interest in theoretical plant physiology. They wanted a practical method of producing healthy seed potatoes in the south, and Lysenko convinced them early in 1935 that he had discovered such a method. He had tried summer planting only twice, on a quarter-hectare experiment plot in 1933, and on 31 hectares scattered in 16 collective farms in 1934; and the results, which he has never published in full, were far from providing clear support for his hopes.[15] Nevertheless, in March 1935 he published his flat assertion that he had discovered how to make the south entirely self-sufficient in the production of seed potatoes.[16] He had thrown aside not merely the complex statistical canons of modern agricultural research, but even the simple wisdom of the Russian proverb, "Measure seven times to cut once." "Do we have the right," he asked, while pushing another innovation on a mass scale,

> when we propose a method that is as yet only theoretically grounded, . . . to lose two to three years in preliminary trial of this method on little plots at several plant-breeding institutions? No, we haven't the right ot lose even a single year.[17]

The Commissariat of Agriculture agreed, for they ordered 500–600 southern farms to begin summer planting on 1600 hectares in 1935, not as a test so much as the start of regular production of seed potatoes by midsummer planting.[18]

In 1936 the Commissariat of Agriculture increased the plan for summer planting to 35,000–40,000 hectares (17,000–18,000 were actually planted).[19] In 1937 summer planting was endorsed by the highest organ of government; the Council of People's Commissars ordered that the method be used on 65,000 hectares (about 20,000 was the result).[20] In 1938 the Council decreed that the success of summer planting justified the cessation, starting the following year, of all rail shipment of potatoes between districts (*oblasti*); the south would assure itself of healthy seed potatoes by midsummer plantings on nearly 50,000 hectares (this time the plan was exceeded; 54,000 were actually planted).[21] In 1939, if we can believe a Lysenkoite report that was made many years after the fact,[22] summer planting rose to 107,000 hectares; in 1940 to 153,000; and in 1941 the German attack cut off a plan for forcing summer planting on a scale of 250,000 hectares.[23] Those figures seem to show that Lysenko had won his audacious gamble, that he had found a truly practical way to make the south self-sufficient in the production of healthy seed potatoes.

But, paradoxical as it may seem, other data published by Lysenko and the potato specialists who joined his cause reveal the opposite. For example, Lysenko wrote that questionnaires were sent to the 500–600 farms that were obliged to try the method in 1935; 420 replied, but Lysenko published the results only for the 50 best.[24] This kind of extremely selective reporting is characteristic of all but two of the 8 or 9 years in which summer planting was pushed on a mass scale. (In those 2 years, Lysenko's chief potato specialist was happy to announce, "not only individual farms but also many counties [*raiony*] and even districts [*oblasti*]" had a successful experience with summer planting.)[25] Of the 17,000–18,000 hectares planted in 1936, to take the year that was crucial for approval on the highest level of government, results were published for only 407 hectares, yet the Council of People's Commissars decreed that the experience of 1936 "fully proved the possibility of obtaining a yield of nondegenerated potato tubers twice as great as the yield of the usual [spring] plantings in the southern part of the Ukraine." [26]

If one stresses "possibility," one begins to understand the Council's paradoxical support of a method that must have been a failure on the overwhelming majority of farms that had tried it. The experience of the unsuccessful farms simply did not count. Lysenko said as much, when he published results for only 10 percent of the farms that tried the method in 1935:

> Positive reports on summer planting of potatoes are given not only by those collective farms that obtained good harvests, but also by those collective farms that obtained poor harvests. To the collective farmers of those collective farms, and also to agricultural officials, it became perfectly clear that the causes of their low harvest from summer plantings, and also the removal of those causes, depended entirely on them. Thus, the collective farm, "Soviet Farmer," in the Bereznigovatskii *raion* of Odessa *oblast'*, characterizes the method of summer planting in the following manner: "Summer plantings are very good but because of poor cultivation of the land we obtained a low harvest." [27]

In the 1930s and 1940s that sort of argument was in complete harmony with the dominant trend of Bolshevik thought on agriculture. Why should the summer planting of potatoes stand or fall according to the average results on all the farms that tried it, when collectivization itself could not pass such a test? It was standard practice to prove the great potential of the new system by citing the spectacular achievements of the best units in it. The failure of the other units to measure up was to be explained only by such subjective factors as laziness, incompetence, poor administration, and "wrecking." That was the burden of Stalin's speeches, hopelessly confounding the possibility of objectivity in agricultural economics.[28]

In the tangled thickets where economics and plant science grew together Lysenko absorbed Stalin's passionate subjectivity and infused it into plant science. Yields, after all, are the complex product of social and natural factors. The economic authorities were unable to analyze the alternative costs of local seed-potato production in the south as against long hauls from specially favored northern areas (Soviet economists are only now coming to such an ability). Intuitively reacting to a severe transportation crisis, they decided for local self-sufficiency. While the potato specialists hemmed and hawed, Lysenko justified this decision by a snap judgment on the causes and cure of degeneration, brushing aside as irrelevant the failure of his cure on most of the farms, which could not be trusted to do the simplest things right. Small wonder then, that the Bolshevik press made him a model for plant scientists, or that Stalin shouted "Bravo, Comrade Lysenko!" toward the end of Lysenko's speech to a Congress of Collective Farm Shock-Brigade Workers.[29]

Against this background the reaction of potato specialists and virologists becomes comprehensible. None of them publicly questioned the practical value of summer planting. A few became enthusiastic supporters of Lysenko. Many uttered a perfunctory endorsement of summer planting, and then dropped the subject of degenerative diseases. A few, granting the practical success of Lysenko's method, diffidently questioned his theoretical explanations. The notion of degenerative diseases as aging was untenable, they suggested; the contagious nature of the diseases had been established beyond doubt, and an assessment of ecological factors must proceed on that basis.[30] Lysenko made short work of such critics. If their theoretical understanding was so great, why did they have no practical solution for degeneration? If his theory was so poor, how had he accomplished a cure? Besides, he did not deny the possibility that viruses might be present in degenerated potatoes; he simply pointed out the main cause of degeneration, which was "stage aging" owing to excessive heat at the time of tuber formation.[31]

Lysenko's position threw doubt on the necessity of virus-control measures in the production of seed potatoes, and at the same time opened a way for virologists to defend such measures. Some virologists tried to square Lysenko's views with their own by attributing the supposed success of summer planting to a decline in aphid infestation during the last part of summer. On this basis some even compared Lysenko with Jenner and Pasteur, who had discovered cures for other virus diseases without knowing about viruses.[32] But such arguments implicitly rejected Lysenko's theory of "stage aging," and gradually such voices fell silent.[33] Lysenko's potato experts, notably Favorov, filled the silence with arguments against the view that degenerative diseases are infectious. Ultimately, the boldest Lysenkoites began to argue against the very existence of plant viruses, picturing them as a metaphysical construct of bourgeois science, comparable to genes.[34] On a practical level, the seed certification law ceased to regard degenerative diseases as contagious. For example, diseased plants were allowed to remain in the seedbeds until harvest, when they could be gathered for food while the plants that still appeared healthy could be certified as seed.[35] In effect, Lysenko had solved the problem of virus diseases of the potato by getting everyone to ignore them. How much the Soviet economy saved in reduced costs of production (notably transportation, labor, and research), how much it lost in reduced potato yields, the public record does not reveal.

As the Second World War ended, some diffident disagreement with Lysenko was expressed,[36] but the famous August Session in 1948 silenced the critics — forever, it seemed at the time. Lysenko not only proclaimed the complete monopoly of his school, he announced that the Central Committee of the Party had approved his monopoly. In the aftermath an intensive drive was launched to reestablish summer planting among southern farmers, who seem to have abandoned it en masse during the war. By 1950 summer plantings were halfway to the peak acreage of the prewar period, but they were doomed to climb no higher. Nearly all the summer plantings of 1949 and 1950 were cut down by drought and disease. Favorov admitted as much in 1952, when he published his magnum opus on summer planting, a strange book that opens with the usual breathtaking arrogance but quickly declines to defensive — and unconvincing — arguments against unpublished criticisms.[37] One gathers from Favorov's book that agricultural officials were quietly tolerating another mass flight from summer planting in the south. Indeed, they allowed works to be published in 1952 which quietly undermined the Lysenkoite program for control of degenerative diseases.[38]

Early in 1953 Stalin died. Within six months Khrushchev made his famous report on the bleak condition of Soviet agriculture. In the aftermath long-pent voices uttered bitter truths. Little more than one-third of all potato plantings were done with certified seed, twenty-five years after the Party had decreed a complete conversion to certified seed within four years.[39] Worse yet, surveys showed that certified seed was often as badly infected with degenerative diseases as noncertified. Vavilov's world famous program of potato breeding had been stalled; new varieties, representing years of labor, had been destroyed by runaway virus infections as soon as they left the plant breeding stations. A potato specialist who had access to the raw data from which the Lysenkoites drew their arguments for summer planting charged Favorov and Lysenko with deliberate distortion and misrepresentation.[40]

When the editorial board that printed such direct attacks on Lysenko was dismissed, potato specialists and virologists moderated their tone. But they had already won crucial changes both in research opportunities and in agricultural practice. In February 1958 a conference of potato specialists had resolved to include virus-control measures in the law on seed certification. The conference also endorsed summer planting as a useful measure in the south, and established an uneasy compromise in the Soviet seed-potato business.[41] Non-Lysenkoite specialists discreetly urged a tightening of the requirements for certification and an increase in regional specialization. They looked forward to the time when the production of seed potatoes would be concentrated in such favored spots that truly rigorous standards of certification would be feasible. Bukasov, a former student of Vavilov's who had become the dean of Soviet potato specialists by kowtowing to Lysenko, was able to give southern farmers the following advice on summer planting, probably the wisest summation of all: If the farmer has irrigated land available for a seedbed; if he uses an early variety whose tubers have a brief enough rest period to make possible summer planting with freshly dug tubers; and if he calculates the midsummer planting date to coincide with the time when the aphid and leafhopper populations drop in his area; then summer planting will give him a tolerable amount of fairly good seed potatoes for next spring's sowing.[42] (It seems likely that the concurrence of such factors gave rise to the tradition of summer planting in scattered spots of Oklahoma, southern France and the Ukraine.)

With the final fall of Lysenkoism in 1965 knowledgeable specialists like Bukasov were able to drop all hypocritical pretense about the value of Lysenko's methods of fighting degenerative diseases.[43] It is still difficult for the Soviet Union to establish the kind of scientific potato production that exists in other advanced countries, but the difficulties are now entirely within the realm of political economy. Whether areas with hot dry summers should import eating potatoes or merely healthy seed tubers from geographically favored areas is now a question that can be faced on its merits as a problem in economics, assuming that Soviet economists and administrators have learned how to measure the alternative costs accurately. It is no longer necessary to pretend that there is a way out of the choice, a cheap way of avoiding any long hauls of potatoes by producing both healthy seed tubers and eating potatoes in every locality. Lysenko's weird substitute for scientific understanding of the potato's viral diseases helped the Stalinist bosses ignore the needs of modern agriculture, while they scrambled frantically to achieve the second military-industrial complex of the world.

CORN

Maize crossbreeds promiscuously. All the tassels thrust up from a field of corn drop pollen for all the silks below; the intervening wind indiscriminately fertilizes all the sticky female parts with the dusty sperm dropped by all the male parts. In this sense "hybrid corn" is as much a tautology as wet water; maize naturally and continuously hybridizes, crossbreeds, as far as the wind carries its pollen. An enormous variety of crosses results, and from them the natural environment and the farmer select those that survive and serve human needs best. The ancient American Indian bent the selective process so strongly toward human needs and away from natural conditions that maize became a purely domesticated creature, entirely dependent on human cultivation for survival. If we recognize that the ancient Indian's thought processes were savage rather than scientific, this initial transformation of maize was a more remarkable accomplishment than the creation of inbred hybrids by twentieth-century geneticists, for the Indian was much further from a genuine understanding of what he was doing.[44]

Yet there are some aspects of the geneticists' achievement — so far their greatest contribution to agriculture — which are hardly less remarkable combinations of a little understanding and much lucky circumstance. Only a few basic insights marked the birth of genetics at the beginning of our century, yet the new science generated great self-confidence, enabling men to reject beliefs based on centuries of practical experience. Forced inbreeding stunts, maims, and sterilizes naturally cross-pollinating plants such as maize. The traditional European cornbreeder therefore followed nature's pattern, as the American

Indian had in his time. The best ears of the best varieties were selected for natural crossing, and some very good open-pollinated varieties were created in that way. But by the end of the nineteenth century, in such highly commercialized countries as the United States, there was growing impatience with the slow rate of varietal improvement achieved by these age-old methods. Breeders began casting about for new methods, but did not seriously consider inbreeding, until the geneticists overturned the ancient belief that it does permanent damage to normally crossbreeding plants. The new distinction between phenotype and genotype made it possible to say that inbreeding damages the former, and not the latter. The cornbreeder was thus encouraged to use inbreeding as a method of sorting out the characteristics that are scrambled together in a natural population of corn. Inbred lines with undesirable characters could be discarded, while those with desirable characters could be recombined. At this point the cornbreeder discovered a great bonus. When he crossed two inbred lines, and then crossed the result with the product of another interlinear cross, the double-cross hybrid was sometimes more vigorous and higher yielding than any natural open-pollinated variety. The average increase in yields was 25 to 35 percent.[45]

Geneticists have still not agreed on an explanation of this hybrid vigor or heterosis, which provides an additional puzzle by falling off sharply after the first generation. One finds repeated grumbling and head-shaking over the "purely empirical," "trial-and-error" development of the new hybrids.[46] But the role of dumb luck has been far greater in the socioeconomic context than in the scientific. A new kind of agriculture was required to make use of these new hybrids, and with it significant changes in the economy as a whole; and with *that,* changes in the mental set of farmers, government officials, scientists, and the general public. Largely without comprehension of the total process, as if guided by Adam Smith's invisible hand, all these people made the intricate rearrangements in their social organization that the new hybrids required.

When the first interlinear hybrids were being developed in the 1920s, even a progressive like Henry Wallace took seriously the traditional rule that yearly purchases of seed corn are the mark of an improvident farmer on the way to ruin. In the first edition of his famous handbook on corn he suggested that the new hybrids would not be commercially successful until breeders learned how to make their hybrid vigor last beyond the first generation.[47] He also expressed the traditional farmer's doubts about the wisdom of buying mineral fertilizers,[48] which proved to be absolutely essential when the new hybrids began to drain the soil of mineral nutrients at a higher rate to produce their higher yields than the old-fashioned varieties of corn. Willingness to make large annual outlays on seed corn and mineral fertilizers depended not only on the biological qualities of these items but even more on a favorable balance between the outlays and the farmer's receipts for his increased output of corn. This balance was in turn dependent on such larger economic forces as the demand for meat and the consequent demand for fodder, which in turn were dependent on the state of the economy as a whole. The role of the government has been crucial. In subsidizing basic research and the production of inbred lines, in maintenance of seed quality, in education and extension services that break in farmers to the new methods, in juggling price supports, taxes, and the other economic levers that synchronize, more or less, the whole complex mechanism, the government coordinates the intricate interaction between science and agriculture. Such success as the United States has achieved in this field has been due in part to the widespread faith in the utility of science and the famous enterprising spirit, but it has also been the product of much blind contingency. One notes, for example, that the great depression and the Second World War facilitated widespread acceptance of the new hybrids. One also notes that the intricate interdependence of farmers, government officials, and scientists is poorly comprehended by the traditional individualistic ideology to which they all swear allegiance.[49]

Marxist-Leninist ideology made that interdependence self-evident to Soviet officials of the 1920s and early 1930s, inspiring extremely self-confident efforts to achieve scientific agriculture at a rush. The transformation of traditional peasant farms into large-scale modern units, whose work would be coordinated by socialist planning, which would be informed by a great network of research institutes, seemed to make possible the achievement of scientific agriculture at a rate that would astound the world. Some scientists shared this official utopianism and reinforced it. "We haven't the slightest doubt," a Soviet cornbreeder wrote in 1931,

> that in our conditions, with all the rigorousness of *khozraschet* [Soviet cost accounting], the problem of using [inbred] hybrid seeds profitably will be solved far more expeditiously than in

> America. . . . [Our price for this seed will be far lower than the prices charged by American profiteers.] Furthermore, in our organized system of seed production hybridization will be carried out far more simply and harmoniously than in the disconnected individual enterprises of the American seedman and farmer.[50]

In painful fact just the opposite occurred. Within the decade following that bold prediction American farmers began to use the new hybrids on a broad scale, while the Soviet program collapsed. Lysenko's critics blame him for the failure, but the historical record shows that they exaggerate his significance. The hybrid corn program was already withering, when he produced a pseudoscientific justification for killing it. Forced collectivization inspired corn specialists with the utopian dream quoted above, and at the same time created the chaos that destroyed the dream.

In 1930 Soviet agricultural officials launched an intensive drive to expand the country's acreage under corn. Corn was to be raised not only south of a June isotherm of 19° C., where existing varieties could ripen into grain, but far to the north of that boundary, where stalks, leaves, and half-ripe ears were supposed to provide fodder for a great increase in livestock.[51] This will have a familiar sound to those who remember Khrushchev's discovery of corn in 1954. Actually he revived a crash program that had been tried — and had crashed — in the early 1930s.[52] Indeed one senses here another of the intriguing continuities in Russian history: once the peasants, then the commissars, turned to corn when in trouble. Repeatedly it has been urged for fields of winter-killed wheat, with the arguments that it often survives droughts that kill spring wheat, and that its yield per acre is much larger than that of wheat. Repeatedly those who have given these arguments have forgotten that corn will not accomplish these wonders unless it has a summer long enough and warm enough, with just the right sequence of sunny and rainy periods.[53]

In the early 1930s, over large areas where corn was thrust upon newly collectivized peasants, the results were very poor. Disappointed officials complained of "a kulak campaign against corn,"[54] when widespread crop failures were the result of meteorological and human resistance to centralized farming by decree. The simultaneous effort to subdue the weather and the peasants made it extremely difficult to tell which was responsible in what degree for the failure. One almost sympathizes with the ardent corn specialists who complained in 1932 that

> the campaign which began so successfully met with dumb but stubborn, deep-rooted resistance. Conservatism, incomprehension, ignorance, prejudices, thoughtlessness, bungling, wrecking, kulak agitation — all this became a powerful obstacle to the introduction of corn in our agriculture. Drought in the latter part of the summer of 1930 found corn in an unprotected agrotechnical condition. . . .[55]

But what is the point of rehearsing their dreary list of slovenly faults in the peasants' use of corn? One author used the simple word, catastrophe, to describe the 1930 results of the corn campaign.[56] Yet here as elsewhere Stalin's government could not distinguish between a setback and a defeat, until a measure had been repeated to the point of extreme pain. Corn was forced on a massive scale in 1930, again in 1931, again in 1932, and still again in 1933, with officials blustering all the while that the slovenly sullen peasants were responsible for the repeated failure, not corn or the weather, and certainly not themselves.[57] Finally, in 1934, they simply dropped the campaign. Corn acreage sank to its traditionally low level, except for the small corn belts in Georgia and the southwest Ukraine.

On the scientific side too the 1930s witnessed a great intensification of work with corn, followed by a slump. At Dniepropetrovsk an All-Union Institute of Corn was created to supervise the development of inbred hybrids and open-pollinated varieties by a network of stations. By 1935 the first double-cross hybrids had been created, with lines imported from America, but perceptive corn specialists realized that the new farm system was hardly ready for them. Even if mass production of inbred lines and double crosses could have been organized in the primitive conditions of Soviet agriculture — and the experts agreed that it could not — the impoverished farms were quite unlikely to buy the expensive seed each year.[58] Somewhat later the Mexican government would show that traditional peasants can be persuaded to use the new hybrids, but the circumstances were drastically different: the Mexican peasants were already very familiar with corn, their traditional social organization was not violently disrupted, they were not obliged to pay the full price of the seed, which has been subsidized by a government that has had foreign assistance and has not been straining to achieve a military-industrial complex at top speed.[59] In the Soviet Union of the 1930s all conditions were unfavorable. The corn campaign collapsed, and in 1934 the ambitious breeding program was severely cut back.

Only two or three institutions continued work with inbred lines, on a reduced basis.[60]

The Bolsheviks did not calmly ask themselves what light the practical failure of the corn campaign shed on the biological and social theories that had inspired it. A rational observer would have exculpated the biological theories simply by glancing at the USA, where the new hybrids were gaining widespread acceptance just as they were being shelved in the USSR. A rational outsider could have seen that the Stalinist social and political outlook, combined with the enthusiasm of some scientific specialists, had inspired a utopian campaign that could not possibly have succeeded. But Bolshevik officials were not rational outsiders; they were hardly prepared to blame themselves or their ideology for the fiascos that accompanied collectivization. The criterion of practice was not to be turned against Marxist-Leninist theory or its chief. After all, Marxism-Leninism was intended to inspire Bolsheviks with self-confidence, not self-doubt. The officials blamed the peasants, and showed a mounting irritation with agricultural scientists, who had originated the dream of salvation through corn, and were still spending much money without noticeable effect on yields.[61]

That was the context within which Lysenko came to attack inbred hybrids, after the modern corn program had already been sharply curtailed. At first his attack on hybrid corn was an incidental by-product of his program of wheat improvement. Lysenko disapproved of the sexual habits of wheat: If inbreeding is universally harmful, as everyone knows it is, then it must also be harmful for such plants as wheat, which inbreeds naturally. Lysenko insisted that the degeneration of good varieties of wheat was not only, or not so much, the result of a poor seed business and backward farming methods, as it was of inbreeding, which always "impoverishes the hereditary foundation." [62] He sent millions of peasants through the wheat fields to strip open the florets and force them to crossbreed. Standard wheat specialists protested that this "intravarietal crossing" was useless, if the fields were planted to truly pure varieties and were truly free of weeds. In most cases it was harmful, for varietal purity and freedom from weeds was only a dream on most Russian fields. But they could not press this argument very hard, for that would be challenging the agricultural officials who backed Lysenko's recipe for "intravarietal crossing" of wheat. So Lysenko's critics argued most about the abstract issue whether inbreeding does "impoverish heredity," which they denied, pointing to modern methods of breeding cross-pollinating plants.[63] Here too the genuine scientists were hamstrung by the criterion of practice. Double-cross hybrids of inbred lines could hardly be offered in defense of genetics, for they had already proved disappointing. Only one corn specialist was sufficiently bold, or naive, to suggest the reason why.

> These varieties [inbred hybrids], for reasons of an organizational-technical nature, do not as yet have a chance of rapid acceptance, in spite of their increased productivity and high quality of grain. With the organization of a harmonious [*stroinoi*] system of seed production these difficulties will naturally be overcome to a significant extent.[64]

To press such an argument would have been to defend genetics by accusing the Soviet agricultural system.

One can therefore understand the weak self-defense of scientists like Vavilov, when they were charged with failure to aid the new farming system. Only rarely did they point to inbred hybrids as a great practical triumph of genetics. The files of the USDA contain a letter from Vavilov, asking if inbreeding had helped their corn specialists to develop any good varieties that could be propagated in the simple old-fashioned way:

> We are now having a serious discussion in our country on the method of inbreeding. . . . This problem is a very serious one for us. Theoretically, it is very clear to me, and to many of us, . . . that by inbreeding it might be possible to get rid of lethals and bad genes, but the practical results have not been very promising. The work requires much time, and in our country, where people like to get results in a short time, it is rather discouraging.[65]

The reply he got must have intensified his discouragement: American cornbreeders had abandoned the effort to develop varieties that could be released to farmers for propagation in the old-fashioned way. Vavilov was told what he must have known already: that first-generation double crosses of inbred lines must be produced by seed companies for annual sale to farmers.[66] At the conference of December 1936, when Vavilov was still trying to placate the Lysenkoites and the agricultural officials who supported them, he defended inbreeding very feebly. He refrained from mentioning the American success with a method that could not be applied in Soviet farms, and Lysenko was able to taunt him for failing to offer any solid evidence that inbreeding had any practical benefit.[67]

But what practical benefit did the Lysenkoites bring to Soviet cornbreeding? They gave "scientific" approval to its backwardness. In fact, they helped push it back still further. When the agronomist Musiiko launched a campaign to check the depressing effect of hot dry winds on fertilization and seed set, Lysenko recognized a kindred soul, and made him his chief corn specialist. Musiiko got peasants to go through the cornfields when the tassels were dropping pollen, with pieces of rabbit fur, touching them from tassels to silks. Musiiko's original explanation was the modest and obvious one: the peasants were replacing pollen killed by the hot dry wind.[68] Lysenko taught him to see profound theoretical significance in this additional pollination. It increased the plant's chances of choosing the pollen most suitable to itself.

> Proceeding from Darwinian theory, one must draw the conclusion that free cross pollination, especially of cross pollinating plants, always, as a rule, leads to the receipt of a greater quantity of seeds from the plants, and also to their greater biological fitness and viability.[69]

Repeating and expanding such arguments, Musiiko became the chief opponent of inbreeding as a method of corn improvement. In its place he and Lysenko justified continuation of old-fashioned methods: mass and individual selection within open pollinated varieties, and efforts to achieve hybrid vigor by crossing such varieties. They argued further that open-pollinated varieties could be grown much closer to each other than existing rules allowed — does not each plant select the pollen best suited for it? The rules were duly changed, simplifying Soviet corn production even more, by further retrogression in seed quality and varietal purity.[70]

In 1939 Musiiko and his friends got a conference of corn breeders to endorse their methods as the chief means of corn improvement.[71] A handful of corn breeders nevertheless continued their quiet work with inbred lines. Within a few areas of the country they had some inbred hybrids certified and used by local collective farms.[72] At the end of the Second World War they seemed to gain some support at the center, when the war's terrible devastation and a severe drought in 1946 caused the Central Committee to repeat the Russian habit of turning to corn in time of trouble. The Committee decreed an increase in corn acreage, stressing the value of hybrid seed but failing to specify either the old-fashioned varietal hybrids, which Lysenko favored, or the new inbred hybrids favored by his critics.[73]

Still, there must have been an unpublished decision at a high level in favor of the inbred hybrids, for Lysenko himself briefly conceded their usefulness in a 1947 speech to the Lenin Academy of Agricultural Sciences.[74] There were other signs too that agrobiology was losing a little ground to standard science in the immediate postwar period. But in 1948 the Lysenkoites not only recovered their losses; they won absolute victory. With the explicit sanction of the Central Committee genetics was totally condemned. Another conference of corn breeders thereupon endorsed the Lysenkoite program,[75] and explicit orders to cease inbreeding were given even to such camplaisant specialists as B. P. Sokolov, who were willing to side with Lysenko if only he would tolerate their work with inbred lines.[76] When toleration was ended, Sokolov and a few other specialists continued inbreeding on the quiet. Dudintsev's fictional revival of the critically thinking lonely rebel was inspired by a revival in fact. In corn research institutes as well as novels honest dedication to technology caused men to defy political authority.[77]

The ban on inbreeding was quietly removed a year or so before Stalin died,[78] one of several signs that Lysenko's imposing power was being undermined by change within the agricultural establishment. The process was greatly accelerated by the death of the great individual in 1953. A pent-up urge to make basic changes was suddenly released. Of course, large parts of the resulting reform were Stalinist in nature; the reformers after all were themselves products of the system they wished to reform. Khrushchev raised a cry for corn, corn, corn all over, with even greater unconcern for weather and economic calculation than similar campaigners had shown in the early 1930s.[79] But this time there was no general fiasco. The Soviet farming system was far stronger both administratively and economically, far more capable of sustaining a hugely wasteful mass experiment. In some places there were very bad results, and in many places, where corn is an uneconomic choice for fodder, farms were obliged to choose it anyhow. Khrushchev had a strange mixture of Stalinist and post-Stalinist attitudes. He insisted that agricultural specialists must be independent of political bosses in their technical judgments. Yet he reversed the corn specialists on the suitability of corn in the Moscow district, with the argument that he got it to grow in his *dacha* garden.[80] Not only did he display preference for intuitive judgments as against the careful statistical calculations of modern agronomy; at times he flaunted his purely political power over the specialists. Once, for example, he fluttered the Central Committee by a brutally frank exchange with a speaker who declared that everyone now understands the importance of corn:

> KHRUSHCHEV: There are still skeptics.
> SPEAKER: Only a few
> KHRUSHCHEV: No, there are many, but they have begun to speak less. (*Animation in the hall.*)[81]

Even without the forceful prodding of the supreme boss, farm managers would have had a hard time making efficient choices of fodder crops, for the science of agricultural economics was only beginning to revive from its Stalinist atrophy. This was not merely a matter of abstract academic research and teaching. At conferences of corn specialists and agricultural managers there were arguments over the prices and bonuses that should or should not be paid for this or that kind of seed corn, indicating that the production of hybrid seed was a profitable enterprise for some farms in some sections, and that some features of bargaining were helping to shape the Soviet mechanism for price determination.[82] For example, many southern farms that were commanded to grow seed corn for northern districts were shirking that order in order to concentrate on the profitable production of varieties for southern districts where nature favors corn.[83]

In 1964, on the eve of Khrushchev's overthrow, an American specialist traveling through the Soviet countryside came to the conclusion that "there is little, if any, true hybrid corn being grown in the Soviet Union in the sense that we use the term."[84] That was an extreme way of noting two peculiarities of the Soviet corn program. One was the continuing popularity of old-fashioned open-pollinated varieties, which was partly the result of lingering Lysenkoite prejudice,[85] and partly the result of general agricultural backwardness. The other was the inadequacy of the seed business. Soviet corn specialists were constantly complaining that proper standards did not prevail in the production of double-cross hybrids of inbred lines.[86] The timely and rigorous detasselling of maternal plants, which is essential to the production of pure seed, was being widely and frequently bungled, even though — or should we say *because* — the Soviet countryside still contained a surplus of peasant labor, poorly motivated and poorly utilized. Soviet corn specialists were therefore more eager than their American colleagues to develop inbred lines with cytoplasmic male sterility, a laborsaving device that ends the necessity of mass detasselling.[87] To be more precise, that technique substitutes the labor of scientific personnel for the unskilled labor of peasants, who could not be counted on to do detasselling properly and on time. Here again one senses the painful persistence of the traditional agricultural problem, exacerbated by the brutal Stalinist mauling of the peasantry. Though scientific specialists have also been harshly treated at some times, they have by and large received far more encouragement than the peasantry, and so have continued the old Russian pattern of government and intelligentsia dragging a recalcitrant peasantry toward modernity. In spite of all the obstacles of climate, Stalinist irrationality, Lysenkoism, and peasant unreliability, the geneticists and corn specialists did get modern hybrids into production. They fell short of American standards of seed purity, but they did achieve an upturn in acreage yields of corn.[88]

What functions and dysfunctions, then, are attributable to Lysenkoism? After the scientists' utopian corn campaign of the early 1930s collapsed, Lysenkoism soothed Bolshevik discontent by denouncing the scientists' basic idea of inbred hybrids. The price paid for that soothing syrup was an excessive slowdown in the development of inbred lines and in the training of competent specialists. Most of the Soviet countryside was too backward for immediate use of inbred hybrids, but it would have cost relatively little to prepare for the competent farmer of the future an array of hybrids specially suited to the various regions where corn is a sensible crop. Lysenkoite hostility to inbreeding caused protracted delay in this preparation — after 1948 it caused acute retrogression[89] — and filled the countryside with ill-trained specialists whose heads were stuffed with nonsense about the value of "additional pollination" and the superiority of open-pollinated hybrids. In the mid-1950s, when a modern corn program was resumed with great vigor, the old sin of excessive enthusiasm was also resumed, and Lysenkoite prejudices were allowed to go on hampering corn improvement until the end of 1964, when the political chiefs finally decided to grant normal autonomy to corn specialists and geneticists, perhaps even to agronomists and farm managers.[90] The bosses did not learn enough science to understand why the criterion of practice required the science of genetics. They simply delimited the sphere within which they would decide what was required by practice. Those who have been saying that inbred hybrids of corn are the greatest practical accomplishment of genetics should now recognize a still more remarkable accomplishment: Geneticists have forced Stalinist bosses to acknowledge some of the limits of their political power.

LAND

According to a famous metaphor, soil is a leaky storage bin of plant nutrients; a cropping system is the farmer's method of systematically replacing the nutrients that his crops take out of the bin; and soil fertility is the balance he maintains, measured by the average crop he extracts each year from a given unit of land. European farming has achieved a rising level of fertility by evolving through four major stages of the cropping system, with science making a contribution only in the recent past:[91]

1. shifting cultivation (the *zalezhnaia* or *perelozhnaia* system in Russian), which restores fertility by allowing land to revert to natural vegetation for ten years or more
2. grain-fallow rotations, often called the two- or three-field system, or the *zernoparovaia* or *parovaia* system
3. rotations of grain with green manure (clover or some other leguminous herbage) and with intertilled (*propashnye*) crops such as potatoes, turnips, sugar beet, or corn
4. diverse schemes that economics and soil science of the industrial age have been multiplying since the nineteenth century

The third system has many variants and names: the Norfolk system, for example, after the region of England that made it famous in the eighteenth century, or the Thaer system, after the agronomist who popularized it in nineteenth-century Germany. It has also been called mixed husbandry or balanced farming, because the farmer abandons grain monoculture and produces two or more field crops along with meat and dairy products. The systematic use of green manure and dung, the control of weeds by cultivation of the intertilled crops, and the reduction of plant diseases and pests through frequent alteration of crops on a given piece of land, raise yields far above the level achieved by the three-field system and provide a better diet for a denser population. The diversity of the farmer's produce also serves as a hedge against drops in market prices. In Russian the usual term for this type of crop rotation is *plodosmen.*

The systems in the fourth category have ranged from the "mechanically elaborate, though . . . agriculturally primitive" system of wheat-fallow farming in the American West,[92] through complex systems of balanced farming, to highly intensive monoculture, such as Roswell Garst showed Khrushchev.[93] (He raises seed corn year after year on the same plots.) Rational farming has ceased to be calm continuation of a pattern established by long use; it has become an endless quest for the elusive "optimum" in combining land, weather, living organisms, machines, chemicals, markets. Government intervention becomes ever more important, not least in organizing the development of science and its applications.

From the establishment of the zemstvos and the Petrovskaia Agricultural Academy in the 1860s to the collectivization of the 1930s a disturbing anomaly was apparent in Russia: agricultural science advanced ever more rapidly than farming methods. The Russian school of soil science became the world's pioneer, while the great bulk of Russian farmers were stuck at the second stage of cropping systems, the three-field or grain-fallow rotation. In sparsely settled parts of the country the *first* stage, the *zalezhnaia* system, survived into the twentieth century.[94] One result of this widening gap between science and agriculture was the occasional eruption of nativist obscurantism, denunciations of "Western" science for its uselessness, and demands for a Russian science that would aid Russian farmers.[95]

At the turn of the century Williams (or Vil'iams) developed a messianic variety of this nativistic revolt against world science. The dominant school of soil science was wrong, he insisted, not only for backward Russia but for the world at large. It stressed the chemical elements of soil processes and plant nutrition, while the crucial factor was actually soil structure, the formation of a crumbly texture that provides roots with an optimum of air and water. He worked out a grand theory to support his insistence on an inevitable, ultimately catastrophic decline in fertility unless farmers systematically restored soil structure by making extended courses of perennial grasses the basis of their cropping systems.[96] That kind of speculation was a relic of a dying age in the development of soil science, an age when so many critical factors were obscure that agronomists vacillated between the rival claims of system builders and narrow reliance on the experience of successful farmers. In the 1880s a series of critical discoveries transformed the study of soil processes and plant nutrition into a genuine science, or perhaps one should say into an amalgamated interdisciplinary science with commonly accepted methods and a growing body of significant results. For farmers the first benefit was reliable advice on the use of mineral fertilizers. West European yields began to climb above the levels attainable through the Norfolk system (*plodosmen*), a climb that is still continuing.

Williams remained aloof from these developments. He dismissed mineral fertilizers as useless, unless their application followed the improvement of soil structure by *travopol'e,* which we may translate as grassland or lea rotations. He refused to admit that the causes of good soil structure and its precise role in fertility were not clearly established. Indeed, they remain obscure to this day. It is doubtful that perennial grasses make such a great improvement in soil structure as Williams thought; artificial soil conditioners developed by chemically oriented soil scientists do the job better.[97] In any case, the best current opinion is that building good soil structure is an important way to raise yields beyond the level of optimum nutrient supply.[98] If that judgment is correct, then Williams was putting the cart before the horse when he deferred improvement in the supply of nutrients until optimum soil structure might be achieved. Of course, Williams should not be condemned for mistaken conclusions. His essential flaw was his method: disdain for the limited real achievements of his science in favor of the limitless triumphs of speculation.

Before the Bolshevik Revolution, Williams' pretentious speculations were considered beneath criticism by his fellow scientists.[99] His classroom eloquence merely captivated some students at the Moscow Agricultural Institute, as the Petrovskaia Academy was renamed following a political upheaval in the 1890s, in which Williams seems to have been the only staff member to enjoy the trust of tsarist authorities.[100] After the Bolshevik Revolution this central institution of agricultural education, renamed the Timiriazev Academy, experienced another, and sharper, clash with an authoritarian government. Williams was the first professor to support the new regime.

> A struggle developed along the line of abolishing the autonomy and independence of the Academy, along the line of struggle with the influence of the peasant parties among the students, among whom the Bolsheviks were then isolated individuals. . . . [This occasioned] a sharp exacerbation of Vasili Robertovich's [Williams'] relations with the majority of professors, and then also with groups of agronomists, who were hostilely inclined against the Soviet regime.[101]

Williams even joined the Party, a very unusual step for a senior scientist in the 1920s.[102] He was lionized by the press, but for almost twenty years Bolshevik agricultural officials were no more willing than their tsarist predecessors to endorse his peculiar agronomy.

The recognized leaders among soil scientists and crop specialists, most notably Prianishnikov and Tulaikov, were slower in coming to terms with the Bolshevik regime, but it was their recommendations that agricultural officials endorsed, not only during the 1920s, when the actual choice of cropping systems was still made by autonomous peasants, but also during the early 1930s, when Bolshevik officials took direct control of farming. To be sure, Williams' political eminence made it impossible for scientists to go on ignoring him. In repeated debates during the 1920s and early 1930s, his windy oratory evoked some Bolshevik sympathy, but the chief officials were not persuaded that a revolution in soil science was required to close the gap between science and agriculture.[103] They were convinced that the main obstacle to the development of scientific farming was the backward social organization of agriculture. One of their important motives in forcing rapid collectivization was the enthusiastic confidence that Russian peasants, once in the new farms, would leap from the medieval to the twentieth-century stage of cropping system. The experience if the first two five year plans shattered this confidence. Yields did not rise approximately 35 percent, as the planners had forecast. They fell; in the all-important case of grain they fell approximately 14 percent.[104] By the mid-1930s Bolshevik officials were ready to heed aged Williams. As the retrospective outside observer sees it, the Bolsheviks were now ready to close the gap between science and agriculture by severe retardation of science.

It is easy for an outside observer to see why the Bolshevik regime blamed science for its disappointment. They could hardly blame the new social organization of agriculture; it was the most advanced system the world had ever known. To doubt this would be to doubt themselves, a mental process that is usually fatal to political leaders, especially Bolsheviks. The difficult question is why the regime, which was so quick to condemn genuine science for failing to produce immediate gains, was then so slow to recognize the uselessness of pseudoscience. The cropping systems recommended by genuine science were tried in the collective and state farms only seven or eight years, 1929 to 1936, before they were condemned. *Travopol'nye* or grassland rotations were given a trial of twenty-four years, 1937 to 1961, before they were finally discarded. Evidently there was something in Williams' doctrine that satisfied the psyche of Bolshevik officials, something that was lacking in genuine science, perhaps also something that met real needs in their agriculture. Let us try to discover what this was by a closer look at Bolshevik experience with the cropping systems recommended by the two schools.

Genuine soil scientists and crop specialists like Prianishnikov and Tulaikov did not insist on a single scheme of crop rotation. They emphasized the need for individual rotations, adapted to the soil, climate, and economic conditions of each firm. Of course they recognized the need for general rules; a single scheme and individualized rotations are not mutually exclusive contradictories but opposite ends of a gradient, and genuine scientists were willing to move toward either end, as the changing situation required. At the beginning of collectivization the emphasis was on regional specialization.[105] The country was divided into regions according to the facts of physical and economic geography. Cropping systems were to depend on the region and its main products but, in any case, were to be worked out on an individual basis by the management of each farm. For example, the southeast was to specialize in grain production. Since it was dry and sparsely populated, Tulaikov recommended a Soviet version of the mechanically elaborate but agriculturally primitive cropping system that had emerged in the American West of the late nineteenth century. Acreage yields were expected to be low, but yields per ruble of investment were to be high. When acreage yields turned out to be much lower than expected, failing to recoup the government's investment, Bolshevik authorities began to denounce Tulaikov's program as a wrecking scheme, designed to give loafers and enemies of the people an excuse for bad work. Poor management was no longer to be helped by simplified agronomy; it was to be whipped into shape by stern insistence on the most advanced agronomy, which alone was appropriate in the most advanced farm organizations the world had ever known.[106]

By 1933 it was clear to Bolshevik officials that the whole emphasis on regional specialization was a scheme designed by "wreckers" to hurt the Soviet economy. The country's transport network was unable to effect large regional exchanges of surpluses, which were mostly nonexistent anyway. The wholesale slaughter of livestock that attended collectivization — also the work of "wreckers" and "enemies of the people"—had created an especially acute shortage not only of meat for consumers but also of manure for the land, while the chemical industry did not produce enough mineral fertilizers to take the place of dung. Clearly a retreat, or rather an advance, from regional specialization to regional self-sufficiency was called for. The specialists thereupon shifted their recommendations from regional systems of crop rotation to regional variations of a single system: *plodosmen* or mixed husbandry. This had been their choice for certain regions from the start. For example, the collective farms of the central nonchernozem region had been instructed to specialize in vegetables and dairy products, while cutting the nation's transportation bill by raising enough grain to feed their own region. Some such rotation as clover, wheat, potatoes, oats, and fallow was to be followed in eight- or ten-field patterns. (Most of Russia is too dry to dispense with fallow altogether, as the classic Norfolk system did.) In 1933 all the regions were ordered to produce their own meat and potatoes, and variants of *plodosmennye* rotations were worked out for all. But this new scheme also failed to bring a sharp upturn in yields, if only because it was never seriously tried. Endorsed in 1933, it was officially discarded in 1937, when only one-third of the nation's farms had started rotations in their fields (another third had rotations on paper, the final third nothing).[107] Indeed, archival research will probably reveal that the crucial decision in favor of Williams' scheme was made as early as 1934, when public endorsement was being given to *plodosmen*.[108]

Standard science had lent itself to the volatile utopianism, the expectation of a quick leap from medieval to twentieth-century cropping systems, on which the Stalinist agricultural revolution was initially based. Once this expectation was disappointed, standard science could only hold out the prospect of following the example of eighteenth-century English farmers, an intolerable affront to Bolshevik pride. "How can one speak of introducing *plodosmennye* rotations in our agriculture, which is the biggest in the world, the most highly mechanized, has the most advanced and progressive structure, in our socialist agriculture?!"[109] Thus the head of the agricultural section of the State Planning Commission in 1949, arguing for an intensification of the twelve-year-old campaign for *travopol'e*. In the peak years of that campaign, 1949 to 1953, the average grain yield per hectare was still slightly below the average for 1925 to 1929, but Bolshevik faith was not broken. In 1950, to be sure, Lysenko called attention to the need for regional variations in *travopol'e*,[110] and in 1954 the stream of decrees imposing it on the nation's farms quietly petered out.[111] But soil science and agronomy were allowed to remain in the grip of Williams' school, which continued to recommend *travopol'e* until 1961, when Khrushchev finally condemned the cropping system and opened the way for criticism of the pseudoscience on which it was based.[112]

The ideological function of this extraordinary persistence is readily apparent in the public record. The volatile utopianism of the Frist Five Year Plan gave way in the mid-1930s to a dogged, sullen utopianism, the characteristic Bolshevik style of resignation to painful reality. Williams' doctrine suited the new attitude very well. He had all along preached that permanently increased yields could be achieved only by massive use of *travopol'nye* rotations over a long period of time. In his view the Soviet state was at war with an age-old, worldwide deterioration of soil and climate. (The famous Stalin plan for altering the climate by massive forest planting derived from Williams.) He made it possible to denounce costly twentieth-century methods of raising yields; they were meretricious, the fever of capitalist agriculture preceding its certain death. Soviet agriculture might seem poorer at the moment, but it was following the one sure way to permanent affluence. Williams' doctrine was a tranquilizer for the Bolshevik psyche, soothing its utopian restlessness during a long period of agricultural backwardness.[113]

Was Williams' doctrine also a harmful drug, as Khrushchev declared in 1961, causing great losses to Soviet agriculture? Until the archives are opened to historians with open minds, the answer to this question can only be surmised. But there is suggestive indirect evidence in the public record. One is struck most of all by the diversity of cropping schemes that acquired the holy name of *travopol'e*. When Williams appropriated the term for his grand scheme, it was generally used to describe the type of dairy and meat farming characteristic of Denmark or Switzerland, where half or two-thirds of the sown area are under grass, and grain is imported.[114] Williams changed the meaning of *travopol'e* to signify a form of the *zalezhnaia* system: wheat was to be the main crop, rotated with long courses of sown grass, which would support livestock as a subsidiary product. In either case *travopol'e* conflicted with the Soviet government's constant effort to overcome the effects of low grain yields by extending the acreage under grain. In the early 1930s there were even charges that *travopol'e* was used by "wreckers" to cut back wheat acreage in favor of grass.[115]

Williams was therefore obliged to allow many alterations of his original scheme. He granted the name of *travopol'nye* rotations to grass courses of only 3- or even 2-years' duration, and therefore reduced his estimate of the amount of land needed for grass to only 22 percent of the sown area.[116] Prianishnikov, the most knowledgeable and persistent critic of Williams, thereupon declared that *travopol'e* would be essentially the same thing as *plodosmen,* if only Williams would recognize the superiority of leguminous herbage to true grass in a country that lacks adequate mineral fertilizer.[117] In 1950 Lysenko placed his imprimatur on further amendments. Not grass but clean fallow must precede grain in dry areas, for grass dries out the soil, leaving inadequate moisture for the grain crop. In regions of adequate moisture but considerable density of population, where farms had little spare land, 1-year courses of grass-legume mixtures were to be used. He even suggested a rough measure of the opportunity cost of keeping land under grass for more than one year: if the first hay cutting produced less than 30–40 centners per hectare, the meadow was too poor to keep; it should be plowed up for a planting of winter grain. In no case, Lysenko insisted, was a farm to reduce grain acreage to make way for grass.[118]

It is therefore hardly surprising that only 10.7 percent of the sown area was under perennial grass in 1953, the peak year of efforts to force *travopol'e* on the farms. 5 percent was under annual grass. (The actual term in the statistical handbooks is *travy,* a term that includes both true grass and leguminous herbage.)[119] One can only guess how much of that grassland was regularly rotated with other crops, for Williams granted the name of *travopol'e* even to rotations that kept meadowland separate from the basic crops. In short, *travopol'e* was an extremely loose term for many ways of combining a bit of meadow and livestock and intertilled crops with the main business of grain production. The changing structure of the sown area suggests that Prianishnikov was right. The actual rotations in most regions were probably varieties of grain-fallow with increasing elements of *plodosmen.* If this tentative conclusion from inadequate evidence is correct, the *travopol'naia* campaign was a great mystification. It was a way of pretending to adopt the most advanced system of crop rotation, while Russia was actually making the transition from the second to the third stage of cropping systems, the transition that Western Europe had made between the seventeenth and nineteenth centuries.

To what extent did the mystification help or hurt the transition in Russia? Such a broad question raises a host of subsidiary questions that can be answered only by archival research and careful economic analysis. One would have to establish what the real alternatives were for Soviet farms, and how the *travopol'shchiki* in various regions bent the farm managers' choices in various directions. Then one could assess the gains and losses of the *travopol'naia* campaign. But some tentative thoughts can be ventured now.

It seems unreasonable to expect that depressed farmers can be highly rational in their cropping systems, can show a lively appreciation of their real alternatives and make carefully calculated choices among them. In collective or private farms, depressed farmers tend to concentrate on traditional crops that are immediately required, whether by the market or the state or their need to keep alive. They tend to neglect such crops as clover or lupine, the green manure that will improve grain yields after a few years, at the immediate cost of a reduction in acreage under cash crops or an increase in labor and funds to till extra land. In 1913, after zemstvo agronomists had been urging the transition to *plodosmen* for decades, less than 3 percent of Russia's sown area was under *travy*. In 1929, on the eve of forced collectivization, that index of the advance of *plodosmen* had risen only to 4 percent, and it remained low during the terrible years of collectivization, partly because the authorities were trying to overleap *plodosmen* in some regions. In the mid-1930s the *travopol'naia* campaign began. By the eve of the Second World War the portion of the sown area under *travy* had jumped to 10.8 percent. By 1953, the peak year of enforced *travopol'e,* it had moved to 15.7 percent.

Prianishnikov did not object to the growing portion of land given to *travy;* he figured that 15–25 percent, depending on the region, should be under green manure.[120] What angered him was the irrational preference that the *travopol'shchiki* showed for true grass. It therefore seems clear that one of the costs of the *travopol'naia* campaign was the loss of the nitrogen that clover and lupine would have fixed in the soil, and the consequent loss of the increased grain yields that would have been accumulated, if the authorities had insisted more on leguminous herbage and less on true grass. At the same time I would suggest that the *travopol'naia* campaign brought some gain: it generated the enthusiasm and justified the force that jolted Soviet farms away from grain-fallow rotations toward some form of *plodosmen.* What can be established only by archival research is how much pushing there was, and how much was really necessary at one time or another. How many farm managers would have made no serious effort to improve rotations, if the crackpots and timeservers of Williams' school had not bombarded them with decrees and fines? [121] The Soviet historian who answers questions of that sort will help us toward an answer to the large question: How much was the advance of modern farming hurt, and how much was it helped, by Stalin's reliance on enthusiasm and force, his neglect of material incentives and local initiative?

It is my guess that the main cost of the Soviet government's long commitment to Williams' school was the resulting distortion of its investment in higher education and research. Williams had only a few disciples before official support came to him in the mid-1930s, and he died soon after. Official support changed things very rapidly. The few disciples were joined by many timeservers, and a real school was soon at work, training young people in their image, bawling at scientists who were bold enough to express doubts. The retardation and distortion of Soviet soil science grew especially marked after the Second World War, when Lysenko threw his support to Williams' school.[122] Lysenko endorsed some sensible criticisms of Williams' agricultural recipes, but he was thoroughly enamored of Williams' grand theory of a single process of soil formation. Indeed, Lysenko persisted in that love even after Khrushchev had condemned Williams' school.[123] He pictured Williams' soil science as a natural partner for his "Michurinist" biology.

One is at first startled by this abrupt discovery of an affinity between two sects that had previously developed separately, but reflection confirms Lysenko's judgment. His school of agrobiologists and Williams' did have a similar method and outlook. Both leaped from speculation to conviction, impatiently brushing aside demands for rigorous proof. Both were ready and willing to use Stalinist methods of suppressing free discussion. Both saw their isolation from world science as proof of their messianic role. (At one point Williams compared himself to persecuted Galileo.)[124] And both showed a persistent vitalism, an insistence that chemical methods of analyzing soil and plants missed the essential quality of living processes.

If Soviet authorities had kept up a modest investment in standard science during the period when their agriculture was unable to make full use of it, they would now be reaping the benefits. Commercial fertilizers, for example, are beginning to flow into Soviet farms in sizable quantities, but there is an acute shortage of the sophisticated soil maps and properly trained agronomists that would achieve maximum efficiency in the use of these fertilizers.[125]

Standard soil science and agronomy have become ever more necessary as, in the post-Stalin era, careful analysis of farm costs and profits (*okupaemost' zatrat*) has become ever more necessary. But in all three fields, soil science, agronomy, and agricultural economics, there has been no smooth and easy progress. Bemused Stalinists have been struggling to overcome the defects of the system that shaped their whole mentality. Shortly after Stalin died they quietly stopped the flow of decrees by which they had been attempting to force *travopol'e* on the nation's farms, but those decrees (133 all told) were not repealed until 1962.[126] In the meantime, soil science and agronomy were allowed to remain in the grip of Williams' school. To be sure, gestures were made in the direction of freedom for scientists and farm managers. Extremists of Williams' school, for the most part unnamed, were criticized for various stupidities in applying the great theories, stupidities that Lysenko had criticized in 1950, at the height of the campaign for *travopol'e*. At the same time, though still addicted to Williams' ideology, the government was seized by fits of impatience with low yields. Envy of the dizzily rising yields in advanced Western countries began to break down the tranquilizing effects of Williams' grand theory. The corn campaign was a major departure from *travopol'e,* but not the end. The devil of the matter was, first, that no one was sure how to measure the relative efficiency of alternatives in cropping systems, and, second, that the regime still doubted the ability of farm managers to make decisions beneficial both to themselves and to the national economy. So appeals for individualized rotations based on cost accounting coexisted with continued preaching of *travopol'e and* with insistence on corn as the main fodder crop all over.

Those contrary pressures generated Khrushchev's thunderclap of 1961. Soil scientists and agronomists were set free from Williams' school, and lectured harshly on the duty of the scientific community to stand up for their convictions as Tulaikov had done.[127] (He was "rehabilitated" posthumously.) At the same time Khrushchev went out of his way to exempt Lysenko from responsibility for the backward state of soil science, and to maintain his immunity to scientific criticism. Farm managers were told that they must choose cropping systems not on the basis of a grandiose single scheme, but on the basis of careful calculation of their real alternatives. At the same time they were harried by a new campaign for a single scheme of crop rotation: the *propashnaia.* This meant a concentration on intertilled crops such as Khrushchev's beloved corn, at the expense of fallow, grass, leguminous herbage, and small-grain fodder crops.

Once again the Soviet government was trying to push farmers at a rush into the fourth stage of cropping systems, this time into a highly intensive variant, shutting its eyes to the limitations imposed by Russian geography and continuing backwardness. The chemical industry was not able to provide the necessary amount of fertilizers, and in many regions and farms the complex of geographical, economic, and human factors is such that a more extensive type of farming is more economical. In the Baltic region, for example, grass and leguminous herbage are a better source of fodder than corn, as the region's specialists proclaimed in the aftermath of Khrushchev's fall. (They called for *"Travy bez travopolia,"* [128] "Grass Without the Grassland System," but their description makes it clear that they are for the Danish or Swiss type of farming, which was the original meaning of *travopol'e.*)

In the spring of 1965 the Minister of Agriculture of the RSFSR boasted that for the first time in more than thirty years no outside authority was trying to impose cropping schemes and methods on farm managers and agronomists.[129] To what extent the boast was justified, and how long Soviet authorities will continue to make it, remain to be seen. It is clear that no government in an advanced country can refrain altogether from interference in the farmer's choice of cropping systems. But it is also clear that the most effective management of farmers rests on official realization that they must have a considerable degree of autonomy within the framework of national controls. The same holds true for management of an advanced nation's scientists, who are not and cannot be the utterly free individualists they often imagine themselves to be, yet cannot be the driven creatures that Stalin, Williams, and Lysenko tried to make of them.

The Soviet authorities show a dawning recognition of these elementary truths, but they are still inhibited from full recognition and appropriate policy by their reluctance to apply to themselves the eloquent appeal that a journalist addressed to a Lysenkoite agronomist:

> Today, at a new stage in our progress, an end has come to subjectivism in the management of agriculture. . . .
>
> Will you have enough strength to argue with yourself every day, and test all you've been doing with great care, to seek with unflagging curiosity to learn what you haven't yet succeeded in learning; will you learn to take a respectful attitude toward the opinion of scientists even if you do not agree with them; will you find your place in science? [130]

When the Soviet press carries similar appeals to the Party chiefs — are they not social scientists? — we will be sure that a proper measure of autonomy has been secured not only by scientists and farm managers but by Soviet citizens at large.

Is that a utopian dream? In 1950 it would have been utopian to imagine that the Soviet press would some day carry such an appeal to the Lysenkoites. At that time it seemed as if intuition and bluster, self-deceiving enthusiasm and force, had permanently suppressed genuine science and free discussion in the Soviet drive for modernization. Evidently there is something in plants, soil, farmers, and scientists that requires rationality and genuine self-criticism of those who would manage them. My hope is that this inherent necessity is limitless in its scope, that it will make itself felt in areas of Soviet life where it now seems utopian to expect it. But I must admit that faith in the necessity of rationality and freedom is not fully sustained by such studies as I have presented here. The modernization of a backward country seems to require some irrational enthusiasm and force, which are very difficult to measure out in usefully moderate doses.

10

Ideologies and Realities

The Stalinists did so many things in such extravagantly brutal and wasteful ways that outside observers came to think of them as mad ideologists whose minds were unhinged by a dream of total power and utopia. The Stalinists regarded themselves as supremely practical people, who subordinated theoretical considerations to practical necessities. And they were in fact notoriously anti-intellectual, though they were equally notorious for their ideological fanaticism. The apparent paradox is easily resolved by definition: ideology need not be intellectual, it need not consist of theoretical doctrines. If it is understood to be unacknowledged dogma that serves a social function, then one finds it frequently embodied in inarticulate beliefs, or in beliefs that are hardly discussed because they seem self-evident. But once definition gets us past the apparent paradox of anti-intellectual ideology, we are enmeshed in problems that cannot be solved without detailed historical study. Above all there is the problem of explaining intense commitment to practicality combined with very impractical behavior. To state the problem is immediately to be involved in another, that of viewpoint. When we judge the Stalinists impractical, maybe we are simply exercising our own ideological beliefs. Where can we find an objective vantage point from which to study them?

The rise and fall of agrobiology presents the problem of Stalinist practicality in an acute form, while affording at the same time an excellent vantage point — world science. Of course the world's scientific communities cannot claim absolute truth, but they can fairly claim that they are closer than anyone else to genuine knowledge concerning their particular fields of inquiry. The Bolshevik leaders, in common with most other rulers of the modern world, have generally acknowledged this claim, but they made a great exception in the sciences that impinge on agriculture. For thirty-five years they showed strong tendencies to dispute many of the truths established by these sciences, preferring the schemes of willful cranks. In this exceptional case the outside observer can identify the ideological elements in Bolshevik thought rather easily, without fear of indulging his own ideological preferences. On the other hand, precisely because he is identifying ideological beliefs by their gross divergence from known truth, he finds it very hard to establish their social functions. There is a great temptation to declare these ideological beliefs purely dysfunctional, for it seems impossible that flouting known truth can ever serve any useful functions.

Still there comes the nagging reminder that the fanatical Bolsheviks did not stumble into permanent disaster. They did carry through a practically successful revolution even in agriculture, their poorest sector. They forced traditional peasants to support rapid industrialization, creating the world's second greatest military-industrial complex; and they were practical enough to start reforming the agricultural system when it showed signs of stagnation. This is hardly Marx's notion of liberating revolutionary *praxis*, but let us put off that delicate dream, as the Bolshevik leaders did within their first decade of rule. Since Stalinism triumphed in the late 1920s the criterion of practice at work in Bolshevik minds has been defined almost exclusively in terms of a military-industrial complex and its many requirements, including a great increase in agricultural productivity. Hope for such an increase was the basis for the original support of genuine agricultural science, for the rapid swing to agrobiology in the time of collectivization, for the protracted wavering over a final choice between the two, and for the ultimate return to unquestioning support of genuine science. If the desire for an increase in agricultural productivity was strong and persistent, as it seems clearly to have been, why did it express itself in such erratic behavior?

The original support of science was based on utopian assumptions about the state's power to accelerate the modernization of agriculture by altering its social organization. When these assumptions took a forceful turn in collectivization and the immediate result was a decline in most indices of productivity, the Bolshevik leaders angrily refused to believe that they had been proved wrong. They had achieved a sharp increase in the state's share of the harvests on a sharply increased acreage, the two indices that did go up, clearly establishing the practical success of the new system. If scientific methods were inapplicable on most of the new farms, the methods must be wrong. Hence the protracted infatuation with agrobiology, which accused science of impracticality, and hence too the ultimate disillusionment with agrobiology, as the critical index failed to go on climbing. The level of low-cost state procurements (*zagotovki*) was the rough-and-ready test of truth in the Bolshevik leaders' anti-intellectual minds. However wasteful, it was an effective test, if only because it finally obliged them to recognize the inefficiency of their farming system. During the long period while they were testing the capacity of the system, using political compulsion as a method of extracting from peasants the agricultural requirements of an industrial-military complex, agrobiology served to reassure them that they were doing things scientifically. As they gave up the test, they gave up the reassuring pseudoscience.

The costs of this narcotic were high, not only in farm labor and materials wasted on nostrums, but even more in misdirected education and research. Peter the Great had bequeathed to his country a peculiar pattern of modernization, from the top down, with advanced learning growing faster than the capacity of society to utilize it. The Bolsheviks started out in this tradition, and then, with collectivization, tried to leap out of it, to force agricultural science quickly out of the experiment stations onto the peasant fields. At the same time they sharply increased the state's share of the peasants' produce. If they had been coldly calculating rulers they would have recognized the incompatibility of these two efforts. They would have continued support of genuine science as a long-term investment, looking forward to the payoff when they could begin to reduce the exploitation of the peasants, and an increasing flow of properly trained specialists could appeal to the usual incentives for the introduction of advanced farming methods. Rulers who saw themselves as slave drivers might even have persisted in the initial effort to force scientific methods on peasants who were denied any profit from them. But Stalinist bosses saw themselves as popular leaders in the creation of an abundant new society; after a few disappointments with agricultural science, they angrily switched their support to pseudoscience. They needed to believe the line they were handing out, that collectivization was creating the most advanced farming system in the world. A great capacity for self-deception has been as crucial to the Stalinist mentality as the famous readiness to deceive others. Perhaps this is a general characteristic of the political species, especially marked in the frenzied Stalinist variety. In any case, a saving limit was provided by the crude Stalinist rule of practicality, a rule of frantic authoritarian trial and painfully protracted recognition of error.

Aside from its psychological uses to the rulers, agrobiology had other latent functions, which can be discovered only by detailed examination of its influence in the experiment station and the farm. In some respects it gave reassurance to lesser people than the rulers. In the case of corn, for example, Lysenkoite opposition to modern methods comforted old-fashioned breeders and farm managers who might otherwise have been ashamed of their open-pollinated varieties and their carelessness about varietal purity. A detailed examination of Lysenko's impact on the small-grain cereals would probably establish a similar pattern: considerable losses resulting from the attacks on modern methods, with psychological satisfaction accruing to breeders and sloppy farm officials who could fancy themselves Michurinists. Occasionally one discerns an economic function in agrobiological recipes, however weird the lingo. For example, vernalization of potatoes was an odd name and ludicrous explanation for sprouting tubers before planting them, which is frequently an effective way to raise yields. And *travopol'e,* as it was actually applied, seems to have been a move not so much toward Williams' crankish universal scheme as toward various forms of mixed husbandry, the stage of cropping systems that advanced European countries had moved through in the eighteenth and nineteenth centuries. In these cases economically useful methods were promoted in the name of pseudoscience, with consequent distortions both of the methods and of the relevant sciences that should have been preparing for further advances. A similar pattern would probably emerge from a detailed study of some agrobiologists' work in fruit growing and stock breeding. But it is noteworthy that one has to look hard for instances of this kind. Most of the most famous agrobiological recipes were simple failures, dropped by Stalinist officials after a few years of massively wasteful trial. Slowly and painfully they learned the necessity of leaving science to the scientist, of restoring government support to normally autonomous communities of genuine specialists.

Perhaps the most valuable function of the Lysenko affair, like a disease that immunizes, was to generate self-restraint and rationality in the ruling class.

One wonders whether the process continues, whether the economic necessity of professional autonomy for specialists entails a general growth of political autonomy for citizens at large. Very tentatively, conscious of the trap of wishful thinking, I would suggest that it does. In the darkest days of terror fortunate disciplines such as physics and chemistry, where professional autonomy was not seriously impaired, produced a handful of Varangians, adventurers who went out of their special fields to defend academic freedom in foreign territory. As the terror has receded into the past, such people have become less rare and more audacious. Together with like-minded artists they have recreated the old Russian intelligentsia with its consciousness of *obshchestvennost'* versus *vlast'* (enlightened public opinion versus political power), pressing for full-fledged constitutionalism, in some cases directly challenging the political bosses to judicial trial of "socialist legality." So far, when the bosses have taken up the challenge, the result has been kangaroo trials that have jailed or banished individual Varangians. But others are left at large, and the larger body of ordinary intelligentsia go on creating a public opinion that is increasingly independent of the officially manufactured one. They do this by talking among themselves, by circulating unpublished manuscripts, by struggling to get nonconformist items into print, sometimes at home, sometimes abroad. Recently a physicist went so far as to publish a straight-forward political critique of the existing system.[1]

What may be more important, producing the milieu within which such bold defiance becomes possible, are quiet acts of disobedience by the bulk of nonheroic specialists. One recalls the refusal of geneticists and plant physiologists to desert their professional standards. Nowadays the outside observer sees indirect signs that analogous professionalism is mounting in the social sciences. For example, after Khrushchev was dismissed for obstructing further agricultural reform, Matskevich returned as Minister of Agriculture with a warning that reform was not to be pushed so far as to question collectivization.[2] The implications for historians were spelled out by S. P. Trapeznikov, a writer on agrarian history who became chief of the Central Committee's Division of Science. He has been dutifully supervising the restoration of biological science, while trying to suppress objective historical inquiry into collectivization. He denounces unnamed

> dilettantes of scholarship, who with a merchant's mentality [*kupecheskim razmakhom,* as opposed to the revolutionary *razmakh* that Stalin celebrated] and petty-bourgeois obstinacy, try to denigrate the celebrated conquests of the Party and the people, to throw a shadow on the great and thorny path traversed by our socialist Nativeland. . . . These knights of nihilism . . . are interested not in the flame of struggle, but the smouldering difficulty, not the steel but the slag, not the cultivated plants but the weeds and thistles.[3]

Trapeznikov warns such historians that scholarship is like an enormous machine in motion, which can crush the careless person who comes too close. Of course, he does not recall Karl Radek, who threatened scholars with a similar metaphor in 1930, a few years before he himself was crushed.[4] Nowadays such talk alarms; it does not terrify. When Trapeznikov was nominated for membership in the Academy of Sciences, the academicians voted him down.[5] In 1929 a similar refusal to elect top Communists triggered a purge of the Academy with numerous arrests. In the late 1960s, Trapeznikov is limited to scolding, to obstructing publication of nonconformist works, and to depriving nonconformists of their jobs.

One hopes that the obstructed historians will be provoked into serious consideration of the questions carved on prison walls by many "enemies of the people": Why? What for? They were asking for the purposes of the Stalinist frenzy, and some had sufficiently detached intellects to wonder also about its causes and functions. These things — purposes, causes, functions — are related, but they should not be confused. To search for purposes is either to ask what were the conscious intentions of the Stalinist leaders, or else to ask what ends or goals they served, if only unwittingly and in the long run. The intentions of political rulers are only part of historical causation, as we can readily appreciate when we reflect how often they achieve quite unintended results. The Stalinists intended to raise farming to the level of science, but they accomplished just the opposite, dragging science down to the primitive level of peasant farming, and ultimately produced a completely unintended restriction of political power. The Stalinist record is so full of similar contrasts between intention and accomplishment; the intentions are such intricate mixtures of public and private, conscious and subconscious elements; the results are always the product of such complex interactions between the leaders' directions and the behavior of millions of other Soviet people, that the search for causation often seems a hopeless enterprise.

Nevertheless, there are ways to understanding. The poet and the creative ideologist use intuition and prophecy. The scholarly inquirer, with a strong determination to know for a fact the whys and wherefores, selects a manageable few of the beliefs and actions of various groups, correlates them with each other and with the external realities they were intended to explain and master, and achieves thus a little understanding of the transitory functions served by the beliefs and the actions. This is not comprehensive history, which at this point can be only poetry or dogma. This limited history does not even establish causal relations in the strict sense of the word. To say that Stalinist irrationality functioned as a wasteful and brutal aid to the modernization of agriculture is not the same as saying that modernization required, could not have occurred without, Stalinist irrationality. This book has tried to prove the first proposition; I am not even sure how one might go about proving the second.

Antiseptic talk of functions and causes is very far from satisfying the urge to know what ends or goals were served by Stalinism, if only unwittingly and in the long run. In 1930 Mayakovsky, on the verge of suicide, shouted to his "comrade descendants digging in this day's petrified shit," commanding future professors to take off their glasses and see the poetry of "built-in-battles socialism." [6] In 1936 his more durable friend Pasternak urged analogous reassurance upon yet another poet in despair: "Believe in revolution as a whole, believe in the future . . . and not in the construction put on things by the Union of Soviet Writers, which will be changed all of a sudden before you have had time to sneeze — believe in the Age and not in the week" of the current bosses, who were hounding Pasternak's friend to his death.[7] The historian is forbidden by his discipline to brush aside all the grubby things done in the week of the Stalinist bosses — the many weeks and continuing years, as it turns out. Can he nevertheless discern what the poets saw in their age? To say that irrational Stalinism helped make Russia powerful is hardly relevant to the Communist vision of Mayakovsky, certainly not to the Christian faith of Pasternak. It is almost a jeer at those who believed in the collaboration of science and the revolution. If Vavilov was allowed to read *Pravda* as he died of malnutrition in jail, he must have groaned to see Lysenko boasting that the plant world had become "clay or plaster for the sculptor: we can easily sculp from them the forms we need." [8] This was a sensationalist perversion of a metaphor that Vavilov had frequently repeated. In the near future, he had said, science would make it possible "to sculp organic forms at will," for the use of prosperous socialist farmers.[9] Tormented in their time by their faith in the liberating power of science and the revolution, Vavilov's generation may yet be justified. Perhaps a very distant future may even see the poetry of Mayakovsky's "built-in-battles socialism."

Appendixes
Bibliography
Notes
Index

Appendix A. Repressed Specialists

The 105 people listed here are only a fraction of all the physicists, philosophers of science, biologists, and agricultural specialists who suffered repression. (For a definition of repression, see p. 112.) Rank-and-file specialists who did their work without publishing suffered repression in obscurity. Even many of those who published and suffered repression are still not in this list, for some published too little to support a judgment of their specialty and position, and in many cases the evidence of repression is not clear. In short, this is a list of specialists who left news of themselves in the public record, and whose repression is established with certainty or high probability in that same record.

C stands for "certainly repressed," as evidenced by contemporaneous testimony, for example, the appellation "wrecker" or "enemy," or by subsequent, often posthumous, "rehabilitation."

P stands for "probably repressed," evidenced by indirect signs. I have tried to be cautious in making such judgments, since I realize that scientists drop out of sight for many other reasons than repression. For example, two Americans on the staff of the Soviet Institute of Genetics in 1936 were forced out of scientific work on their return to the United States, because American administrators of science were afraid to hire them. (Carlos Offerman and Daniel Raffel — their cases were described to me by H. J. Muller.) As Soviet people understand repression and as I have used the term, these Americans were not repressed.

The reader should note that sixty-eight of the cases listed here are C, certainly repressed; thirty-seven are P, probably repressed.

The sources of information were too diverse to permit footnoting. I would greatly appreciate corrections and additions. I do not pretend to have searched the public record exhaustively. I searched the record until my patience was exhausted.

A.1 Physicists and Philosophers of Physics*

Name	Dates	Specialty and post	Public Position on modern physics	Fate, year of arrest
Berg, Aksel Ivanovich	1893–	Radio specialist; Leningrad Electrotech. Inst.	Took it for granted	P, 1937; subsequently released
Bronshtein, Matvei Petrovich	1906–1938	Theorist; worked at Physico-Tech. Inst., Leningrad	Defender	C, 1937
Bursiian, Viktor Robertovich	1886–?	Theorist; Leningrad Univ.	Defender	P, 1937?
Frederiks Vsevolod Konstantinovich	1885–1944	Theorist	Defender	C, 1937?
Gessen (or Hessen) Boris Mikhailovich	1883–1937?	Philosopher, popularizer, administrator of physics; dir. of Inst. of Physics at Moscow Univ.	Defender	C, 1936
Houtermanns, Friedrich		Ger. physicist; worked at Physico-Tech. Inst. Kharkov	Took it for granted	P, 1937; deported to Germany, 1940
Ivanenko, Dmitrii Dmitrievich	1904–	Theorist	Defender	P, 1938?; subsequently released
Krutkov, Iurii Aleksandrovich	1890–1952	Theorist	Defender	C, 1937?
Landau, Lev Davidovich	1908–1969	Theorist; Inst. of Physical Problems of Acad. of Sciences at time of arrest	Defender	C, 1938; subsequently released
Leipunskii, Aleksandr Il'ich	1903–	Experimentalist; dir. of Physico-Tech. Inst., Kharkov	Took it for granted	P, 1938; subsequently released
Lukirskii, Peter Ivanovich	1894–1954	Experimentalist; Physico-Tech. Inst. of Academy of Sciences	Took it for granted	P, 1938? subsequently released
Obreimov, Ivan Vasil'evich	1894–	Chief crystallographer (former dir.); at Physico-Tech. Inst., Kharkov	Took it for granted	P, 1937; subsequently released

A.1 (*continued*)

Name	Dates	Specialty and post	Public position on modern physics	Fate, year of arrest
Rozenkevich, Lev Viktorovich	?–1937	Theorist; working at Physico-Tech. Inst., Kharkov at time of arrest	Took it for granted	P, 1937
Rumer, Iulii Borisovich		Theorist	Defender	P, 1938?; subsequently released
Semkovskii Semen Iulevich	1882–	Philosopher, special interest in theory of relativity; leading philosopher of science in the Ukraine	Defender	P, 1937?
Shpil'rein, Ian Nikolaevich	1887–?	Theorist; eminent prof. at Physico-Tech. Inst., Leningrad	Defender	P, 1938?
Shubin, Semen Petrovich	1908–1938	Theorist	Defender	C, 1938
Shubnikov, Lev Vasil'evich		Experimentalist; chief of low-temperature lab. at Physico-Tech. Inst., Kharkov	Took it for granted	P, 1938
Uranovskii, Iakov Markovich		Philosopher of science; admin. at Leningrad Univ.	Defender	C, 1937
Vasil'ev, Sergei Fedorovich	1898–	Philosopher of science; worked in Acad. of Sciences	Defender	C, 1936
Vitt, Aleksandr Adolfovich		Experimentalist; at Moscow Univ.	Took it for granted	P, 1937?
Weissberg, Alexander		Austrian physicist; at Physico-Tech. Inst., Kharkov.	Took it for granted	C, 1937; deported to Germany, 1940

* It is probable that some obscurantist critics of modern physics will be on the complete list, when it is dug out of the archives. See note 77, Chapter 5.

A.2 Biologists, Philosophers of Biology, and Agricultural Specialists

Name	Dates	Specialty and post	Public position on agrobiology	Fate, year of arrest
Abolin, Robert Ivanovich	1886–"after 1939"	Botanist, ecologist, soil scientist; head of section of VIR studying deserts	None, except for tribute to Williams, 1935	P, 1938 or 1939
Agol, Israel Iosifovich	1891–1937	Philosopher of biology, geneticist; various posts as researcher, teacher, and editor	None; leader in Communist discussions of biology in 1920s and early 1930s	C, 1936
Arnol'd, Boris Mikailovich	1891–?	Plant breeder, botanist; chief of millet studies at Saratov Inst. of Grain Culture	Not applicable; disappeared too early	P, 1931?
Artemov, Petr Ksenofont	unknown	Crop specialist; chief of variety testing and certification in early 1930s	None	P, 1937?
Avdulov, Nikolai	1899–?	Cytogeneticist; at VIR and then at Saratov Inst. of Grain Culture	None	P, 1937?
Barykin, Vladimir Aleksandrovich	1901–?	Microbiologist; director of State Scientific Microbiological Inst.	None	C, 1937?
Bauer, Erwin (Hungarian refugee)	1890–1942	Biophysicist, philosophical interpreter of biology; Communist Acad. and Inst. of Experimental Med.	None	C, 1937
Berg, Viktor Romual'dovich	1883–?	Plant breeder; scientific dir. of Siberian Inst. of Grain Culture	Not applicable; disappeared too early	C, 1933 or 1934
Bordakov, Leonid Petrovich	1885–?	Plant breeder; head of cotton section of VIR	None	C, 1937
Bykhovskaia Anna Markovna	1901–?	Zoologist; dean of biology faculty and dir. of Inst. of Zoology, Moscow State Univ.	None	C, 1937

A.2 (*continued*)

Name	Dates	Specialty and post	Public position on agrobiology	Fate, year of arrest
Chaianov, Aleksandr Vasil'evich	1888–?	Agricultural economist; dir., Inst. of Agri. Economics, Timiriazev Acad.	Not applicable	C, 1930
Chelintsev, Aleksandr Nikolaevich		Agricultural economist	Not applicable, though he published a very respectful critique of Michurin in 1912	C, 1930
Chetverikov, Sergei Sergeevich	1880–1959	Geneticist; prof. at Moscow State Univ.; after release, in 1935 prof. at Gorky Univ.	None after release; in 1927 ridiculed cult of Kammerer	C, 1931; subsequently released
Chizhevskii, Aleksandr Leonidovich	1897–?	Philosopher and crank; dir. of Central Lab. of Ionization, est. for him, 1931, by Comm. of Agri.	None, beyond defense of his own version of agrobiology	C, 1936
David, Rudol'f Eduardovich	1887–1939	Meteorologist; chief of agri. meteorology at Saratov Inst. of Grain Culture	None	C, 1937
Doiarenko, Aleksei Grigor'evich	1874–1958	Agronomist; prof. and dept. chrmn. at Timiriazev Acad.; after release worked at Saratov Inst. of Grain Culture	Forthright opponent of Williams' schemes and critic of hasty plans for "grain factories"	C, 1930; subsequently released
Efroimson, Vladimir Pavlovich		Geneticist; *Dotsent* at Kharkov U. at time of arrest	Anti-Lysenkoite	C, 1936–1937 C, 1949–1955
Ermakov, Grigorii Efremovich	1900–?	Stockbreeder; dir. All-Union Inst. of Stockbreeding	Lysenkoite	C, 1937
Favorskii, N. V.		Cytologist; chief of cytological lab. at Saratov Inst. of Grain Culture	None	P, 1937
Ferri, L. V.		Geneticist	None	P, 1936

A.2 (*continued*)

Name	Dates	Specialty and post	Public position on agrobiology	Fate, year of arrest
Filip'ev, Ivan Nikolaevich	1889–?	Entomologist; Leningrad Univ.	None	C, 1937?
Fliaksberger, Konstantin Andreevich	1880–?	Wheat specialist; chief of wheat studies at VIR	None	C, 1941?
Gartokh, Oskar Oskarovich	1881–?	Microbiologist; chief of bacteriology at All-Union Inst. of Exp. Medicine	None	C, 1938?
Gogol'-Ianovskii, Georgii Ivanovich	1868–?	Horticulturist; dir. of exp. div. of Com. of Agri., early 1920s	None, except for restrained praise of Michurin in 1919	P, 1933?
Govorov, Leonid Ipat'evich	1885–?	Plant breeder; chief of Div. of Grains and Legumes at VIR	Anti-Lysenkoite	C, 1940
Gredeskul, Nikolai Andreevich	1865–?	Philosopher of biology, anthropology	None	P, 1933?
Ianata, Aleksandr Aloizovich	1888–?	Crop specialist, soil scientist; dir. of Ukrainian Inst. of Applied Botany	None	C, 1933?
Ignat'ev, M. V.		Mathematician, specializing in population genetics; in charge of such studies at Inst. of Medical Genetics	None	P, 1937?
Il'in, Nikolai Alekseevich	1903–?	Geneticist, stockbreeder; dir. of Wool Inst.	None	P, 1939?
Ivanov, Il'ia Ivanovich	1870–1932	Zoologist, pioneer in artificial insemination of livestock; head of lab. of artif. insem. at All-Union Inst. of Stockbreeding	Not applicable	C, imprisoned Dec. 1930–June 1931
Karpechenko, Georgii Dmitrovich	1899–1942	Geneticist; head of dept. at Leningrad Univ.	Anti-Lysenkoite	C, 1941

A.2 (*continued*)

Name	Dates	Specialty and post	Public position on agrobiology	Fate, year of arrest
Kol', Aleksandr Karlovich	1877–?	Botanist; head of plant introduction at VIR	Lysenkoite	P, 1938?
Kondrat'ev, Nikolai Dmitrovich	1892–?	Agri. economist; prof. at Timiriazev Acad.	Not applicable	C, 1930
Korolev, Sergei Ivanovich	1894–1946	Plant breeder; official in variety testing before arrest, prof. at Gorky Agri. Inst. after release	None	C, 1934? subsequently released
Koval' Aleksei Pavlovich		Horticulturist; dir. of Mysovskaia Sta. of fruit breeding	Offended Michurinists by refusing to endorse their varieties and methods	P, 1936?
Kovalev, Nikolai Vasil'evich	1888–?	Plant breeder; various posts at VIR before arrest, head of fruit breeding at Cent. Asian Sta. of VIR after arrest	Compromiser, failed to satisfy Lysenko	C, 1941?; subsequently released
Krichevskii Il'ia Levich	1885–?	Microbiologist; director Microbio. Inst. of Commissariat of Ed.	None	C, 1937?
Kuz'min, Valentin Petrovich	1893–	Plant breeder; before arrest worked in VIR, after release at a plant breeding sta. in Kazakhstan	None	C, 1936?; subsequently released
Lepin, Denis Karlovich	1895–1964	Geneticist; Inst. of Genetics, Acad. of Sci. before arrest	Anti-Lysenkoite	P, 1938? subsequently released
Levin, Max (Liudvigovich added when he moved to Russ.)	1885–1937?	Philosopher of biology; worked at Timiriazev Inst., Comm. Acad.	None	C, 1936
Levit, Solomon Grigor'evich	1894–1938?	Geneticist; dir. of Inst. of Medical Genetics	None	C, 1936 or 1937

A.2 (*continued*)

Name	Dates	Specialty and post	Public position on agrobiology	Fate, year of arrest
Levitskii, Grigorii Andreevich	1878–1947	Cytogenetics; prof. at Leningrad Univ.	Anti-Lysenkoite	C, 1941
Makarov, Nikolai Pavlovich	1886–?	Agri. economist; prof. at Timiriazev Acad.	Critic of scheme for "grain factories"	C, 1930
Margolin, Lev Solomonovich		Crop specialist; academic sec'y of Lenin Acad. of Agri. Sciences	Vigorous opponent of Williams' schemes, no public stand on Lysenkoism	C, 1937
Meister, Georgii Karlovich	1873–?	Plant breeder; chief of breeding at Saratov Inst. of Grain Culture	Major proponent of compromised coexistence between genetics and Lysenkoism	C, 1937
Mikheev, Aleksandr Aleksandrovich	1884–?	Botanist, forest specialist; prof. at Polytechnic Inst., Baku	None	C, 1937?
Nadson, Georgii Adamovich	1867–1940	Geneticist; dir. of Inst. of Microbio., Acad. of Sciences	None	C, 1937
Nurinov, Aleksandr Ageevich		Stockbreeder; director of Inst. of Hybridization at Ascania Nova	Lysenkoite	P, 1937?
Odintsov, Vasilii Alekseevich	1899–	Horticulturist; dir. of Michurin Inst. of Fruit Culture, after release worked in Horticulture Inst. of Non-Black-Earth Zone of RSFSR	Leading Michurinist, helped the fusion of Michurinism with Lysenkoism	C, 1937; subsequently released
Orlov, Aleksandr Alekseevich		Plant breeder; worked at VIR	None	C, 1937?
Pan'shin, Igor Viktorovich		Geneticist; worked at Inst. of Genetics, then at Inst. of Exp. Bio.	None	C, 1946? (accused of desertion from Army); subsequently released

A.2 (*continued*)

Name	Dates	Specialty and post	Public position on agrobiology	Fate, year of arrest
Pisarev, Viktor Evgrafovich	1883–	Plant breeder; major official at VIR before arrest, worked at Moscow Reg. Exp. Sta. after release	Compromiser, won approval of Lysenkoites	C, 1933?; subsequently released
Plachek, Evgeniia Mikhailovna	1878–?	Plant breeder; chief of sunflower work at Saratov Inst. of Grain Cul.	Anti-Lysenkoite	C, 1937?
Poretskii, Artemii Sergeevich	1901–42	Botanist; worked at Far Eastern Branch of Acad. of Sciences and at Bot. Inst., Leningrad	None	C, 1937
Pozdni..kov, L. V		Horticulturist	Offended Michurinists by failing to endorse their varieties and methods	P, 1933?
Raikov, Boris Evgen'ich	1884–1966	Specialist in biological educ.; prof. at Leningrad Pedagogical Inst.	None	C, 1931?; subsequently released
Ravich-Cherkasskii, M.		Philosopher of biology	None	C, 1931?
Romashov, Dmitrii Dmitrevich	1899–	Geneticist; Inst. of Exp. Bio.	Anti-Lysenkoite	C, 1948?; released 1955
Rubtsov, Grigorii Aleksandrovich		Breeder of fruit trees; worked at VIR	Compromiser	P, 1937?
Samarin, Nikodim Gavrilovich		Agri. economist; worked at Saratov Inst. of Grain Culture	Not applicable, except for opposition to Williams' scheme	C, 1933
Sambuk, Feodosii Viktor	1900–42	Botanist; worked at Bot. Inst., Leningrad	None	C, 1937
Shlykov, Grigorii Nikolaevich	1903–	Botanist; head of new crops research at VIR, after release worked in fruit sec. of VIR	Lysenkoite	C, 1940?; subsequently released

A.2 (*continued*)

Name	Dates	Specialty and post	Public position on agrobiology	Fate, year of arrest
Shtutser, Mikhail Ivanovich	1879–?	Microbiologist; director of Microbio. Inst. in Rostov (Don)	None	C, 1937?
Shul'meister, Konstantin Georgevich		Crop specialist; before arrest dir. of Krasnokut Exp. Sta., since release prof. at Stalingrad Agri. Inst.	None	P, 1937?; subsequently released
Simirenko, Vladimir L'vovich		Fruit specialist; scientific dir. of All-Union Fruit Inst., Mleev	Offended Michurinists by refusing to endorse their varieties and methods	C, 1933
Slepkov, Vasilii N.		Philosopher of biology; in 1920s worked at Sverdlov Comm. Univ. and Comm. Acad.	None	P, 1935?
Talanov, Viktor Viktorovich	1871–1936	Plant breeder and crop specialist; many imp. posts in 1920s incl. chief of variety testing and certification, last known post at Gorky Dist. Exp. Sta.	None	C, 1936?
Timofeev-Ressovskii, Nikolai Vladimirovich	1901–	Geneticist; worked in Berlin until arrest in 1945, after release worked at Inst. of Med. Radiology, Obninsk	None	C, 1945; released 1955?
Troitskii, Nikolai Nikolaevich	1887–?	Entomologist; All-Union Inst. of Plant Protection	None	C, 1937?
Tsinzerling, Iurii Dmitr.	1894–1939	Botanist; acting dir. of Bot. Inst., Leningrad	None	C, 1938

A.2 (*continued*)

Name	Dates	Specialty and post	Public position on agrobiology	Fate, year of arrest
Tulaikov, Nikolai Maksimovich	1875–1938	Crop specialist; dir. of Saratov Inst. of Grain Cult., until 1933 one of the govt's two principal advisers on problems of agri. techniques	Compromiser with respect to Lysenkoites; anti-Williams	C, 1937
Uspenskii, Evgenii Evgen'evich	1889–?	Plant physiologist; head of work in that field at Biol. Inst. of Comm. Acad. (Timiriazev Inst.), also head of Dept. of Microbio. at Moscow Univ.	Compromiser	P, 1938?
Vasil'ev, Boris Ivanovich	1904–?	Cytogeneticist; *Dotsent* at Leningrad Univ.	None	P, 1940?
Vavilov, Nikolai Ivanovich	1887–1943	Botanist; many imp. posts, incl. dir. of VIR; until 1935 one of gov't's two principal advisers on problems of agri. techniques	Compromiser until 1937, then militant anti-Lysenkoite	C, 1940
Vermel', Iulii Matveevich	1906–?	Zoologist, worked at Comm. Acad. and Moscow Univ.	Lamarckist in 1920s; no record of public position in debates of 1930s	P, 1936?
Verushkin, Sergei Makarovich	1895–?	Plant breeder and geneticist; worked at Saratov Inst. of Grain Culture	None, except for offense given to Tsitsin on issue of hybrids of wheat and couch grass	P, 1937?
Voitkevich, I. I.		Plant breeder; dir. of exp. sta. working on improvement of sugar beet	None	P, 1933?

A.2 (*continued*)

Name	Dates	Specialty and post	Public position on agrobiology	Fate, year of arrest
Vol'f, Moisei Mikhailovich		Agricultural economist; head of agri. sec. of State Planning Commission	None	C, 1933
Zaporozhets, Anton Kuzmich	1895–?	Agronomist and economist; dir. of All-Union Inst. of Fertilizer	None	C, 1937
Zarkevich, A. V.		Stockbreeder; worked at All-Union Inst. of Stockbreeding	Lysenkoite	C, 1938
Zdrodovskii, Pavel Feliksovich	1890–	Microbiologist; chief of epidemiology at All-Union Inst. of Exp. Med.	None	C, 1938?
Zil'ber, Lev Aleksandrov	1905–	Virologist; chief of virology in Inst. of Microbio. at Acad. of Sciences	None	P, 1938?; subsequently released
Znamenskii, Aleksandr Vasil'evich	1891–?	Entomologist; All-Union Inst. of Plant Protection	None	C, 1937?

Appendix B. For Kremlinologists

In this book the behavior of politicians has been correlated with the development of policies and of institutions. Of course I am aware that individual politicians are primarily concerned with personal relations among themselves. I sympathize with the effort "to lay bare factional alignments within the [Soviet] regime mainly by identifying the personal followings of various leaders. . . ." (Carl Linden, *Khrushchev and the Soviet Leadership.* Baltimore, 1966, p. 6.) Such alignments and followings are undoubtedly the primary concern of politicians, and therefore of conventional political historians. The trouble is that available sources are inadequate for conventional political history of the Soviet regime, except for its first five or at most ten years. Refusal to recognize this simple truth gives rise to Kremlinology, the effort to do archival research in high Soviet politics without access to the archives.

As a challenge to those who may wish to prove me wrong, I append a list of politicians who were associated with the Soviet campaign for scientific agriculture. The voluminous public record of the campaign allows these men to be associated at particular times with particular institutions and with particular policies, and these associations enable one to correlate the interacting development of institutions and policies. That is what I have done in this book. The public record does *not* permit the kind of personal correlations that the Kremlinologist seeks. He is obliged to use a great deal of guesswork and to ignore most of the public information about most of these men, for that information associates them with policies and institutions. It contains only tantalizing, inadequate hints of their personal groupings and maneuverings.

The list is somewhat arbitrary. I have tried to make it a representative sample of key politicians involved in the campaign for scientific agriculture from the 1920s into the 1960s. No doubt other investigators would strike some of the names and add others, but I doubt that any list of comparable length and representative quality would be amenable to Kremlinological analysis.

The list is also somewhat arbitrary in its assignment of each individual to one of three simple categories. Some of these men straddled the categories, and some moved from one to another in the course of time. Furthermore, a proper respect for the complexity of Soviet institutions would require more than these three over-simplified categories. I hope I have shown such respect in the main text of the book. The purpose of this appendix is to present the Kremlinologist with a challenge, a simplified list of politicians who would have to be included in a conventional political history of the rise and fall of agrobiology.

NKZ stands for People's Commissar of Agriculture; MZ for Minister of Agriculture; NKP for People's Commissar of Education; Ts.K. for Central Committee; Ts.I.K. for Central Executive Committee.

B.1 Agricultural Bosses

Name	Dates	Positions	Fate
Basov, A.		MZ RSFSR 1965	
Beliaev, Nikolai Il'ich	1903–	Party official for agri., 1950s	
Benediktov, Ivan Aleksandr.	1902–	NKZ and MZ USSR and MZ RSFSR at various times, 1937–1959	
Butenko, P. P.		NKZ Ukr. SSR, late 1930s	
Chekmenev, E. M.		Various posts in agri., 1940s and 1950s	
Chernov, Mikhail Aleksandr.		NKZ, 1934–1938	Repressed
Iakovenko, Vasilii Grigor.	1889–1938	NKZ, 1922–1923; worked in NKZ again in mid-1930s	Repressed?
Iakovlev, Iakov Arkad'evich	1896–1939	Chief of agri. journalism in 1920s; NKZ and Party chief of agri., 1930s	Repressed
Iurkin, Tikhon Aleksandr.	1898–	Imp. agri. posts, 1930s and again in 1954–1961	Repressed in the period between?
Kalmanovich, Moisei Iosifovich		NKSovkhozov, 1930s	Repressed

B.1 (*continued*)

Name	Dates	Positions	Fate
Kozlov, A. I.		Various agri. positions in 1940s and 1950s	
Kubiak, N. A.		NKZ RSFSR, late 1920s	
Laptev, I. D.		Party official in agri. economics, 1930s to 1960s	
Lisitsyn, N. V.	1891–1939	NKZ RSFSR, 1930s	Repressed?
Lobanov, Pavel Pavl.	1902–	Various posts in agri. and agri. science, 1930s to 1960s	
Matskevich, Vladimir Vladim.	1909–	Various posts in agri. science and agri., 1930s to 1960s	
Mikhailov, Mikhail Lazarevich		Head of agri. pub. house, 1930s	
Morozov, P. I.		Deputy MZ USSR, 1960s	
Ol'shanskii, Mikhail Aleksandr.	1908–	MZ USSR, 1960s	
Poliakov, Vasilii Ivanovich	1913–	Chief of Ts.K. Div. of of Agri. in 1960s	
Pysin, Konstantin Georgievich	1910–	MZ USSR, 1960s	
Sereda, Semen Pafnut'evich	1871–1933	NKZ 1918–1921, and then various posts in agri. and planning	
Shlikhter, Aleksandr Grigor.	1868–1940	NKZ Ukr. SSR, 1920s and 1930s	
Skvortsov, Nikolai Aleksandr.		Min. of State Farms, 1940s	
Smirnov, Alexsandr Petrov.	1877–	NKZ, 1920s	Repressed
Soms, Karl Petrovich		Member of Central Control Commission and State Farm supervisor, 1930s	Repressed?
Sosnovskii, Lev Semenovich	1886–1937	Leading agri. journalist, 1920s; minor role, 1930s	Repressed
Spivak, M. S.		MZ Ukr. SSR, 1950s and 1960s	

B.1 (*continued*)

Name	Dates	Positions	Fate
Utekhin, A. A.		Chief of Ts.K. Div. of Agri., 1962–63	
Vol'f, Moisei Mikhail.	?–1933	Head of agri. section of Gosplan, late 1920s to execution	Repressed
Volovchenko, Ivan Platon.	1917–	MZ and Deputy MZ USSR, 1960s	

B.2 Bosses of Higher Learning and Theoretical Ideology

Name	Dates	Positions	Fate
Bauman, Karl Ianovich	1892–1937	Chief of Ts.K. Div. of Science in mid-1930s	Repressed
Bondarenko, Aleksandr Stepan.	1893–	Party official at Lenin Acad. 1930s	
Bubnov, Andrei Sergeevich	1883–1939	Old Bolshevik, imp. leader in 1917, winding up as NKP RSFSR in 1930s	Repressed
Cherviakov, Alexsandr Grigor.	1892–	Pres. Ts.I.K. in mid-1920s, supervising ed. and science	Repressed
Doroshev, I. A.		Party official for science and theoretical ideology, 1930s to 1950s	
Eliutin, Viacheslav Petrovich	1907–	Min. of Higher Ed., USSR, 1950s and 1960s	
Enukidze, Avel' Safronovich	1877–1937	Party leader involved with science, 1930s	Repressed
Gorbunov, Nikolai Petrov.	1892–	Party official involved with science, 1920s and 1930s	Repressed
Il'ichev, Leonid Fedorovich	1906–	Various posts on ideological front, 1930s to 1960s	
Kaftanov, Sergei Vasil'evich	1905–	Admin. of higher education, 1930s to 1950s	
Kaminskii, G. N.		Comm. of Health, 1930s	Repressed?
Kirillin, Vladimir Alekseev.	1913–	Chief of Ts.K. Div. of Science, 1950s and 1960s	
Kol'man, Ernst Iaromirov.		Party official involved with science from late 1920s to 1960s	
Krzhizhanovskii, Gleb Maksimil.	1872–	Various posts in charge of science and technology, 1920s and 1930s	
Lunacharskii, Anatolii Vasil.	1875–1933	NKP, 1920s	
Mezhlauk, I. I.	1891–1941	Party official in field of higher learning, 1930s	Repressed

B.2 (*continued*)

Name	Dates	Positions	Fate
Miliutin, Vladimir Pavl.	1881–1938	Old Bolshevik, imp. leader in 1917; Various posts in adm. of science, 1920s and 1930s	Repressed
Muralov, Aleksandr Ivan.	1886–	Admin. of agri. and agri. science, 1920s and 1930s	Repressed
Muralov, Nikolai Ivan.	1877–1937	Imp. leader in 1917; in mid-1920s and again in mid-1930s rector of Timiriazev Acad.	Repressed
Osinskii, N. (pseudonym of Obolenskii, Valerian Valerianov.)	1887–1938	Old Bolshevik; various posts in charge of agri. and agri. science, 1920s and 1930s	Repressed
Petrov, Fedor Nikol.	1876–	Admin. of science, 1920s	
Rumiantsev, Aleksei Matveevich	1905–	Party official in higher learning and theoretical ideology, 1940s to 1960s	
Semashko, Nikolai Aleksandrovich	1874–1949	Comm. of Health in 1920s, then other posts in admin. of med. and higher learning	
Stetskii, Aleksei Ivan.		Party official of science and culture, 1930s	Repressed?
Svetlov, V. O.		Dep. Min. of Higher Ed., 1940s	
Tal', B.		Head of Ts.K. Div. of Press and Pub. 1930s	
Targul'ian, O. M.		Party official in agri. science, 1920s and 1930s	
Teodorovich, Ivan Adol'fov.	1875–1940	Old Bolshevik; admin. of agri. and agri. science, 1920s and 1930s	Repressed?
Trapeznikov, Sergei Pavl.	1912–	Chief of Ts.K. Div. of Science, 1965–	
Zhdanov, Iurii Andreev.		Chief of Ts.K. Div. of Science, 1940s and 1950s	

B.3 Bosses at Large

Name	Dates	Fate
Andreev, Andrei Andreevich	1895–	
Brezhnev, Leonid Il'ich	1906–	
Bukharin, Nikolai Ivan.	1888–1938	Repressed
Chubar', Vlas Iakovlevich	1891–1939	Repressed
Demichev, Petr Nilovich	1918–	
Egorychev, Nikolai Grigor.	1920–	
Eikhe, Robert Indrikovich	1890–1940	Repressed
Ignatov, Nikolai Grigor.	1901–	
Kalinin, Mikhail Ivan.	1875–1946	
Khrushchev, Nikita Serg.	1894–	
Kiselev, A. S.	1879–1938	Repressed?
Kosior, Stanislav Vikent'evich	1889–1939	Repressed
Kuibyshev, Valerian Vladim.	1888–1935	
Lenin, Vladimir Il'ich	1870–1924	
Malenkov, Georgii Maksimilian.	1902–	
Molotov, Viacheslav Mikhailovich	1890–	
Podgornyi, Nikolai Viktor.	1903–	
Polianskii, Dmitrii Stepan.	1917–	
Postyshev, Pavel Petrov.	1888–1940	Repressed
Rudzutak, Ian Ernestovich	1887–1938	Repressed
Stalin, Iosif Vissarion.	1879–1953	
Suslov, Mikhail Andre.	1902–	
Tsiurupa, Aleksandr Dmitr.	1870–1928	
Zhdanov, Andrei Aleksandr.	1896–1948	

Bibliography

This bibliography is limited to books, pamphlets, newspapers, and other serials that have been cited in the notes, usually in abbreviated form. It does not include articles, which have been cited in full. The purpose of this list is to give the reader the bibliographical data he needs to find any citation in a library. The list is thus only a rough guide to source materials, and an even rougher guide to the scholarly studies of the many large topics that are touched upon in this book.

NEWSPAPERS

Bednota.
Ekonomicheskaia gazeta.
Ekonomicheskaia zhizn'.
Izvestiia.
Komsomolskaia pravda.
Kul'tura i zhizn'.
Leningradskii universitet.
Literaturnaia gazeta.
Meditsinskii rabotnik.
Moskovskii universitet.
Pravda.
Sel'skokhoziaistvennaia gazeta.
Sotsialisticheskoe zemledelie. See SZ.
Sovetskaia Rossiia.
SZ. Abbreviation for *Sotsialisticheskoe zemledelie.* In 1953 it was renamed *Sel'skoe khoziaistvo;* in 1960 *Sel'skaia zhizn'.*
Timiriazevka.
Za proletarskie kadry. Published at Moscow University.

OTHER SERIALS

Agrobiologiia.
Agronom.
Agrotekhpropaganda.
Akademiia Nauk SSSR. *Doklady.*
———. *Izvestiia, seriia biologicheskaia.*
———. *Izvestiia, seriia fizicheskaia.*
Akademiia Nauk SSSR, Biuro po Genetike i Evgenike. *Izvestiia.* See Akademiia Nauk SSSR, Institut Genetiki.

Akademiia Nauk SSSR, Institut Etnografii. *Trudy, novaia seriia.*
Akademiia Nauk SSSR, Institut Fiziologii Rastenii. *Trudy.*
Akademiia Nauk SSSR, Institut Genetiki. *Trudy,* 1933–65.
Superseded Laboratoriia Genetiki, 1930–33, which superseded Biuro po Genetiki i Evgenike, *Izvestiia,* 1925–29, which superseded Biuro po Evgenike, 1922–24. At the outset it was subordinated to the Komissiia po izucheniiu estestvennykh proizvodstvennykh sil SSSR.
Akademiia Nauk SSSR, Institut Lesa. *Trudy.*
Akademiia Nauk Ukrainskoi SSR, Institut Entomologii i Fitopatologii. *Trudy.*
American Naturalist.
Annual Review of Plant Physiology.
Antropologicheskii zhurnal.
Archiv für Rassen- und Gesellschaftsbiologie.
Biologicheskie nauki. Sometimes known as *Nauchnye doklady vysshei shkoly, Biologicheskie nauki.*
Biologicheskii zhurnal.
Biologiia v shkole.
Biulleten' iarovizatsii.
BMOIP, otdel biol. Abbreviation for *Biulleten' moskovskogo obshchestva ispytatelei prirody, otdel biologicheskii.*
Bol'shevik. 1924–52, when its title was changed to *Kommunist.*
Botanicheskii zhurnal.
BZ. Abbreviation for *Botanicheskii zhurnal.*
Current Sociology.
Discovery.
Dnepropetrovsk, Institut Kukuruzy. *Biulleten'.*
Estestvoznanie i marksizm.
Estestvoznanie v shkole.
Fiziologiia rastenii.
FNIT. Abbreviation for *Front nauki i tekhniki.*
Foreign Agriculture.
Front nauki i tekhniki.
Genetika.
Gossortoset', Informatsionnyi i metodologicheskii sbornik. Published by Vsesoiuznyi Inst. Rastenievodstva.
History and Theory.
Iarovizatsiia.
Imperial Bureau of Plant Genetics, Aberystwyth, Wales. *Bulletin.*
Issledovaniia po genetike. Sborniki, published by Leningrad University, M. E. Lobashev, ed.
Istoricheskii arkhiv.
Izvestiia Akademii Nauk SSSR, seriia biologicheskaia.
———, *seriia fizicheskaia.*
Journal of Forestry.
The Journal of Heredity.
Journal of the History of Ideas.
Der Kampf.
Kartofel'.
Kartofel' i ovoshchi.
Khimizatsiia sotsialisticheskogo zemledeliia.
Kniga i proletarskaia revoliutsiia.

Kommunist. See *Bol'shevik.*
Kraevedenie.
Kukuruza.
Die Kulturpflanze.
Lesnoe khoziaistvo.
The Listener.
Materialy k biobibliografii uchenykh SSSR. Seriia biologicheskukh nauk.
Mediko-biologicheskii zhurnal.
Mezhdunarodnyi sel'skokhoziaistvennyi zhurnal.
Mikrobiologiia.
Na agrarnom fronte.
Narodnyi uchitel'.
Nashi dostizheniia.
Die Naturwissenschaften.
Nauchnoe slovo.
Nauka i zhizn'.
Neva.
New York Review of Books.
Novaia Petrovka.
Novyi mir.
Oktiabr'.
Partiinaia zhizn'.
Partrabotnik. Leningrad.
Pechat' i revoliutsiia.
Pflanzenschutz-Nachrichten "Bayer".
Philosophy of Science.
Planovoe khoziaistvo.
Plodoovoshchnoe khoziaistvo.
Pochvovedenie.
Pod markso-leninskim znamenem.
Pod znamenem marksizma.
Priroda.
Problemy botaniki. Published by Vsesoiuznoe botanicheskoe obshchestvo.
Problemy mira i sotsializma.
Problemy zhivotnovodstva.
Proceedings of the American Philosophical Society.
Progressivnoe sadovodstvo i ogorodnichestvo.
Prostor.
Puti sel'skogo khoziaistva.
PZM. Abbreviation for *Pod znamenem marksizma.*
Russkii evgenicheskii zhurnal.
Russkoe sadovodstvo i ogorodnichestvo.
Sad i ogorod.
Sadovodstvo i ogorodnichestvo.
St. Antony's Papers. Publication of St. Antony's College, Oxford U.
Science.
Sel'skoe i lesnoe khoziaistvo.
Sel'skokhoziaistvennaia literatura SSSR.
Sel'skokhoziaistvennaia zhizn'. Ofiitsial'nyi organ Narkomzema RSFSR.
Semenovodstvo.
Slavic Review.
Sobranie postanovlenii pravitel'stva SSSR.

Sobranie uzakonenii i rasporiazhenii rabochego i krest'ianskogo pravitel'stva RSFSR.
Sobranie zakonov i rasporiazhenii raboche-krest'ianskogo pravitel'stva SSSR.
Sorena. Sotsialisticheskaia rekonstruktsiia i nauka.
Sotsialisticheskaia rekonstruktsiia sel'skogo khoziaistva.
Sotsialisticheskoe plodo-ovoshchnoe khoziaistvo.
Sotsialisticheskoe zernovoe khoziaistvo.
Sotsial'naia gigiena.
Sovetskaia agronomiia.
Sovetskaia botanika.
Sovetskaia etnografiia.
Sovetskaia nauka.
Sovetskoe studenchestvo.
Soviet Studies.
Soviet Studies in Philosophy.
SRSKh. Abbreviation for *Sotsialisticheskaia rekonstruktsiia sel'skogo khoziaistva.*
Studies in Soviet Thought.
Tekhnika sotsialisticheskogo zemledeliia.
Theoria.
TPBGS. Abbreviation for *Trudy po prikladnoi botanike, genetike, i selektsii.* For the complex publishing history of this journal, which has had many changes of name and numeration, see *Periodicheskaia pechat' SSSR, 1917–49.*
Tridtsat' dnei.
Trudy Instituta fiziologii rastenii. See under Akademiia Nauk SSSR.
Trudy po prikladnoi botanike, genetike, i selektsii. See *TPBGS.*
Trudy po sel'skokhoziaistvennoi meteorologii.
Uchenye zapiski Tomskogo gosudarstvennogo universiteta.
U.S. Department of Agriculture. *Yearbook.* Washington, D.C.
Unter dem Banner des Marxismus.
Uspekhi eksperiemental'noi biologii.
Uspekhi fizicheskikh nauk.
Uspekhi sovremennoi biologii.
VAN. Abbreviation for *Vestnik Akademii Nauk SSSR.*
VASKhNIL. Abbreviation for Vsesoiuznaia Akademiia Sel'skokhoziaistvennykh Nauk imeni V. I. Lenina. *Trudy.*
Vestnik kommunisticheskoi akademii.
Vestnik leningradskogo universiteta.
Vestnik sel'skokhoziaistvennoi literatury.
Vestnik sel'skokhoziaistvennoi nauki.
Vestnik sotsialisticheskoi akademii.
Vestnik statistiki.
Vestnik vysshei shkoly.
Vestnik znaniia.
VF. Abbreviation for *Voprosy filosofii.*
VKA. Abbreviation for *Vestnik kommunisticheskoi akademii.*
Voprosy antropologii.
Voprosy ekonomiki.
Voprosy filosofii.
Voprosy istorii.

Vsesoiuznyi issledovatel'skii institut plodovogo i iagodnogo khoziastva. Kiev. *Trudy.*
Za marksistsko-leninskoe estestvoznanie.
Za michurinskoe plodovodstvo.
Zeitschrift fur induktive Abstammungs- und Vererbungslehre.
Zemledelie.
Zhivotnovodstvo.
Zhurnal eksperimental'noi biologii, seriia A.
Zhurnal obshchei biologii.
Zhurnal opytnoi agronomii iugo-vostoka.

BOOKS AND PAMPHLETS

Agriculture in the Twentieth Century; Essays on Research, Practice and Organization to be Presented to Sir Daniel Hall. Oxford, 1939.
Agronomicheskaia pomoshch' v Rossi. Petrograd, 1914. Published by Department zemledeliia, V. V. Morachevskii, ed.
Akademiia Nauk Belorusskoi SSR, Institut Filosofii i Prava. *Dialekticheskii materializm kak metodologiia estestvennonauchnogo poznaniia.* Minsk, 1965.
———. *Rol' dialektiki v izuchenii biologicheskikh iavlenii.* Minsk, 1967.
Akademiia Nauk SSSR. *220 let Akademii Nauk SSSR. Spravochnaia Kniga.* M., 1945.
———. *Metodologicheskie problemy nauki. Materialy zasedaniia Prezidiuma AN SSSR 18 oktiabria 1963.* M., 1964.
———. *Pochvovedenie i agrokhimiia. Trudy maiskoi sessii 1935* g. M., 1936.
Akademiia Nauk SSSR, Institut Filosofii. *Istoriia filosofii.* 6 vols. M., 1957–65.
———. *Marksistsko-leninskaia filosofiia i sotsiologiia v SSSR i evropeiskikh sotsialisticheskikh stranakh. Istoriko-filosofskie ocherki.* M., 1965.
———. *Osnovy marksistskoi filosofii.* M., 1958.
———. *Problema prichinnosti v sovremennoi biologii.* M., 1961.
Akademiia Nauk SSSR, Institut Genetiki. *Genetika — sel'skomu khoziaistvu.* M., 1963.
———. *Nasledstvennost' i izmenchivost' rastenii, zhivotnykh, i mikroorganizmov.* 2 vols. M. 1959.
———. *Protiv reaktsionnogo mendelizma-morganizma.* M., 1950.
Akademiia Nauk SSSR, Vsesoiuznoe botanicheskoe obshchestvo. *Voprosy evoliutsii, biogeografii, genetiki i selektsii. Sbornik posviashchennyi 70-letiiu so dnia rozhdeniia Akademika N. I. Vavilova.* M., 1960.
Akademiia Nauk Ukrainskoi SSR. *Filosofskie voprosy sovremennoi biologii (Materialy ukrainskogo soveshchaniia po filosofskim vorprosam biologii).* Kiev, 1962.
———. *Modelirovanie v biologii i meditsine.* Kiev, 1965.
Akademiia obshchestvennykh nauk pri TsK KPSS, Kafedra istorii sovetskogo obshchestva. *Iz istorii sovetskoi intelligentsii.* M., 1966.
Akademik Vasilii Robertovich Vil'iams. Iubileinyi sbornik. M., 1935.
Akademiku V. N. Sukachevu k 75-letiiu so dnia rozhdeniia. Sbornik rabot po geobotanike, lesovedeniiu, paleografii i floristike. M., 1956.
Alikhanian, S. I. *Teoreticheskie osnovy ucheniia Michurina o peredelke rastenii.* M., 1966.

Andics, Hellmut. *Der grosse Terror*. Vienna, 1967. English translation as *Rule of Terror*. N.Y., 1969.
Anuchin, D. N. *Proiskhozhdenie cheloveka*, 3rd ed. M., 1927.
Ashby, Eric. *Scientist in Russia*. Harmondsworth, 1947.
Avakian, A. A. *Biologiia razvitiia sel'skokhoziaistvennykh rastenii*. M., 1960.
Azimov, G. I. *Proiskhozhdenie zhizni na zemle*. M., 1925.
Bakharev, A. N. *I .V. Michurin; k 60-letiiu deialtel'nosti i 80-letiiu zhizni*. M., 1934.
———. *Michurin, Ivan Vladimirovich (1855–1935). Opis' dokumental'nykh materialov lichnogo fonda No. 6856. Krainye daty dokumental'nykh materialov 1883–1941* gg. M., 1952.
———. *Selektsionno-geneticheskaia stantsiia I. V. Michurina. Istoricheskii ocherk s prilozheniem riada statei I. V. Michurina i programmy dlia kruzhkov im. Michurina*. M., 1933.
Barghoorn, F. C. *Soviet Russian Nationalism*. N.Y., 1956.
Barnett, S. A., ed. *A Century of Darwinism*. London, 1958.
Barth, Hans. *Wahrheit und Ideologie*. Erlenbach-Zurich und Stuttgart, 1961.
Baskova, A. A. and Lysenko, T. D. *Iarovizatsiia i glazkovanie kartofelia*. Kharkov, 1933.
Batkis, G. A. *1933/34 uch. god. Programma po sotsial'noi gigiene dlia medvuzov*. M., 1934.
Bauer, Raymond. *The New Man in Soviet Psychology*. Cambridge, Mass., 1952.
Beckner, Morton. *The Biological Way of Thought*. N.Y., 1959.
Berg, L. S. *Nomogenez, ili evoliutsiia na osnovy zakonomernostei*. Petrograd, 1922. English translation: *Nomogenesis, or Evolution by Law*. London, 1926.
Bergson, Abram. *The Economics of Soviet Planning*. New Haven and London, 1964.
Berman, Harold. *Justice in the U.S.S.R. An Interpretation of Soviet Law*. Cambridge, Mass., 1963.
Bernal, J. D. *The Origin of Life*. Cleveland, 1967.
———. *Science in History*, 1st and 3rd editions. London, 1954 and 1965.
Bertalanffy, Ludwig von. *Problems of Life. An Evaluation of Modern Biological and Scientific Thought*. N.Y., 1952 and 1960.
Bibliografiia po raionirovaniiu i razmeshcheniiu sel'skogo khoziaistva SSSR (1818–1860). M., 1961. Published by AN SSSR, Sovet po izucheniiu proizvoditel'nykh sil, Sektor seti spetsial'nykh bibliotek. E. N. Morachevskaia, compiler, and N. N. Pel't, ed.
Bibliographie der sowjetischen Philosophie, 1947–60. 5 vols. Dordrecht, 1954–64. Veröffentlichungen des Ost-Europa Instituts, Universität Freiburg/Schweiz. J. M. Bochenski, ed.
Biograficheskii slovar' deiatelei estestvoznaniia i tekhniki. 2 vols. M., 1958. A. A. Zvorykin, ed.
Blake, William. *The Poetry and Prose*. New York, 1965.
Bliakher, L. Ia. et al., eds. *Istoriia estestvoznaniia v Rossii*, III. *Geologo-geograficheskie i biologicheskie nauki*. M., 1962. Published by AN SSSR, Institut Istoriia Estestvoznaniia i Tekhniki.
BME. Abbreviation for *Bol'shaia meditsinskaia entsiklopediia*. 1st ed. 35 vols. M., 1928–36.

Bogdanov, A. A. *Filosofiia zhivogo opyta. Populiarnye ocherki. Materializm, empiriokrititsizm, dialekticheskii materializm, empiriomonizm, nauka budushchago*. St. Petersburg, 1913.
Bogdenko, M. L. *Stroitel'stvo zernovykh sovkhozov na tselinnykh i zalezhnykh zemliakh (1929–31 gg.)*. M., 1958.
Bolgov, A. V. *Differentsial'naia zemel'naia renta v usloviiakh sotsializma (ocherk teorii)*. M., 1963.
Bondarenko, A. S., ed. *Sel'skokhoziaistvennaia nauka v SSSR. Sbornik statei*. M., 1934. Published by VASKhNIL.
Bor'ba s bolezniami sel'skokhoziaistvennykh kul'tur. Kiev, 1953. Vol. IV of AN Ukr. SSR, Institut Entomologii i Fitopatologii, *Trudy*.
Bosh'ian, G. M. *O prirode virusov i mikrobov*, 2nd ed. M., 1950.
Brecht, Bertolt. *Gedichte*, 9 vols. Frankfurt/Main, 1960–1966.
Brzezinski, A. *The Permanent Purge*. Cambridge, Mass., 1956.
BSE. Abbreviation for *Bol'shaia sovetskaia entsiklopediia*. 1st ed. 65 vols. M., 1928–46. 2nd ed. 51 vols. M., 1949–58.
Bukasov, S. M. and Sharina, N. E. *Istoriia kartofelia*. M., 1938.
Bukasov, S. M., ed. *Kartofel'*. L., 1965. Vol. XXXVII, No. 3, of *TPBGS*.
———. *Kul'tura kartofelia*. L., 1948.
Bukasov, S. M. and Lebedeva, N. A. *Michurinskie metody v selekstii kartofelia*. L., 1949.
Bukasov, S. M. and Kameraz, A. Ia. *Osnovy selektsii kartofelia*. M., 1959.
Bukasov, S. M. *Revoliutsiia v selektsii kartofelia*. L., 1933.
Bunak, V. V. and Nesturkh, M. F. and Roginskii, Ia. Ia. *Antropologiia; kratkii kurs*. M., 1941.
Burt, Olive. *Luther Burbank, Boy Wizard*. Indianapolis, 1962.
Burton, W. G. *Kartofel'*. M., 1952. Translation of Burton, *The Potato; a Survey of Its History, and of Factors Influencing Its Yield, Nutritive Value and Storage*. London, 1948.
Carr, E. H. *Bolshevik Revolution*, 3 vols. N.Y., 1951–53.
———. *A History of Soviet Russia*. 7 vols. N.Y., 1951–60.
Chaianov, A. V. *Petrovsko-Razumovskoe v ego proshlom i nastoiashchem; putevoditel' po Timiriazevskoi sel'skokhoziaistvennoi akademii*. M., 1925.
———. *The Theory of Peasant Economy*. Homewood, Ill., 1966. Translation of his *Organizatsiia krest'ianskogo khoziaistva*.
Chaianov, S. K. et al. *Sel'skokhoziaistvennoe opytnoe delo Narkomzema RSFSR v programmakh na 1927/28–1931/32 gg*. M., 1928.
Chambers, J. D. and Mingay, G. E. *The Agricultural Revolution, 1750–1880*. London, 1966.
Chambers, Whittaker. *Witness*. N.Y., 1952.
Chekotillo, A. M. and Kagan, B. O. *Kukuruza i ee rol' v sotsialisticheskoi rekonstruktsii narodnogo khoziaistva v SSSR*. 2 eds. M., 1930 and 1932.
Cherdantsev, G. N. *Osnovy sel'skokhoziaistvennoi geografii*. M., 1931.
Chesnokov, P. G. *Bolezni vyrozhdeniia kartofelia v SSSR i bor'ba s nimi*. L., 1961.
Chizhevskii, A. L. *Fizicheskie faktory istoricheskogo protessa*. Kaluga, 1924.
Comparisons of the United States and Soviet Economies, vol. I. Washington, D.C., 1959. Published by U.S. Congress, Joint Economic Committee.
Conquest, Robert. *The Great Terror*. N.Y., 1968.

Corbett, Patrick. *Ideologies.* London, 1965.
Danilevskii, N. Ia. *Darvinizm; Kriticheskoe izsledovanie,* 2 vols. S. Petersburg, 1885–9.
Darlington, C. D. *Chromosome Botany and the Origin of Cultivated Plants.* N.Y., 1963.
Deiateli Soiuza Sovetskikh Sotsialisticheskikh Respublik i Oktiabr'skoi Revoliutsii (Avtobiografii i biografii). 3 vols. bound as one. Photographic reproduction of appendices in *Entsiklopedicheskii slovar' Granata,* XLI, 3 parts. M., 1925?
Delone, L. N. and Grishko, N. N. *Kurs genetiki.* M., 1938.
Dickinson, Emily. *The Complete Poems.* Boston, 1960.
Dmitriev, V. S. *Akad. Vil'iams, osnovopolozhnik ucheniia o travopol'noi sisteme zemledeliia.* M., 1949.
Dobzhansky, Theodosius. *Mankind Evolving; The Evolution of the Human Species.* New Haven, 1962.
Doiarenko, A. G. *Iz agronomicheskogo proshlogo.* M., 1965.
Dostizheniia sovetskoi selektsii, December, 1936. See VASKhNIL, *Dostizheniia,* etc.
Dubinin, N. P. *Teoreticheskie osnovy i metody rabot I. V. Michurina.* M., 1966.
Dudintsev, V. *Not By Bread Alone.* N.Y., 1957.
Dunin, M. S. *Stimulirovanie (vymachivanie) semian po polevym opytam krest'ian-korrespondentov laboratorii 'Bednoty'.* M., 1927.
Dunin, M. S., ed. *Virusnye bolezni rastenii.* M., 1938.
Dunn, L. C., ed. *Genetics in the Twentieth Century; Essays on the Progress of Genetics during the First 50 Years.* N.Y., 1951.
Dunn, L. C. *Heredity and Evolution in Human Populations.* N.Y., 1965.
———. *A Short History of Genetics; The Development of Some of the Main Lines of Thought, 1864–1939.* N.Y., 1965.
Efroimson, V. P. *Vvedenie v meditsinskuiu genetiku.* M., 1964.
Ehrenburg, Ilya. *Trest D. E.* Berlin, 1923.
———. *Sobranie sochinenii,* 9 vols. M., 1962–66.
Emel'ianov, I. E., ed. *Kukuruza; bibliograficheskii ukazatel' otechestvennoi literatury za 1794–1959 gg.* M., 1961.
Engels, F. *Anti-Dühring.* N.Y., 1939.
———. *Dialectics of Nature.* N.Y., 1940.
———, and Marx, K. *Pis'ma Marksa i Engel'sa.* M., 1922 and 1923.
Entsiklopedicheskii slovar' Granata. 45 vols.
Erevanskii simpozium po ontogenezu vysshikh rastenii. Doklady. Erevan, 1966.
Erlich, A. *The Soviet Industrialization Debate, 1924–28.* Cambridge, Mass., 1960.
Evolution: Symposia of the Society for Experimental Biology, VII. N.Y., 1953.
Fainsod, M. *Smolensk under Soviet Rule.* Cambridge, Mass., 1958.
Favorov, A. M. and Kotov, A. F. *Letniaia posadka kartofelia.* M., 1952.
Fediukin, S. A. *Sovetskaia vlast' i burzhuaznye spetsialisty.* M., 1965.
Fedorenko, E. G., ed. *Filosofskie voprosy meditsiny i biologii.* Kiev, 1965.
Feiginson, N. I. *Chego dobilis' krest'iane opytniki.* M., 1929.
———, and Terent'ev, N. *Komsomol'tsy labkory; kak provodit' opytnicheskuiu rabotu komsomol'skoi iacheike.* L., 1930.
———. *Korpuskuliarnaia genetika (kriticheskii obzor).* M., 1963.

Filipchenko (or Filippchenko), Iu. A. *Chto takoe evgenika?* Petrograd, 1921.
———. *Eksperimental'naia zoologiia.* L., 1932.
———. *Puti uluchsheniia chelovecheskogo roda (evgenika).* L., 1924.
Filosofskie problemy sovremennogo estestvoznaniia. Trudy Vsesoiuznogo soveshchaniia po filosofskim voprosam estestvoznaniia. M., 1959.
Fischer, George, ed. *Science and Ideology in Soviet Society.* N.Y., 1967.
Fisher, A. (Fischer, Alfons). *Osnovy sotsial'noi gigieny.* M., 1929. Translation of *Grundriss der sozialen Hygiene,* 2nd ed. Karlsruhe, 1925.
Fogg, G. E. *The Growth of Plants.* Harmondsworth, 1963.
Frolov, I. T. *Ocherki metodologii biologicheskogo issledovaniia (Sistema metodov biologii).* M., 1965.
Frost, Robert. *Complete Poems.* N.Y., 1961.
Gabel, Joseph. *La fausse conscience.* Paris, 1962.
Gaidukov, Iu. G. *Rol' praktiki v protsesse poznaniia.* M., 1964.
Gaisinovich, A. E. *Zarozhdenie genetiki.* M., 1967.
Garvud, V. S. *Obnovlennaia zemlia.* M., 1909. (Russian translation of William S. Harwood, *The New Earth.*)
Gibbons, P. A. *Ideas of Political Representation in Parliament, 1660–1832.* Oxford, 1914.
Gillispie, C. C. *The Edge of Objectivity. An Essay in the History of Scientific Ideas.* Princeton, 1960.
Glinka, K. D. *The Great Soil Groups of the World and Their Development.* Ann Arbor, 1927.
Glushchenko, I. E. *Nasledstvennost' i izmenchivost' kul'turnykh rastenii.* M., 1961.
———. *Strany, vstrechi, uchenye (Zapiski biologa).* M., 1963.
———. *U zarubezhnykh druzei.* M., 1957.
———. *Vegetativnaia gibridizatsiia rastenii.* M., 1948.
Gorbunov, N. P. *Kak rabotal Vladimir Il'ich. Sbornik statei i vospominanii.* Kharkov, 1925 and M., 1933.
Gordeev, G. S. *Sel'skoe khoziaistvo v voine i revoliutsii.* M., 1925.
Gorshkov, I. S. *I. V. Michurin, ego zhizn' i rabota. Ko dniu 50–letiiu nauchnoprakticheskoi deiatel'nosti.* M., 1925.
Gowen, John W., ed. *Heterosis. A Record of Researches Directed Toward Explaining and Utilizing the Vigor of Hybrids.* Ames, Iowa, 1952.
Grashchenkov, N. I. *O polozhenii v biologicheskoi nauke i zadachi biologicheskoi nauki v BSSSR.* Minsk, 1948.
Gray, Peter, ed. *Encyclopedia of the Biological Sciences.* N.Y., 1961.
Gremiatskii, M. A., ed. *Evoliutsiia cheloveka. Sbornik.* M., Sverdlov Communist Univ., 1925.
Grishko, N. N. (or Mikola Mikolaiovich) and Delone, L. N. *Kurs genetiki.* M., 1938.
Gurvitch, Georges, ed. *Traité de sociologie,* 2 vols. Paris, 1960.
Haldane, J. B. S. *Biochemistry of Genetics.* London, 1954.
———. *Heredity and Politics.* N.Y., 1938.
Harcave, Sidney. *Russia, a History.* Philadelphia, 1956.
Harris, Marvin. *The Rise of Anthropological Theory; A History of Theories of Culture.* N.Y., 1968.
Harwood, William S. *The New Earth: A Recital of the Triumphs of Modern Agriculture in America.* N.Y., 1906.
Hazard, John N. *The Soviet System of Government,* 3rd ed. Chicago, 1964.

Howard, W. L. *Luther Burbank, A Victim of Hero Worship.* Waltham, Mass., 1946, vol. IX of the series, *Chronica Botanica.*
———. *Luther Burbank's Plant Contributions.* Berkeley, 1945. Bulletin 691 of Univ. of Calif., College of Agriculture, Agricultural Experiment Station.
Hudson, P. S. and Richens, R. H. *The New Genetics in the Soviet Union.* Cambridge, 1946. Published by Imperial Bureau of Plant Breeding and Genetics.
Huxley, Julian. *Soviet Genetics and World Science, Lysenko and the Meaning of Heredity.* London, 1949.
I. V. Michurin v vospominaniiakh sovremennikov. Tambov, 1963. I. S. Gorshkov, ed.
Iadov, V. A. *Ideologiia kak forma dukhovnoi deiatel'nosti obshchestva.* L., 1961.
Iakovlev, Ia. A. *Voprosy organizatsii sotsialisticheskogo sel'skogo khoziaistva.* M., 1933.
Iakushkin, I. V. *Rastenievodstvo,* 1st ed. M., 1947; 2nd ed. M., 1953.
Ihde, A. J. *The Development of Modern Chemistry.* N.Y., 1964.
Il'in, M. (pen name of I. Ia. Marshak) *Gory i liudi. Rasskazy o perestroike prirody.* 3rd ed. L., 1936.
Istoriia estestvoznaniia; bibliograficheskii ukazatel'. Literatura, opublikovannaia v SSSR (1917–56), 3 vols. M., 1949–63. Published by AN SSSR, Inst. Istorii Estestvoznaniia, various editors and compilers.
Itogi i perspektivy issledovanii razvitiia rastenii; sbornik rabot po materialam II delegatskogo s'ezda Vsesoiuznogo botanicheskogo obshchestva, 9–15 maia 1957 g. L., 1959.
Iur'ev, V. Ia., *et al. Obshchaia selektsiia i semnovodstvo polevykh kul'tur,* 1st–3rd eds. M., 1940, '50, '58.
Jasny, Naum. *Khrushchev's Crop Policy.* Glasgow, 1965.
———. *The Socialized Agriculture of the USSR; Plans and Performance.* Stanford, 1949.
Johansson, Ivar, ed. *Handbuch der Tierzüchtung.* Vol. II trans. by Kh. F. Kushner, as Iogansson, I. *Rukovodstvo po razvedeniiu zhivotnykh. Geneticheskie osnovy produktivnosti i selektsii.* M., 1963.
Joravsky, David. *Soviet Marxism and Natural Science, 1917–32.* N.Y., 1961.
Jordan, Z. A. *Philosophy and Ideology; The Development of Philosophy and Marxism-Leninism in Poland since the Second World War.* Dordrecht, 1963.
Kaftanov, S. V. *Za bezrazdel'noe gospodstvo michurinskoi biologicheskoi nauki.* M., 1948.
Kanaev, I. I. *Nasledstvennost'. Nauchno-populiarnyi ocherk.* L., 1925.
Karcz, Jerzy F. *Soviet Agricultural Marketings and Prices, 1928–1954.* Santa Monica, California, 1957.
———, ed. *Soviet and East European Agriculture. Berkeley,* 1967.
Kautsky, Karl. *Der Einfluss der Volksvermehrung auf den Fortschritt der Gesellschaft.* Wien, 1880.
———. *Ethics and the Materialist Conception of History.* Chicago, 1907.
———. *Materialistische Geschichsauffassung, 2 vols. Berlin,* 1927.
———. *Nauka, zhizn' i etika.* M., 1918.
———. *Obshchestvennye instinkty v mire zhivotnykh i u liudei.* M., 1922.
———. *Sochineniia,* 13 vols.? M., 1923–28?

———. *Vermehrung und Entwicklung in Natur and Gesellschaft.* Stuttgart, 1910.
———. *Vozniknovenie braka i sem'i.* Petrograd, 1923.
Kelle, V., and Koval'zon, M. *Formy obshchestvennogo soznaniia.* M., 1959.
Keller, B. A. *Genetika. Kratkii ocherk.* M., 1933. This is not to be confused with his *Botanika s osnovami fiziologii,* volume III of which also appeared in 1933 with the title *Genetika.*
Khaies, B. (B. Chajes). *Kratkii kurs sotsial'noi gigieny.* M., 1923.
Khrushchev, N. S. *Stroitel'stvo kommunizma v SSSR i razvitie sel'skogo khoziaistva,* 8 vols. M., 1962–64.
Klebs, G. A. *Willkürliche Entwickelungsänderungen bei Pflanzen.* Jena, 1903.
Koestler, Arthur. *The Ghost in the Machine.* N.Y., 1967.
Kol'tsov, N. K. *Les molécules héréditaires. Paris,* 1938.
———. *Uluchshenie chelovecheskoi porody.* Petrograd, 1923.
Komarov, V. L. *Izbrannye sochineniia,* 12 vols. M., 1945–58.
———. *Lamark.* M., 1925.
———. *Proiskhozhdenie kul'turnykh rastenii.* M., 1931. The 2nd ed. of 1938 is reprinted in vol. XII of his *Izbrannye sochineniia.*
———. *Proiskhozhdenie rastenii,* 1st–5th eds. M., 1933–36.
Kommunisticheskaia Akademiia, Biologicheskii Institut. *Pamiati K. A. Timiriazeva. Sbornik dokladov i materialov sessii Biologicheskogo Instituta im. K. A. Timiriazeva, posviashchennoi 15-letiiu so dnia smerti K. A. Timiriazeva,* 1920–35. M., 1936. P. P. Bondarenko, *et al.,* eds.
———. *Probleme der theoretischen Biologie. Arbeiten aus dem Timirjaseff-institut für Biologie, Moskau, herausgegeben zum 15. Todestag von K. A. Timirjaseff.* M., 1935.
———. *Protiv mekhanisticheskogo materializma i menshevistvuiushchego idealizma v biologii.* M., 1931.
Kompleksnaia nauchnaia ekspeditsiia [*Akademii Nauk SSSR*] *po voprosam polezashchitnogo lesorazvedeniia. Trudy,* 2 vols. M., 1951–53.
Kononova, M. M. *Organicheskoe veshchestvo pochvy, ego priroda, svoistva i metody izucheniia.* M., 1963.
Korolenko, V. G. *Istoriia moego sovremmenika.* 2 vols. M., 1948.
Koval'zon, M. and Kelle, V. *Formy obshchestvennogo soznaniia.* M., 1959.
Kozodoev, I. I. *Zemel'nye otnosheniia v sotsialisticheskikh stranakh.* M., 1960.
KPSS (Kommunisticheskaia Partiia Sovetskogo Soiuza). *KPSS v rezoliutsiiakh i resheniiakh s'ezdov, konferentsii i plenumov TsK,* 7th ed. 4 vols. M., 1953–60.
———. *Protokoly i stenograficheskie otchety s'ezdov i konferentsii.* 2nd ed. Number and date indicate the particular meeting.
———. TsK. (Tsentral'nyi Komitet). *Plenum.* Date indicates the particular meeting.
Kraevoi, S. Ia. *Vosmozhna li vegetativnaia gibridizatsiia rastenii posredstvom privivok?* M., 1967.
Kruzhilin, A. S. *Kartofel' na iugo-vostoke SSSR.* Saratov, 1941.
Kuhn, Thomas S. *The Structure of Scientific Revolutions.* Chicago, 1962.
Kukol, V. V. *Posev iarovykh kul'tur s oseni (pod zimu).* L., 1932.
Kul'tura pshenitsy. M., 1936. vol. VIII of VASKhNIL, *Trudy.*
Kuperman, F. M. and Rzhanova, E. I. *Biologiia razvitiia rastenii.* M., 1963.

Kushner, Kh. F. *Nasledstvennost' sel'skokhoziaistvennykh zhivotnykh s elementami selektsii.* M., 1964.
Kuznetsov, A. I. *Timiriazevka i deiateli kul'tury.* M., 1965.
Lamarck, J. B. *Filosofia zoologii,* I. M., 1935.
Lasswell, H. D. and Kaplan, A. *Power and Society: A Framework for Political Inquiry.* New Haven, 1950
Lebedev, V. A. *Obnovitel' sadov; provest' o Michurine.* M., 1936.
Lebner (Löbner), Max. *Sortovodstvo dlia sadovodov.* S. Petersburg, 1912. Translated from German.
Lenin. V. I. *Polnoe sobranie sochinenii,* 5th ed. 55 vols. M., 1958–65.
———. *Sochineniia,* 4th ed. 44 vols. M. 1941–67.
Lenin Academy of Agricultural Sciences. See under VASKhNIL.
Lenk, Kurt. *Ideologie: Ideologiekritik und Wissenssoziologie.* Neuwied, 1961.
Leonov Leonid. *Russkii les,* M., 1955.
Lepeshinskaia, Olga B. *Proiskhozhdenie kletok iz zhivogo veshchestva i rol' zhivogo veshchestva v organizme.* M., 1945.
Lévi-Strauss, Claude. *The Savage Mind.* Chicago, 1966.
Lewin, Moshe. *La paysannerie et le pouvoir soviétique, 1928–1930.* 's-Gravenhage, 1966. Translated as *Russian Peasants and Soviet Power.* Evanston, 1968.
Lewytskyj, Boris. *Vom roten Terror zur sozialistischen Gesetzlichkeit.* Munich, 1961.
Lichtheim, George. *Marxism. An Historical and Critical Study.* N.Y., 1962.
Lipshits, S. Iu, compiler. *Russkie botaniki; Biografo-bibliograficheskii slovar',* 4 vols. (A-Kiuz). 1947–52.
Lobanov, P. P. *Sel'skokhoziaistvennaia nauka v semiletke.* M., 1959.
Lobashev, M. E. *Genetika; Kurs lektsii.* 1st and 2nd ed. L. 1963 and 1967.
———. *Ocherki po istorii russkogo zhivotonovodstva.* M., 1954.
Löbner (see Lebner).
Luk'ianiuk, V. I., ed. *Kul'tura kukuruzy v SSSR.* M., 1957.
Luss, A. I. *Tsitrusovye kul'tury v* SSSR. M., 1947.
Lysenko, T. D. *Agrobiologiia: raboty po voprosam genetiki, selektsii i semenovodstva.* 1st–6th eds. M., 1943–52. English translation: *Agrobiology.* M., 1954.
———, and Babak, M. K. *Bor'ba s vyrozhdeniem kartofelia na iuge SSSR; instruktivnye ukazaniia.* M., 1936.
——— and Baskova, A. A. *Iarovizatsiia i glazkovanie kartofelia.* Kharkov, 1933.
——— and Favorov, A. M. *Letnie posadki kartofelia.* M., 1938.
———. *Nekotorye voprosy agrotekhniki vesennego seva.* M., 1944.
———. *Novoe v nauke o biologicheskom vide.* M., 1952.
———. *Ob agronomicheskom uchenii V. R. Vil'iams.* M., 1950.
———. *Pochvennoe pitanie rastenii — korennoi vopros nauki zemledeliia.* 1st-3rd eds. M., 1955–62.
———. *Roboty v dni velikoi otechestvennoi voiny.* M., 1943.
——— and Prezent, I. I. *Selektsiia i teoriia stadiinogo razvitiia rastenii.* M., 1935.
———. *Vliianie termicheskogo faktora na prodolzhitel'nost' faz razvitiia rastenii; opyt so zlakami i khlopchatnikom.* Baku, 1928. Published by Azerbaidzhanskaia tsentral'naia opytnoselektsionnaia stantsiia imeni tov. Ordzhonikidze v Gandzhe, *Trudy,* vypusk 3.

——— and Nuzhdin, N. I. *Za materializm v biologii.* M., 1958.
Macrae, Donald G. *Ideology and Society.* London. 1961.
Maiakovskii, V. V. (or Mayakovsky). *The Bedbug; and Selected Poetry.* N.Y., 1960.
———. *Polnoe sobranie sochinenii,* 13 vols. M., 1955–61.
Maksimov, N. A. *Kratkii kurs fiziologii rastenii.* 3rd–9th eds. M., 1931–58. The 2nd ed., M., 1929, is translated as *A Textbook of Plant Physiology.* N.Y., 1930.
———. *The Theoretical Significance of Vernalization.* Aberystwyth, 1934. Published by Imperial Bureau of Plant Genetics, Herbage Plants, *Bulletin* No. 16.
Malenkov, G. M. *Otchetnyi doklad XIX s'ezdu partii o rabote tsentral'nogo komiteta VKP(b).* M., 1952.
Malthus, T. R. *An Essay on the Principle of Population,* 6th ed. London, 1826. 2 vols.
Mangelsdorf, P. C. *The Origin of Indian Corn and Its Relatives.* College Station, Texas. Bulletin 574 of Texas Agri. Exp. Sta.
Mariagin, G. A. *Postyshev.* M., 1965.
Marx, Karl and Engels, Friedrich. *Correspondence, 1846–1895.* N.Y., 1936.
Marx, Engels, and Lenin. *Marks, Engel's, Lenin o biologii.* M., 1933. Published by Kommunisticheskaia Akademiia, Biologicheskii Institut.
Marx, K. and Engels, F. *Pis'ma Marksa i Engel'sa.* M., 1922 and 1923.
———.*Werke.* 38 vols. Berlin, 1960–68.
Materialy K vsesoiuznoi konferentsii po planirovaniiu genetiko-selektsionnykh issledovanii. L., 1932.
Matson, Floyd W. *The Broken Image: Man, Science and Society.* Garden City, N.Y., 1966.
Medvedev, Zhores A. *The Rise and Fall of T. D. Lysenko.* N.Y., 1969.
Meister, G. K. *Kriticheskii ocherk osnovnykh poniatii genetiki.* M., 1934.
Mel'nikov, M. I., ed. *Iz opyta prepodavaniia biologii v srednei shkole.* M., 1950.
——— *et al. Osnovy darvinizma: uchebnoe posobie dlia srednei shkoly.* 1st–7th eds. M., 1941–56.
Merton, Robert K. *Social Theory and Social Structure.* N.Y., 1965.
Metodologicheskie problemy nauki. M., 1964.
Meyer, Alfred G. *The Soviet Political System.* N.Y., 1965.
Michurin, I. V. *Itogi ego deiatel'nosti v oblasti gibridizatsii po plodovodstvu.* M., 1924.
———. *Itogi poluvekovykh rabot po vyvedeniiu novykh sortov plodovykh rastenii,* 2 vols. M., 1929–32.
———. *Itogi shestidesiatiletnikh rabot.* 3rd–5th eds. M., 1934–49.
———. *Sochineniia.* 1st ed., 4 vols. M., 1939–41.
———. *Sochineniia.* 2nd ed., 4 vols. M., 1948.
———. *Vyvedenie iz semian novykh kul'turnykh sortov plodovykh derev'ev i kustarnikov.* M., 1921.
Mishnev, Vasilii. *Uchenaia stepen'.* Minsk, 1963.
Moody, P. A. *Genetics of Man.* N.Y., 1967.
Moscow, Nauchno-issledovatel'skii Institut Kartefel'nogo Khoziaistva, *Semenovodstvo i aprobatsiia kartofelia.* M., 1946.
Müller-Markus, S. *Einstein und die Sowjet-Philosophie: Krisis einer Lehre.* 2 vols. Dordrecht, Holland, 1960–66.

Murneek, A. E. *et al. Vernalization and Photoperiodism; a Symposium.* Waltham, Mass., 1948. Vol. I of Chronica Botanica Lotsya.

Naess, Arne and associates. *Democracy, Ideology, and Objectivity: Studies in the Semantics and Cognitive Analysis of Ideological Controversy.* Oxford, 1956.

Nagel, Ernest. *The Structure of Science: Problems in the Logic of Scientific Explanation.* N.Y., 1961.

Narodnoe khoziaistvo SSSR v 1960 godu. M., 1961.

Natali, V. F. *Nauchnye osnovy selektsii i evoliutsionnoe uchenie.* M., 1931.

Nauchnaia sessiia po voprosam biologii i sel'skogo khoziaistva. Riga, 22–26 oktiabria 1951 g. M., 1953.

Nauka XX veka, Biologiia. 2 vols. M., 1928–29.

Nauka i nauchnye rabotniki SSSR. 6 vols. M., 1925–34.

Neel, J. V. and Schull, W. J. *Nasledstvennost' cheloveka,* M., 1958. Translation of their *Human Heredity.*

Nove, Alec. *Economic Rationality and Soviet Politics, or Was Stalin Really Necessary?* N.Y., 1964.

O polozhenii v biologicheskoi nauke. Stenograficheskii otchet sessii VASKhNILa 31 iiulia-7 avgusta 1948 g. M., 1948. There are translations into English and other languages.

Ocherki po istorii russkoi botaniki. M., 1947. Published by Moskovskoe obshchestvo ispytatelei prirody. N. A. Komarnitskii, ed.

Odintsov, V. A. *Itogi raboty n.-i. instituta plodovogo i iagodnogo khoziaistva RSFSR i BSSR im. I. V. Michurina za 3-letnii period (1931–34).* Voronezh, 1935.

Olby, Robert C. *Origins of Mendelism.* London, 1966.

Oparin, A. I. *Aleksandr Ivanovich Oparin.* M., 1964. Biobibliografiia uchenykh SSSR, Seriia biokhimii, No. 6.

———. *Fermenty, ikh rol' i znachenie v zhizni organizmov.* M., 1923.

———. *The Origin of Life.* N.Y., 1953.

———. *Proiskhozhdenie zhizni.* M., 1924 and 1935.

———. *Vozniknovenie zhizni na zemle.* 1st–3rd eds. M., 1936–57.

Osobo opasnye gosudarstvennye prestupleniia. M., 1963. V. I. Kurlianskii and M. P. Mikhailov, eds.

Osoboe soveshchanie o nuzhdakh sel'skokhoziaistvennoi promyshlennosti, *Svod trudov,* 4 vols. St. Petersburg, 1902–05.

———. *Trudy,* 58 vols. St. Petersburg, 1902–05.

Pamiati Akademika N. A. Maksimova; sbornik statei. M., 1957. A. L. Kursanov *et al.,* eds.

Pamiati K. A. Timiriazeva; sbornik dokladov i materialov sessii biologicheskogo instituta im. K. A. Timiriazeva, posviashchennoi 15-letiiu so dnia smerti K. A. Timiriazeva, 1920–35. M., 1936. P. P. Bondarenko *et al.,* eds.

Pamiati V. I. Lenina, 1924–34. M., 1934.

Pannikov, V. D. *Pochvy, udobreniia i urozhai.* M., 1964.

Pashkevich, V. V., ed. *K standartizatsiiu sortov plodovykh derev'ev i iagodnykh kul'tur.* L., 1931.

Pasternak, Boris. *Letters to Georgian Friends.* N.Y., 1968.

Pavlenko, S. M. *Estestvennonauchnye osnovy meditsinskoi genetiki.* M., 1963.

Peters, James A., ed. *Classic Papers in Genetics.* Englewood Cliffs, New Jersey, 1959.

Pilipenko, N. V. *Neobkhodimost' i sluchainost'*. M., 1965.
Pisarev, D. I. *Sochineniia*. 6 vols. St. Petersburg, 1894–1905.
———. *Sochineniia*. 4 vols. M., 1955–56.
Platonov, G. V. *Dialekticheskii materializm i voprosy genetiki*. M., 1961.
———, ed. *Ocherk dialektiki zhivoi prirody*. M., 1963.
Pledge, H. T. *Science since 1500. A Short History of Mathematics, Physics, Chemistry, Biology*. London, 1939.
Plekhanov, G. V. *Izbrannye filosofskie proizvedeniia*. 5 vols. M., 1956–58.
Plenum TsK KPSS . . . , with date indicating the particular meeting.
Ploss, Sidney. *Conflict and Decision-Making in Soviet Russia. A Case Study of Agricultural Policy, 1953–63*. Princeton, 1965.
Poliakov, I. M. *Kurs darvinizma*. M., 1941.
———. *Zh.-B. Lamark i uchenie ob evoliutsii organicheskogo mira*. M., 1962.
Polianskii, Iu. I., ed. *Obshchaia biologiia*. M., 1966.
Popovskii, A. *Iskusstvo tvoreniia*. M., 1948. This is a 2nd ed. of his *Zakony rozhdeniia*. M., 1947.
Prezent, I. I. *Biolog-materialist Zhan Batist Lamark*. M., 1960.
———. *I. V. Michurin i ego uchenie*. M., 1961.
———. *Klassovaia bor'ba na estestvenno-nauchnom fronte*. M., 1932.
——— and Lysenko, T. D. *Selektsiia i teoriia stadiinogo ravitiia rastenii*. M., 1935.
———. *Teoriia Darvina v svete dialekticheskogo materializma. Tezisy k 50-letiiu so dnia smerti Ch. Darvina*. L., 1932.
———. *V sodruzhestve s prirodoi: I. V. Michurin i ego uchenie*. 1st and 2nd eds. L., 1946 and 1948.
Prianishnikov, D. N. *Azot v zhizni rastenii i v zemledelii SSSR*. M., 1945. There is an English translation, *Nitrogen in the Life of Plants*. Madison, Wisconsin, 1951.
———. *Moi vospominaniia*. 1st and 2nd eds. M., 1957 and 1961.
Problemy biokhimii v michurinskoi biologii. M., 1949. Published by AN SSSR, Institut Biokhimii.
Professor Vasilii Robertovich Vil'iams. K 25-letiiu nauchnoi, pedagogicheskoi i s.-kh.-oi obshchestvennoi deiatel'nosti. Izdanie studencheskoi iubileinoi kommissii. M., 1914.
Ral'tsevich, V., ed. *Materializm i empiriokrititsizm V. I. Lenina*. M., 1935. Published by Kommunisticheskaia Akademiia, Institut Filosofii, Leningradskoe otdelenie.
Rasovaia teoriia na sluzhbe fashizma. Sbornik statei. Kiev, 1935. Published by Ukrainskaia Assotsiatsiia markso-leninskikh n.-i. institutov, Institut Filosofii.
Ravich-Cherkasskii, M., ed. *Darvinizm i marksizm*. 1st and 2nd eds. Kharkov, 1923 and 1925.
Raznoglasia na filosofskom fronte. M., 1931.
Razumov, V. I. *Sreda i osobennosti razvitiia rastenii*. M., 1954.
Razvitie biologii v SSSR. M., 1967. Published by AN SSSR, Inst. Istorii Estestvoznaniia i Tekhniki.
Revenkova, A. I. *Akademik N. I. Vavilov. 1887–1943*. M., 1962.
Riadom s N. I. Vavilovym. Sbornik vospominanii. M., 1963. Iu. N. Vavilov, compiler.
Richey, F. D. *Selektsiia kukuruzy*. M., 1931. Translation of USDA Bulletin No. 1489.

Rieger, R. and Michaelis, A. *Genetisches und cytogenetisches Wörterbuch.* 2nd ed. Berlin, 1958.
Robinson, G. T. *Rural Russia under the Old Regime.* N.Y., 1931.
Rokitskii, P. F. *Genetika.* 2nd ed. M., 1934.
Rozental', M. M. and Iudin, P. F., eds. *Kratkii filosofskii slovar'.* 1st–4th eds. M., 1939–54. Two further eds., 1963 and 1968, entitled *Filosofskii slovar'.*
Rudenko, A. I. *Opredelenie faz razvitiia s.-kh. rastenii.* M., 1950.
Ruhland, W., ed. *Handbuch der Pflanzenphysiologie.* 18 vols. Berlin, 1955–62.
Russkaia intelligentsiia i krest'ianstvo; kriticheskii analiz trudov mestnykh komitetov o nuzhdakh sel'skokhoziaistvennoi promyshlennosti. M., 1904. Published by Osoboe soveshchanie o nuzhdakh s.-kh. promyshlennosti.
Rutkevich, M. N. *Dialekticheskii materializm. Kurs lektsii dlia estestvennykh fakul'tetov.* M., 1959.
———. *Praktika—osnova poznaniia i kriterii istiny.* M., 1952.
Ryzhkov, V. L., ed. *Virusnye bolezni rastenii i mery bor'by s nimi. Trudy soveshchaniia po virusnym bolezniam rastenii. Moskva, 4–7/II 1940 g.* M., 1941.
Sabinin, D. A. *Fiziologiia razvitiia rastenii.* M., 1963.
Sakharov, A. D. *Progress, Coexistence, and Intellectual Freedom.* N.Y., 1968.
Salaman, R. N. *The History and Social Influence of the Potato.* Cambridge, Mass., 1949.
Samoilov, A. *Detskaia bolezn' "levizny" v materializme (pokhod protiv filosofii, ideologii, mirovozzreniia).* L., 1926.
Sbornik diskussionnykh statei po voprosam genetiki i selektsii. M., 1936. Published by VASKhNIL, O. M. Targul'ian, ed.
Science, Medicine, and History: Essays on the Evolution of Scientific Thought and Medical Practice Written in Honour of Charles Singer, collected and edited by E. Ashworth Underwood. vol. II. London, 1953.
Selektsionnaia rabota po plodovo-iagodym kul'turam; materialy IV plenuma Sektsii plodovo-ovoshchnykh kul'tur, 25 iiunia-1 iiulia 1936 g. M., 1937. Published by VASKhNIL.
Selo Viriatino v proshlom i nastoiashchem; opyt etnograficheskogo izucheniia russkoi kolkhoznoi derevni. Vol. XLI of AN SSSR, Inst. Etnografii, *Trudy,* novaia seriia.
Sel'skoe khoziaistvo SSSR; statisticheskii sbornik. M., 1960.
Sel'skokhoziaistvennaia entsiklopediia. 4 vols. 1st–3rd eds. M., 1932–56.
Semashko, N. A. *Nauka o zdorov'e obshchestva (sotsial'naia gigiena).* 1st and 2nd eds. M., 1922 and 1926.
———. *Vvedenie v sotsial'nuiu gigienu.* M., 1927.
Semenov, Iu. I. *Kak vozniklo chelovechestvo.* M., 1966.
Shekhurdin, A. P. *Izbrannye sochineniia.* M., 1961.
Shennikov, A. P. *Ekologiia rastenii.* M., 1950.
Shevchenko, A. S. *Kukuruza; dlia obmena opytom dveri shiroko otkryty.* M., 1960.
Shlykov, G. N. *Introduktsiia rastenii.* M., 1936.
Shmerling, V. G. *Michurin; ocherki o zhizni i rabote.* L., 1934.
Shpil'man (see Spielmann).
Singer, Charles *et al.,* eds. *A History of Technology.* 5 vols. N.Y. and London, 1954–58.

Slepkov, V. N. *Evgenika. Uluchshenie chelovecheskoi prirody.* M., 1927.
Slicher van Bath, B. H. *The Agrarian History of Western Europe, A. D. 500–1850.* London, 1963.
Slusser, R. M. and Wolin, Simon, eds. *The Soviet Secret Police.* N.Y., 1957.
Smith, Jay W. *The Mind and the Sword.* N.Y., 1961.
Soveshchanie peredovikov urozhainosti po zernu, traktoristov i mashinistov molotilok s rukovoditeliami partii i pravitel'stva. Moskva, 1935. Stenograficheskii otchet. M., 1936.
Soveshchanie peredovikov zhivotnovodstva s rukovoditeliami partii i pravitel'stva. 13–16 fev. 1936 g. M., 1936.
Soveshchanie po probleme zhivogo veshchestva i razvitiia kletok. 22–24 maia 1950 g. M., 1951.
The Soviet Linguistics Controversy. N.Y., 1951. Published by Columbia University, E. J. Simmons, ed.
Speranskii, A. D. *Elementy postroeniia teorii meditsiny.* M., 1935. Translated as *A Basis for the Theory of Medicine.* N.Y., 1944.
Spielmann, T. I. (Shpil'man). *Teoriia i praktika biontizatsii i mutatsii.* M., 1935.
Spornye voprosy genetiki i selektsii. Raboty IV sessii VASKhNILa 19–27 dek. 1936 g. M., 1937.
Sprague, G. F., ed. *Corn and Corn Improvement.* N.Y., 1955. Translated as *Kukuruza i ee uluchshenie.* M., 1957.
SSSR. M., 1957. Separate printing of vol. L of *BSE*, 2nd ed.
Stalin, J. V. *Sochineniia.* 13 vols. M., 1946–51. vols. XIV [1]–XVI [3] have been published by the Hoover Institution. Stanford, California, 1967. R. H. McNeal, ed.
The State of Soviet Science. Cambridge, Mass., 1965. Edited by the editors of *Survey*, published by Mass. Inst. of Tech.
Stauffer, R. C., ed. *Science and Civilization.* Madison, Wisconsin, 1949.
Stoletov, V. N. *Vnutrividovye prevrashcheniia i ikh kharakter.* M., 1957.
Stuart, William. *The Potato. Its Culture, Uses, History, and Classification.* Philadelphia, 1923.
Sukachev, V. N., ed. *Nauchnaia sessiia po voprosam biologii i sel'skogo khoziaistva, Riga, 22–26 okt. 1951 g.* M., 1953.
———. *Stalinskii plan preobrazovaniia prirody.* M., 1950.
Talanov, V. V. et al. *Kukuruza i nailuchshie priemy ee vozdelyvaniia po dannym opytnykh stantsii SSSR i SASSh.* M., 1931.
———. ed. *Selektsiia i semonovodstvo v SSSR. Obzor rezul'tatov deiatel'nosti selektsionnykh i semenovodstvennykh organizatsii k 1923.* M., 1924.
Taliev, V. I. *Osnovy botaniki v evoliutsionnom izlozhenii.* 7th ed. M., 1933.
Taton, René, ed. *Science in the Twentieth Century.* N.Y., 1966.
Tezisy dokladov konferentsii po problemam darvinizma (3–8 II 1948). M., 1948.
Thomas, Lawrence L. *The Linguistic Theories of N. Ja. Marr.* Berkeley, California, 1957. Univ. of California publications in linguistics, vol. XIV.
Timiriazev, K. A. *Sochineniia.* 10 vols. M., 1937–40.
Tokin, B. P. *Voprosy biologii.* Tashkent, 1935.
Tomilin, S. A. *Sotsial'no-meditsinskaia profilaktika; teoreticheskoe obosnovanie i prakticheskaia postanovka.* Kharkov, 1931.

Trapeznikov, S. P. *Istoricheskii opyt KPSS v sotsialisticheskom preobrazovanii sel'skogo khoziaistva.* M., 1959.
———. *Leninizm i agrarno-krest'ianskii vopros.* 2 vols. M., 1967.
Triska, Jan F., ed. *Soviet Communism: Programs and Rules. Official Texts of 1919, 1952 (1956), 1961.* San Francisco, 1962.
Troepol'skii, G. N. *Kandidat nauk.* M., 1959.
Trotsky, L. D. *Sochineniia.* 14 vols. M., 1925–27.
Trudy Vserossiiskogo s'ezda po selektsii i semenovodstvu v g. Saratove, iiun' 4–13, 1920. Place? 1920.
Trudy Vserossiiskogo s'ezda po sel'skokhoziaistvennomu opytnomu delu, VII, 15–25 iiun' 1921. M., 1922.
Trudy [I] vsesoiuznogo s'ezda po genetike, selektsii, semenovodstvu i plemennomu zhivotnovodstvu, v Leningrade 10–16 ianvaria 1929 g. 3 vols. L., 1929–30.
Tsitsin, N. V. *Bol'shoi kolos.* M., 1960.
———, ed. *Otdalennaia gibridizatsiia v semeistve zlakovykh.* M., 1958.
———. *Problema ozimykh i mnogoletnikh pshenits.* Omsk, 1933.
———, ed. *Sorta polevykh kul'tur.* M., 1944.
Tulaikov, N. M. *Izbrannye proizvedeniia. Kritika travopol'noi sistemy zemledeliia.* M., 1963.
———. *Organizatsiia rasprostraneniia s.-kh. znanii sredi naseleniia soedinennykh shtatov.* M., 1923.
Turbin, N. V., ed. *Genetika i tsitologiia rastenii.* Minsk, 1962.
———. *Genetika s osnovami selektsii.* M., 1950.
———, ed. *Geterozis (teoriia i metody prakticheskogo ispol'zovaniia).* Minsk, 1961.
Ul'ianovskaia, V. A. *Formirovanie nauchnoi intelligentsii v SSSR, 1917–37.* M., 1966.
UNESCO. United Nations Economic Social and Cultural Organization. *The Race Concept: Results of an Inquiry.* Paris, 1952.
U.S. Congress, Joint Economic Committee. *Comparisons of the United States and Soviet Economies,* Part I. Washington, D.C., 1959.
U.S. Department of Agriculture, Soil Conservation Service. *Soil and Water Use in the Soviet Union.* Washington, D.C., 1959.
Ushakov, D. N., ed. *Tolkovyi slovar' russkogo iazyka.* 4 vols. M., 1935–40.
Vakar, B. A. *Vazhneishie khlebnye zlaki.* Novosibrisk, 1929.
Vasil'chenko, I. T. *I. V. Michurin.* 2 eds. M., 1950 and 1963.
VASKhNIL. Abbreviation for Vsesoiuznaia Akademiia Sel'skokhoziaistvennykh Nauk imeni V. I. Lenina.
———. *Dostizheniia sovetskoi selektsii; raboty IV sessii 19–26 dek. 1936.* M., 1937.
———. *Kalendar' spravochnik.* 3 eds. M., 1936, 1937, 1958.
———. *Rastenievodstvo SSSR; materialy k sostavleniiu gosudarstvennogo plana po sel'skomu khoziaistvu na 1931 i blizhaishie gody.* 2 eds. L., 1930 and 1933.
———. *Selektsionnaia rabota po plodovo-iagodnym kul'turam; materialy IV plenuma Sektsii plodovo-ovoshchnykh kul'tur, 25 iiunia-1 iiulia 1936 g.* M., 1937.
———. *Trudy sessii VASKhNILa posviashchennoi voprosam podniatiia urozhainosti i razvitiia zhivotnovodstva (g. Voronezh, 15–20 fevralia, 1933 g.).* Voronezh, 1933/34.

VASKhNIL, Laboratoriia po biontizatsii semian. *Biontizatsiia semian — novyi faktor povysheniia urozhainosti.* M., 1931.
Vavilov, N. I. *Genetika na sluzhbe sotsialisticheskogo zemledeliia.* M., 1932.
———. *Izbrannye trudy.* 5 vols. M., 1959–65.
———. *Sovremmenye zadachi sel'skokhoziaistvennago rastenievodstva.* Saratov, 1917.
———, ed. *Teoreticheskie osnovy selektsii rastenii.* 3 vols. M., 1935–37.
Veselovskii, B. B. *Istoriia zemstva za sorok let.* 2 vols. St. Petersburg, 1909.
Veselovskii, I. A. *Kartofel' semenami severnym, gornym i otdalennym raionam SSSR.* L., 1933.
Vil'iams, V. R. *Izbrannye sochineniia.* 3 vols. M., 1950–55.
———. *Sobranie sochinenii.* 12 vols. M., 1948–53.
Vishnevskii, B. N. *Proiskhozhdenie cheloveka.* 1st and 2nd eds. M., 1934 and 1935.
Vladimirskii, A. P. *Peredaiutsia li po nasledstvu priobretennye priznaki?* M., 1927.
Volotskoi, M. V. *Podniatie zhiznennykh sil rasy.* 1st and 2nd eds. M., 1923 and 1926.
Vorob'ev, A. I. *Osnovy michurinskoi genetiki.* 1st and 2nd eds. M., 1950 and 1953.
Vsesoiuznaia Akademiia Sel'skokhoziaistvennykh Nauk imeni Lenina. See under VASKhNIL.
Vsesoiuznaia konferentsiia po bor'be s zasukhoi, Oktiabr', 1931. *Biulleten'.* Nos. 1–7.
———. *Tezisy i materialy.* Nos. 1–3.
Vsesoiuznaia konferentsiia po planirovaniiu genetiko-selektsionnykh issledovanii. *Materialy.* L., 1932.
———. *Trudy.* L., 1933.
Vsesoiuznoe soveshchanie nauchno-issledovatel'skikh uchrezhdenii po sel'skomu khoziaistvu. Moskva, 1945. *Materialy.* M., 1948.
Vsesoiuznoe soveshchanie po proizvodstvu gibridnykh semian kukuruzy, Dnepropetrovsk, 28–30 marta 1956 g. M., 1956.
Vsesoiuznoe soveshchanie po proizvodstvu kukuruzy. Krasnodar', 9–13 fev. 1960 g. M., 1961.
Vsesoiuznoe soveshchanie rabotnikov sel'skokhoziaistvennoi nauki, 19–23 iiunia 1956 g. *Materialy.* M., 1957.
Wallace, Alfred Russell. *On Miracles and Modern Spiritualism. Three Essays.* London, 1875.
———. *Studies Scientific and Social,* 2 vols. London, 1900.
Wallace, Henry A. and Bressman, E. N. *Corn and Corn-Growing.* 1st–5th eds. Des Moines, 1923–N.Y., 1949. Translated as *Kukuruza i ee vozdelyvanie.* M., 1954.
Weatherwax, Paul. *Indian Corn in Old America.* N.Y., 1954.
Weir, W. W. *Soil Science.* Chicago, 1949.
Weissberg, Alexander. *The Accused.* N.Y., 1951.
Wetter, G. A. *Dialectical Materialism. A Historical and Systematic Survey of Philosophy in the Soviet Union.* N.Y., 1958.
Williams, V. R. See Vil'iams, V. R.
Zavadovskii, B. M. *O korobke konservov i proiskhozhdenii zhizni na zemle.* M., 1926.
Zemel'naia renta v sotsialisticheskom sel'skom khoziaistve, M., 1960.

Zhdanov, A. A. *Essays on Literature, Philosophy, and Music.* N.Y., 1950.
Zhdanov, Iu. A. *Lenin i estestvoznanie.* M., 1959.
———. *Vozdeistvie cheloveka na prirodnye protsessy.* M., 1952.
Zhegalov, S. I. *Vvedenie v selektskiiu sel'skokhoziaistvennykh rastenii.* 3rd ed. M., 1930.
Zhizn' i tekhnika budushchego (sotsial'nye i nauchno-tekhnicheskie utopii). M., 1928.
Zhuravlev, V. V., *Marksizm-leninizm ob otnositel'noi samostoiatel'nosti obshchestvennogo soznaniia.* M., 1961.
Zirkle, Conway. *Death of a Science in Russia.* Philadelphia, 1949.
———. *Evolution, Marxian Biology, and the Social Scene.* Philadelphia, 1959.
Zvorykin, A. A., ed. *Biograficheskii slovar' deiatelei estestvoznaniia i tekhniki.* 2 vols. M., 1959.

Notes

References to books, pamphlets, journals, and newspapers have been shortened to the barest minimum necessary for identification of the full citation in the Bibliography. References to articles are given in full, since articles are not listed in the Bibliography.

ABBREVIATIONS

BME *Bol'shaia meditsinskaia entsiklopediia.* 1st ed.
BMOIP, otdel biol. *Biulleten' moskovskogo obshchestva ispytatelei prirody, otdel biologicheskii.*
BSE *Bol'shaia sovetskaia entsiklopediia.* 1st or 2nd ed., as specified.
BZ *Botanicheskii zhurnal.*
FNIT *Front nauki i tekhniki.*
PZM *Pod znamenem marksizma.*
SRSKh *Sotsialisticheskaia rekonstruktsiia sel'skogo khoziaistva.*
SSSR Separate printing of vol. L of *BSE,* 2nd ed. (M., 1957).
SZ *Sotsialisticheskoe zemledelie.* A newspaper that was renamed *Sel'skoe khoziaistvo* in 1953, and renamed *Sel'skaia zhizn'* in 1960.
TPBGS *Trudy po prikladnoi botanike, genetike, i selektsii.*
VAN *Vestnik Akademii Nauk SSSR.*
VF *Voprosy filosofii.*
VKA *Vestnik Kommunisticheskoi Akademii.*

PREFACE

1. Brecht, VII, 64–80. The poem, "Die Erziehung der Hirse," was originally published in 1950.

1. SOVIET IDEOLOGY AS A PROBLEM

1. Alfred G. Meyer, "The Functions of Ideology in the Soviet Party System," *Soviet Studies,* 17:273–285 (Jan. 1966).
2. Naess and associates, pp. 171–2.
3. *Ibid.* For another effort to avoid pejorative statements in the definition of ideology, see Corbett, pp. 11–13.
4. See Merton, pp. 439–508; Gurvitch, pp. 103–136; George Lichtheim, "The Concept of Ideology," *History and Theory,* 4:164–195 (1965). For an annotated bibliography see N. Birnbaum, "The Sociological Study of Ideology (1940–60)," *Current Sociology,* 9:91–172 (No. 2, 1960). An historical anthology is provided by Lenk, *Ideologie.*

5. On the difference between functional correlations and causal connections, see Nagel, pp. 522–6 *et passim.*

6. Kuhn, *passim.*

7. That ambiguous question provides an excellent illustration of the difference between rational discourse, which is usually understood to prohibit *ad hominem* arguments, and ideological analysis, which cannot avoid them. If the question is purely rhetorical, it means: This is an inconsistent pattern of thought. If it is read as a genuine question, it invites the answer: This inconsistent pattern of thought is characteristic of a group (Western students of Soviet affairs), not because it satisfies the criteria of truthful thought, but because it satisfies the group's interest in denigrating the Communists.

8. The prerevolutionary philosophical controversies of the Russian Marxists were concerned in part with this issue. See, for example, Lenin, *Sochineniia,* XIV, 123, for his objection to Bogdanov's view of ideology. He and Bogdanov both understood the term nonpejoratively, to include all forms of thought, but Bogdanov held that ideology can never claim absolute truth, only social usefulness. For an example of Bogdanov's repeated argument that Lenin's and Plekhanov's insistence on absolute truth is inevitably authoritarian, see Bogdanov, pp. 216–17.

V. V. Adoratsky, in his Preface to the one-volume edition of *Pis'ma Marksa i Engel'sa* (Moscow, 1922 and 1923), revived Marx's pejorative use of the term ideology, arguing that the proletariat needs no ideology but only natural science and Marxism, which is social science. A controversy ensued, which covered a wide range of issues, from Marx's use of the term (see, e.g., I. P. Razumovskii, "Sushchnost' ideologicheskogo vozzreniia," *Vestnik sotsialisticheskoi akademii,* 1923, no. 4) to the psychological sources of irrationality (see, e.g., M. A. Reisner, "Sotsial'naia psikhologiia i uchenie Freida," *Pechat' i revoliutsiia,* 1925, nos. 3, 4, 5–6, and "Ideologiia i politika" *Vestnik kommunisticheskoi akademii,* 1929, no. 33/34). For Bogdanov's restatement of his view — this time distinguishing two senses of the term ideology — see *Vestnik kommunisticheskoi akademii,* 1924, no. 9, pp. 318–21. For one of the conventional efforts to prove that there are no serious difficulties in believing Marxism to be science and philosophy and ideology, see A. Samoilov. These controversies were frozen in 1930–31 by the rigid subordination of philosophy and ideology to politics, that is, to direct control by the Central Committee. Natural science was assigned in principle to ideology; in practice much of it remained autonomous.

9. See Stalin's famous articles on linguistics, which excluded language from the "ideological superstructure" and suggested that science too might be excluded. They appeared originally in *Pravda,* 20 June, 4 July, and 2 August 1950, and were widely reprinted. Although philosophers showed great reluctance to develop Stalin's suggestion, natural scientists began to press the distinction between science and ideology while the old man was still alive.

10. See Helmut Fleischer, "The Limits of 'Party-Mindedness': A Selection of Texts (Recent Soviet Discussions on the Concept of Ideology," *Studies in Soviet Thought,* 2:119–31 (June 1962). The quotation from Kelle and Koval'zon at the end of this perceptive survey is strikingly similar to the courageous protests of Karev and Sten in 1930, when Mitin and the other Stalinists reduced philosophy and ideology to politics. See

Raznoglasiia, pp. 125, 159. (Karev and Sten were subsequently condemned as "enemies of the people.") In Poland, Czechoslovakia, and Yugoslavia efforts to disentangle science and philosophy from ideology have been much bolder than in the Soviet Union. They were a major cause of the official campaign against revisionism in the late 1950s. See, e.g., Vladimir Ruml, "Ideologiia i nauka," *Problemy mira i sotsializma,* 1960, no. 7, pp. 39–45. What is at stake in these quarrels is the Central Committee's control of the intelligentsia.

11. See, e.g., J. M. Bochenski, "Toward a Systematic Logic of Communist Ideology," *Studies in Soviet Thought,* 4:185–205 (September 1964).

12. *Pravda,* June 20, 1950.

13. See, for example, Karl Nielsen, "On Speaking of God," *Theoria,* 28:110–137 (1962). Cf. Gabel, for an effort to define ideology as a form of collective insanity.

14. Sometimes an opposite argument was used against universal suffrage: It would give wealthy men inordinate power, since they could buy the votes of the poor. See Gibbons, *passim.*

15. For the origin of the concept of ideology, and Napoleon's reaction to it, see Jay W. Smith. Cf. also Barth, pp. 13–31.

16. For definitions of ideology that include the confusion of moral and factual judgment as an essential characteristic, see Nielsen, as cited in n. 13, and Walter P. Metzger, "Ideology and the Intellectual: A Study of Thorstein Veblen," *Philosophy of Science,* 16 (April, 1949), p. 125. For a laborious effort to turn political "realism" into political science, see Lasswell and Kaplan, especially pp. 119–121, where they agree with Bertrand Russell's acid comment: "Beliefs which have been successful in inspiring respect for the existing distribution of power have usually been such as cannot stand against intellectual criticism." But then they take the rational acid from this comment, and turn it into "realism," by arguing that intellectual criticism is inappropriate to such beliefs. Only political effect is worth considering.

17. The revisionists in Poland, Czechoslovakia, and Yugoslavia made this one of the main points of their rebellion in the post-Stalin era. See especially the writings of L. Kolakowski. Even the orthodox philosophers in those countries tended to agree. See, e.g., N. Lobkowicz, "Philosophical Revisionism in Post-War Czechoslovakia," *Studies in Soviet Thought,* 4:89–101 (June 1964). The stubborn refusal of official Soviet philosophers to recognise the fact that Stalin made ideological "theory" an irrelevant comment on intuitive "practice" is one of the symptoms — and one of the causes — of the continuing irrelevance of theoretical ideology in the Soviet Union. See, for example, Akademiia Nauk, Institut Filosofii, *Istoriia filosofii,* VI, Part I, 151–4 *et passim.*

18. See, for example, M. Sulkovskii, "Zemel'naia renta," *BSE,* Ist ed., XXVI (1933), column 583.

19. See Bauer, pp. 123 *et seq.*

20. Hazard, p. 69. Shortly thereafter (p. 71) he frames a conflicting hypothesis: The people "appear, in the main, to have accepted terror during Stalin's time as a necessary evil. In short, one must understand that terrorizing as a technique of government was rationalized by Stalin for most of the rank and file of Soviet citizens." In that final shift to an insistent indicative mood — "one must understand that terrorizing . . .

was rationalized" — even Hazard, one of our most careful students of the Soviet political process, betrays the common tendency to transmute favored hypotheses into proven theories.

21. "Moreover, the men in the Kremlin may also have convinced themselves that mass terror endangers and damages the political system more than it benefits the ruling elite, perhaps . . . because legitimacy achieved through dread rests on too insecure a basis." Meyer, p. 331. Of course, Meyer may not have meant the last clause to be a universal principle. (Luther's brutal aphorism, "Frogs must be ruled by storks," may generalize a larger portion of historical experience.) He may have meant it to generalize the experience of modern governments, or simply of the Soviet government, in which case we are back at the locked doors of the archives.

22. This is not to say that framing hypotheses in this area is futile. For example, Meyer uses two valid methods in disputing Brzezinski's hypothetical explanation of Soviet terror. He finds it more contrived, less "economical" than his own (a logical method of ranking hypotheses), and he doubts that totalitarian "systems have no other method [than terror] of ensuring social mobility and a turnover of elites" (*ibid.* p. 328). The second argument refers the analysis of terror to a kind of sociological study that can be done, however imperfectly, with data available in the public record. For example, my own accumulation of biographical facts concerning a few hundred Soviet philosophers and scientists makes me share Meyer's doubt. The overwhelming majority did not obtain or lose posts as a result of terror.

23. See Frank Durgin, "Monetization and Policy in Soviet Agriculture Since 1952," *Soviet Studies,* 15:375–407 (April 1964); Karcz, *Soviet Agricultural Marketings*; Nancy Nimitz, "Soviet Agricultural Prices and Costs," in *Comparisons of the United States and Soviet Economies,* I, pp. 239–84.

24. See Stalin, XII, 151–52. As Bolgov points out, in this speech at the Conference on Agricultural Economics Stalin's actual words were "extremely vague," but he was understood to be denying the existence of any form of rent. See Bolgov, pp. 8–10, for a brief review of the previous discussions of rent. The main articles appeared in the journals *Na agrarnom fronte* and *PZM,* 1925–29.

25. Cf. Bergson, pp. 192–3, 196, 210, for a discussion of the intricate relationships between distribution of these proceeds and the promotion of long-run efficiency. Bergson considers it probable that, before collectivization, the peasants were receiving most of the proceeds from rent. It would follow that Stalin's denial of rent functioned in part to obscure the unpleasant fact that collectivization was sharply increasing the state's take. It would not necessarily follow that Stalin was fully aware of this function. Indeed, if he had been an astute liar rather than an ideologist, he would probably have arranged the extraction of rent more efficiently.

26. I have run together two quotations from the symposium *Zemel'naia renta,* pp. 48 and 187. Such economists are a vanishing breed. This particular slogan, which derives from V. R. Williams, and was once mandatory, is now most frequently quoted for derogatory purposes. See, e.g., Bolgov, pp. 114 *et passim.*

27. See I. D. Laptev, "Kolkhoznye dokhody i differentsial'naia renta," *Bol'shevik,* 1944, no. 16, pp. 18–22, and "Zemel'naia renta," *BSE,* 2nd ed., XVI (1952), pp. 631–3, for late Stalinist acknowledgment of rent and

analysis of its distribution. Cf. Kozodoev, p. 179. Mr. Durgin, as cited in n. 23, suggests that the acknowledgment of rent has been a result of the declining abundance of unused land. He argues that the scarcity of capital, being more acute than the scarcity of land, brought about an earlier recognition of interest, while Soviet ideology *helped* the leaders to perceive the necessity of measuring the opportunity costs of labor.

For other perceptive studies of the acknowledgment of economic problems that were once denied, see R. W. Campbell, "Marx, Kantorovich, and Novozhilov: *Stoimost'* versus Reality," *Slavic Review,* 20:402–18 (October, 1961); and Alfred Zauberman, "New Winds in Soviet Planning," *Soviet Studies,* 12:1–13 (July 1960).

28. Cf. Macrae, p. 65: ". . . Society itself is only rendered possible by people having fairly accurate assessments of their response systems to different circumstances, and similar assessments of the response systems of others — that is, in part, of their ideologies."

29. See, for example, Nagel, pp. 473–502. Kelle and Koval'zon, pp. 11 *et seq.,* created a stir among Soviet philosophers by making a sharp distinction between scientific cognition and ideology, which may be intricately combined "in real life," but must and can be distinguished "by means of theoretical analysis, in abstraction." If they were to develop this distinction, it seems to me that they would arrive at conclusions similar to Nagel's.

30. Evidence abounds. It has even become a cliché of Soviet ideologists to concede that there is some merit in the works of such philosophers as Wittgenstein, Carnap, or Bertrand Russell. See, for example, Zhuravlev, p. 57. More significant evidence can be found in the publications of philosophers like Ernst Kol'man, who has been transformed beyond recognition. (In the 1930s and 1940s he was one of the most savage Stalinists on the front of science and philosophy.)

31. The *de facto* narrowing of the extension of ideology is much more impressive than the theorists' efforts to redefine the concept, but a few of the latter are worth noting. See, for example, Kelle and Koval'zon, and Iadov. Most impressive is the withering of the tradition that Marxist political leaders deliver new truths of theoretical ideology to their followers. The leaders have virtually ceased to pretend that they are the prophetic successors of Marx and Lenin.

32. For a poignant case of two excellent minds straining at the limits of rational thought about power politics, see the exchange between D. G. Brennan and Anatol Rapport, "Strategy and Conscience," *Bulletin of the Atomic Scientists,* 21:25–36 (December 1965). Each author argues that the other's position on Machiavelli's basic rule is untenable, yet each admits, in effect, that his own position is basically ideological, i.e., not justifiable on rational grounds.

2. A CRISIS OF FAITH IN SCIENCE

1. For a convenient summary, see Singer et al., IV, Chapter 1. For more extensive accounts, see Chambers and Mingay, and Slicher van Bath.

2. See Ihde, pp. 424–6.

3. These estimates were made by Prianishnikov, *Azot v zhizni rastenii,* pp. 134–6. For somewhat different estimates, see Weir, pp. 17–18. It goes without saying that undue importance should not be assigned to such

global figures. Cf. the comments on increases in yields in Chambers and Mingay, *passim*.

4. See *Agriculture in the Twentieth Century*, pp. 246–8.

5. For a history of efforts to improve Russian agriculture, see *Agronomicheskaia pomoshch' v Rossii*. See pp. 350–1 for the anonymous author's comment on the Russian reversal of the Western pattern.

6. See USDA, *Yearbook*, 1938, pp. 881–2. Cf. Marbut's introduction to Glinka.

7. Vavilov's famous collection was begun by other scientists before the Revolution. See his essay in Talanov, ed., *Selektsiia i semenovodstvo SSSR*, pp. 31–46.

8. See G. T. Robinson.

9. Concerning the Academy, see Prianishnikov, *Moi vospominaniia*; also Doiarenko, Korolenko, A. V. Chaianov, and Kuznetsov, as listed in the bibliography. The major work on the zemstvos is still B. B. Veselovskii, *Istoriia zemstv*.

10. *Agronomicheskaia pomoshch' v Rossii*, p. 88.

11. The incident is told by Prianishnikov, *Moi vospominaniia*, pp. 116–117. Before the Revolution of 1905 the most extensive discussion of the causes of Russian agricultural backwardness was organized by the *Osoboe soveshchanie o nuzhdakh sel'skokhoziaistvennoi promyshlennosti* (Special Conference on the Needs of Agricultural Production), convened and supervised by Witte. See its *Trudy*, 58 vols., and *Svod trudov*, 4 vols. Cf. also *Russkaia intelligentsiia i krest'ianstvo*.

12. For the transformation of the Academy into the Moscow Agricultural Institute, see the works cited in note 9. Note Prianishnikov's remark, p. 154, that Williams "was shown special trust by the [tsarist] government." The adulatory Soviet biographies of Williams make no effort to explain why the tsarist officials who were closing the Academy for its political sins chose Williams to be curator of the buildings, or why they allowed him to continue teaching in the new Institute. Oral tradition apparently kept alive knowledge of Williams' political opportunism. Cf. Khrushchev's brief, and unconvincing, rebuttal of this tradition, in Khrushchev, VI, 56–7.

13. The words are K. A. Timiriazev's, in his preface to the Russian translation of Harwood, *The New Earth*. See Garvud, p. iii. This romantical exaggeration of the feats of the USDA had a great impact on the Russian intelligentsia, including Lenin. A second edition of the Russian translation appeared in 1918, and throughout the 1920s men like Gorbunov, Lenin's assistant for science and technology, attested to the book's influence on their chief. For direct evidence, see Lenin, *Polnoe sobranie*, LIV, 300–301.

14. See again Robinson, *Rural Russia*.

15. See *Agronomicheskaia pomoshch' v Rossii*, *passim*, for a telling contrast between the indices of modest gains in agricultural productivity and the impressive expansion of scientific and educational institutions.

16. See Vakar, pp. 274–284.

17. The argument continued into the Soviet period. See Prianishnikov's brief summary in Akademiia Nauk SSSR, *Pochvovedenie*, pp. 356–360.

18. See *Pochvovedenie*, 1907, No. 1, pp. 57–78, for two scientists replying to Prianishnikov's criticism.

19. Timiriazev, III, 18–19.

20. For an attempt to give an objective account of this controversy, see Bliakher et al., III, 216–38.

21. Iarilov, " 'Istinnaia,' 'russkaia,' 'samobytnaia' nauka i eia universitetskie predstaviteli," *Pochvovedenie*, 1905, No. 3, pp. 232–3. The first two installments of this long polemic appeared in Nos. 1 and 2.

22. See Iarilov's celebration of Williams, in *Akademik V.R. Vil'iams*, pp. 199–227. See also his articles in *Pochvovedenie*, 1936, No. 3, 1937, No. 10, and 1938, No. 4.

23. Burbank remains the hero of many writers for children. See, for example Olive Burt, *Luther Burbank, Boy Wizard.* The most damning account of Burbank's life and works is given inadvertently by an author who tries hard to rescue Burbank from the scorn of scientists. See W. L. Howard, *Luther Burbank, A Victim of Hero Worship*, and *Luther Burbank's Plant Contributions.* Niels Hansen was a self-advertising plant breeder and collector who, in the decade before the First World War, won political support to overcome the well-merited hostility of genuine scientists. See File 346 of the USDA, National Archives, Washington, D.C.

24. *Professor Vasilii Robertovich Vil'iams.*

25. V. N. Sukachev, "O 'teorii dernovogo protsessa' prof. Vil'iamsa," *Pochvovedenie*, 1916, No. 2, pp. 1–26. The only previous notice of Williams' ideas in this official journal of Russian soil scientists may be found in the issue of 1902, No. 4, pp. 434–6, where Williams' scheme is briefly described, and regret is expressed that he has offered no proof of it, "for it completely contradicts the reigning scientific views of the present time." For Williams' part in academic politics, see the works cited in note 9, especially the reminiscences of Prianishnikov.

26. See Sukachev's speech, *VAN*, 1948, No. 9, pp. 44–48. In fairness to Sukachev it must be pointed out that this speech was probably insincere, a concession to the overwhelming political power mobilized on behalf of pseudoscience in 1948. Subsequently he proved to be one of the leaders in the defense of genuine science.

27. See Chapter 3, "Michurin."

28. See *Agronom*, 1927, No. 11, pp. 79–80.

29. *Novaia Petrovka*, 1922, No. 2, pp. 26–7.

30. *Ibid.*, 1923, No. 3–4, p. 7.

31. Quoted by Commissar of Education A. V. Lunacharskii, "Lenin v ego otnoshenii k nauke i iskusstvu," *Nauchnyi rabotnik*, 1926, No. 1, p. 13. Cf. Fediukin.

32. Prianishnikov was the elected rector who was forced to resign by conflicts with the new regime. See his cryptic report in *Moi vospominaniia*, pp. 216–17. For frank accounts of the conflicts, see the student magazine, *Novaia Petrovka*, 1922–23. See also the reminiscence by S. Molchanov in the newspaper *Timiriazevka*, 1926, No. 7. For another frank account, see *Akademik V. R. Vil'iams*, pp. 24–7.

33. *Izvestiia*, Nov. 24, 1923.

34. Trotsky, XXI, 283.

35. *Agronom*, 1924, No. 1, pp. 18–21.

36. See Joravsky, p. 65.

37. See below, Chapter 9, "Land."

38. See below, Chapter 3.

39. For the most part, Lenin turned over problems of science and technology to his administrative assistant, N. P. Gorbunov. See Gorbunov's

reminiscences, *Kak rabotal Vladimir Il'ich.* For agricultural science see his speech in *Trudy vsesoiuznogo s'ezda po genetike . . .* , I, 121–2, and the documents published in *Istoricheskii arkhiv,* 1962, No. 1. Cf. the reminiscences of the eminent plant breeder, P. I. Lisitsyn, in *Nashi dostizheniia,* 1929, No. 4.

40. A professor at the Agricultural Academy told the graduating class that they were going out to battle "the darkness and ignorance of the people." *Novaia Petrovka,* 1923, No. 7, p. 5.

41. *Agronom,* 1927, No. 11, pp. 83–4.

42. See *Agronomicheskaia pomoshch' v Rossii, passim,* for evidence of American influence on Russian agricultural specialists before the Revolution. For Lenin's sharing in this common attitude, see the references in note 13. Note especially *Istoricheskii arkhiv,* 1962, No. 1, p. 61, for Osinskii, one of the first Commissars of Agriculture, twitting Lenin for an excessive interest in the American model.

43. Tulaikov, *Organizatsiia.*

44. Vavilov, *Sovremennye zadachi,* pp. 10, 18.

45. Vavilov, "Zakon gomologicheskikh riadov v nasledstvennoi izmenchivosti," in *Trudy Vserossiiskogo s'ezda po selektsii . . . , vypusk* I, pp. 41–56.

46. *Riadom s Vavilovym,* p. 70.

47. Recalled by Baranov, among others, in Akademiia Nauk SSSR, Vsesoiuznoe Botanicheskoe Obshchetvo, *Voprosy evoliutsii . . .* , p. 6.

48. Quoted in Revenkova, p. 46.

49. Iu. A. Filipchenko, "O parallelizme v zhivoi prirode," *Uspekhi eksperimental'noi biologii,* 1924, No. 3–4, pp. 242–258.

50. See *Nauka XX veka,* II (1929), p. 8. Cf. the comment of the eminent plant breeder, G. K. Meister, in *Zhurnal opytnoi agronomii iugo-vostoka,* 1927, p. 58.

51. See Darlington for a sympathetic appraisal of Vavilov's theories in the light of contemporary knowledge.

52. See the founding statute, in *Sobranie zakonov, otdel* 1, 1929, *No.* 42, *stat'ia* 375.

53. *Novaia Petrovka,* 1922, No. 1, p. 4. Usually Soviet writers acknowledged the retrogressive aspects of the agrarian revolution with euphemistic phrases. See, e.g., Gordeev, p. 120. But cf. N. D. Kondrat'ev, "K vorprosu ob izmenenii polevodstva v krest'ianskom khoziaistve za period 1916–20," *Sel'skoe i lesnoe khoziaistvo* 1921, No. 1–3, pp. 50–78, for a masterful analysis.

54. *Trudy Vserossiiskogo s'ezda po selesktsii . . .* , vypusk I, p. 64. Cf. Bushuev's argument in *Planovoe khoziaistve,* 1928, No. 9, pp. 275–8.

55. *Trudy Vserossiiskogo s'ezda po sel'skokhoziaistvennomu opytnomu delu,* vypusk I, pp. 3–19.

56. *TPBGS,* XIV (1925), pp. 1–16. This was a speech that Vavilov delivered to a meeting in the Kremlin, celebrating the founding of a new institute.

57. There is a large body of writing on the background to collectivization, all of it crippled by lack of objectivity, or by lack of archival material, or by both. For some of the more substantial writing, see Carr, *A History of Soviet Russia;* Erlich; V. P. Danilov, "K itogam izucheniia istorii sovetskogo krest'ianstva i kolkhoznogo stroitel'stva v SSSR," *Voprosy istorii,*

1960, No. 8, pp. 34–64; Moshe Lewin; and a number of articles in *Soviet Studies,* which cite the Soviet literature of the mid-1960s.

58. See the founding statute, cited in note 52. During 1929 *Sel'skokhoziaistvennaia gazeta* was full of articles describing the creation and expansion of institutes and experiment stations. For the detailed, extremely ambitious plan of the specialists, see S. K. Chaianov et al.

59. *Agronom,* 1927, No. 11, p. 84.

60. See his speech in KPSS, *15-yi s'ezd,* II, 1353–1363.

61. For the economists' forecast, see N. P. Oganovskii, "Perspektivnyi plan rekonstruktsii sel'skogo khoziastva na piatiletie 1927/28–1931/32 gg.," *Planovoe khoziaistvo,* 1928, No. 1, pp. 33–56. For Iakovlev's response see *Ekonomicheskaia zhizn',* July 3, 1928.

62. For Kol'tsov's articles on this theme, see *Sel'skokhoziaistvennaia gazeta,* 1929, April 16, May 14 and 22. See Serebrovskii's article, *ibid.,* Oct. 27, 1929, and see especially *Materialy k vsesoiuznoi konferentsii.* For Iakovlev's endorsement of Serebrovskii's work, see *SRSKh,* 1931, No. 11, p. 11.

63. Serebrovskii became a candidate-member in 1930; his promotion to full membership seems to have been stopped by the Lysenkoite attack of the mid-1930s. See VASKhNIL, *Kalendar'* (1937). Tulaikov became a member in 1930. VASKhNIL, *Kalendar'* (1936).

64. KPSS, *16-aia konferentsiia,* p. 343.

65. A handwritten letter of January 1, 1930, on the stationery of a Cleveland hotel. Professor Dobzhansky, who kindly allowed me to examine his correspondence, was then in California. He recalls that Vavilov expressed such sentiments in private conversation as well as letters — until 1932, when Vavilov cryptically advised Dobzhansky not to return. This final conversation occurred at the International Congress of Genetics in Ithaca, after which Vavilov went back to the Soviet Union, never to leave again.

66. See *Trudy vsesoiuznogo s'ezda po genetike.*

67. Gorbunov was the *upravdel* or chief secretary of the central government's Council of People's Commissars with special responsibility for problems of science and technology. Such agencies as *Glavnauki* and *Glavprofobr* seem to have been subject to his supervision. In the 1930s ultimate supervision was vested in a newly created section of the Party's Central Committee.

68. See *Trudy vsesoiuznogo s'ezda po genetike,* I, 121. For Kirov's speech, see pp. 117–118; for Gorbunov's pp. 121–22.

69. *Ibid.,* pp. 136–7, cf. N. I. Vavilov et al., "Opytnoe delo i rekonstruktsiia sel'skogo khoziaistva," *Sel'skokhoziaistvennaia gazeta,* April 24, 1929. See *ibid.,* July 11, 1929, for the announcement of additional appropriations, including two million rubles for *nauchnye komandirovki.* For the background of these financial difficulties, see S. K. Chaianov, pp. 6–7.

70. See Joravsky, Chapter 16.

71. *Timiriazevka,* 1929, Nos. 23 (44) and 24 (45). At this time, May–June, 1929, Doiarenko was merely dismissed. See the issue of 1930, No. 29–30 (75–76), for the news that he had been convicted of "wrecking."

72. *Ibid.,* 1929, No. 1 (47), for the number of those dismissed, as of Sept. 19, 1929. See 1930, No. 29–30 (75–76) for the five "wreckers":

Doiarenko, Kondrat'ev, Chaianov, Makarov, and Rybnikov. All but Doiarenko were agricultural economists.

73. Stalin, XII, 141–172. Cf. Joravsky, Chapters 16–17.

74. Stalin, IV, 261.

75. See especially V. V. Matiukhin, "K rekonstruktsii nauchnogo i opytnogo dela v sel'skom khoziaistve," *Puti sel'skogo khoziaistva,* 1929, No. 12 (54), pp. 26–39. Cf. *Agronom,* 1929, No. 11–12, pp. 96–109.

3. HARMLESS CRANKS

1. S. V. Pokrovskii, "O michurinskom pitomnike," *Narodnyi uchitel',* 1933, No. 5, p. 102.

2. There are many biographies of Michurin. See, e.g., A. N. Bakharev, "I. V. Michurin, biograficheskii ocherk," in I. V. Michurin, *Sochineniia* (2nd ed.), I, 3–108; and I. T. Vasil'chenko. These accounts do not agree, on certain details of Michurin's early life, with Lebedev, a biographical novel based on conversations with Michurin as well as documentary material. Though Lebedev's tale is obviously overdrawn, on such matters as Michurin's mental health Lebedev seems closer to original sources than the recent biographers.

3. See, e.g., the spirit expressed in Michurin, *Sochineniia* (2nd ed.), I, 173–74. For Michurin's explanations of his motives in taking up plant breeding, see IV, 3–4, 9–12.

4. See Michurin, *ibid.*, I, 124–27. This article was originally published in 1905.

5. For the mixing of pollen, see *ibid.*, pp. 122–24. For his methods as a whole, see the essays in this same volume I.

6. That Michurin's methods were derived from the ancient folklore of gardening is made clear by A. I. Luss, "Vzaimootnoshenie podvoia i privoia," in N. I. Vavilov, ed., *Teoreticheskie osnovy,* I, 689–752.

7. Michurin, *Sochineniia* (2nd ed.), I, 463.

8. *Ibid.*, p. 496. Cf. also p. 550, III, 452, and IV, 136.

9. *Ibid.*, I, 304.

10. Quoted in A. A. Zvorykin, II, 43. Cf. Michurin, *Sochineniia* (2nd ed.), IV, 488–90.

11. *Ibid.*, p. 5.

12. See *Istoricheskii arkhiv,* 1955, No. 4, p. 132, for Prezent's very brief summary of this document. See *ibid.*, p. 99, for Michurin, in his communication of June 12, 1908, summarizing the one of November 15, 1905. The quotations above are taken from the communication of 1908. Nationalist sentiments were a recurrent theme in Michurin's prerevolutionary writings. See, e.g., Michurin, *Sochineniia* (2nd ed.), I, 285, 293.

13. These points are made in Michurin's communication of June 12, 1908. See *Istoricheskii arkhiv,* 1955, No. 4, pp. 99–104.

14. *Ibid.*, pp. 98–99.

15. *Ibid.*, pp. 99–104.

16. See K. I. Pangalo, "Selektsiia, ee razvitie i znachenie v narodnom khoziaistve," *Priroda,* 1931, No. 1, pp. 40–42; and N. M. Tulaikov, "Glavnye etapy istorii sel'skokhoziaistvennogo opytnogo dela v SSSR," *Agronom,* 1926, No. 6, pp. 35–43.

17. See Michurin, *Sochineniia* (2nd ed.), IV, 466–67, 470, 477–78, 483–89. See also *Istoricheskii arkhiv,* 1955, No. 4, pp. 105–106.

18. Michurin, *Sochineniia* (2nd ed.), IV, 3–8.
19. *Ibid.*
20. See *Letters of Frank N. Meyer,* National Archives, Washington, D.C. From the letter of Dec. 6, 1911 it is evident that Meyer had then learned of Michurin for the first time. There is an outside possibility that Michurin had been visited earlier by someone else from the United States. He complained to Meyer that there had been an earlier visitor who had taken plant stock of great value and given no credit to the originator, Michurin. For Meyer's accounts of his two visits with Michurin, see the letters of Dec. 29, 1911 and Jan. 31, 1913. For Michurin's exchange with David Fairchild, see Michurin, *Sochineniia* (1st ed.), I, 37, and IV, 235. (These letters are omitted in the second edition.) See Meyer's letter to Fairchild, Sept. 24, 1913, for evidence that the effort to buy cuttings from Michurin's nursery had been dropped.
21. For evidence that these tales came from Michurin, see his *Sochineniia* (2nd ed.), I, 429 and IV, 539 and 491. Note also that the most exaggerated tales were based on interviews with Michurin. For example, the salary of $32,000 was reported by Shmerling, p. 56. Keller, p. 113, tells how he got the story from Michurin, this time with a salary of $8,000. In the post-Stalin period the legend of the American invitation has been expiring, but it is still not quite dead. See, for example, *I. V. Michurin v vospominaniiakh,* p. 154, for another acquaintance of Michurin's repeating the legend and explicitly attributing it to the old man himself. (Once again the salary is $8,000.) I am grateful to David Comey for calling my attention to this last book, and for lending it to me.
22. USDA, *Yearbook* (1913), p. 12.
23. See D. Kashkarov, "Eshche o deiatel'nosti L. Burbanka," *Progressivnoe sadovostvo i ogorodnichestvo,* 1912, No. 26, p. 747.
24. *Ibid.* For Meyer's words at first hand, see citations in note 20. On Burbank see again Howard.
25. Michurin, *Sochineniia* (2nd ed.), IV, 5. As given by Michurin at various times and repeated by his disciples, this estimate has varied greatly, from a low near one hundred to a high of "several hundred." The most disturbing feature of the estimates is that no effort is usually made to distinguish between hybrids of proven commercial value and those of experimental interest only.
26. See Pashkevich. Cf. the angry response of a Michurinist in *Sotsialisticheskoe plodo-ovoshchnoe khoziaistvo,* 1932, No. 5, p. 47, which reminds the committee that prepared this book that the Party Control Commission had issued a decree on Michurin's behalf. See No. 6, p. 10 for a letter of apology from the responsible specialist. As a result of such pressure eight of Michurin's varieties were certified by a conference of fruit specialists in Dec., 1931.
27. See below, p. 73.
28. For the agrarian revolution in Tambov, see references in Carr, *Bolshevik Revolution,* II, 38, 170.
29. See Michurin's report of June 24, 1918, in *Istoricheskii arkhiv,* 1955, No. 4, pp. 109–110. At the end of 1922 all but the household plot was still in a nearly wild condition. See the report in *Istoricheskii arkhiv,* 1955, No. 4, p. 121.
30. *Ibid.*, pp. 110–113.

31. See Bakharev, *Selektsionno-geneticheskaia stantsiia,* pp. 40–42; and *Oktiabr',* 1930, No. 7, p. 182; and Michurin, *Sochineniia* (2nd ed.), IV, 507–8.

32. Quoted in *I. V. Michurin v vospominaniiakh,* p. 70. I have mistranslated *verba* as pussy willow.

33. See Vavilov's account in Michurin, *Itogi* (1924), p. 4, and his somewhat different version in *Novyi mir,* 1934, No. 11, p. 142. In either version Vavilov slides past critical issues with tactful silences and bland ambiguities.

34. See the fictional version in Lebedev, pp. 263–5. (Lebedev used archival materials that have not been published.) Already in 1912 a generally favorable appraisal of Michurin's work noted as a major drawback the lack of widespread trials of his varieties. See A. N. Chelintsev, "Kratkie svedeniia iz istorii rabot po plodovomu sortovodstvu," in Löbner (or Lebner), *Sortovodstvo,* p. 154.

35. Only bits and hints of the investigations and decisions in these first years of the Soviet regime have been published. For Michurin's allusions, see his *Sochineniia* (2nd ed.), I, 429–30, and IV, 594.

36. See *ibid.,* IV, 504–5.

37. In 1934 Kalinin recalled 1919 as the year of his visit. See *Istoricheskii arkhiv,* 1955, No. 4, p. 129. Bakharev, the local journalist who became Michurin's press agent, is more reliable in his memory of September 1922 as the time of the President's visit. "From that time," Bakharev wrote in 1933, "Michurin's nursery has enjoyed special attention from the government." See Bakharev, *Selektsionno-geneticheskaia stantsiia,* p. 18.

38. *Sel'skokhoziaistvennaia zhizn',* 1922, No. 5, p. 17. Cf. No. 6, p. 7, for a story of Michurin, based on an interview with him, trustfully repeating his inflated version of his achievements. This journal was an official organ of the Commissariat of Agriculture.

39. For the telegram see *Istoricheskii arkhiv,* 1955, No. 4, p. 118. Cf. Michurin, *Sochineniia* (2nd ed.), IV, 508–13, for Michurin's part in another round of investigations and conflicting reports, which seems to have been started by Gorbunov's wire.

40. See *Istoricheskii arkhiv,* 1955, No. 4, p. 122.

41. Maiakovskii, *Polnoe sobranie,* II, 298–300. I have mistranslated *ukrop* as parsley.

42. For evidence on Kalinin, see above, p. 26. See also note 37, and *Istoricheskii arkhiv,* 1955, No. 4, p. 125, for Kalinin early in 1923 giving his approval to the illegal steps that a local official had taken to increase support for Michurin.

L. S. Sosnovskii was an early chief of Agitprop, the founder and for six years the chief editor of *Bednota,* and, most important for present purposes, the originator of "militant feature articles in the Party press on such themes as stockbreeding, the cultivation of root crops, new varieties of wheat, or a new crop — *kenaf.*" The quotation is from his autobiography in *Deiateli,* XLI, pt. 3, p. 107.

43. See *Pravda,* September 26, 1923, for the brief mention of Michurin. Cf. B. A. Keller, "Blestiashchie dostizheniia russkoi nauki," *Pravda,* September 11, 1923, which does not mention Michurin at all.

44. Quoted by Gorshkov, in *I. V. Michurin v vospominaniiakh,* p. 22.

45. Michurin's letters to Gorshkov, heavily censored, are in Michurin, *Sochineniia* (2nd ed.), IV, 517–22.

46. A. Bragin, "Kozlov ili Vashington?" *Izvestiia,* October 14, 1923.
47. See the photograph of the watch in *I. V. Michurin v vospominaniiakh,* p. 22.
48. See *Izvestiia,* November 24, 1923, and *Istoricheskii arkhiv,* 1955, No. 4, pp. 125–26.
49. See *Pravda,* October 25, 1925, for the highpoint of the celebration of Michurin in the 1920s.
50. *Ibid.*
51. See Michurin, *Sochineniia* (2nd ed.), IV, p. 22.
52. Michurin, *Sochineniia* (1st ed.), III, 309.
53. Michurin, *Sochineniia* (2nd ed.), I, 459.
54. V. V. Pashkevich, "Russkii originator-plodovod I. V. Michurin," in Michurin, *Itogi* (1924), pp. 16–61. In calling this little collection the first edition of Michurin's "works" I am following the convention of Michurin's disciples, who disdain to count the still smaller pamphlet, Michurin, *Vyvedenie iz semian* (1921). That had no notes or commentary.
55. Michurin, *Sochineniia* (1st ed.), IV, 264.
56. By 1929 Michurin's press agent was calling the publisher's delay "a flagrant crime." Quoted in *Oktiabr',* 1930, No. 7, p. 186, from the Kozlov (Michurinsk) newspaper, which I was unable to consult.
57. N. I. Vavilov, "O mezhdurodovykh gibridakh dyn', arbuzov i tykv," *TPBGS,* XIV (1924), No. 2, pp. 3–35.
58. P. N. Shteinberg, "O 'chudesakh' i 'charodeiakh' v sel'skom khoziaistve," *Vestnik znaniia,* 1926, No. 117, pp. 1129–1138.
59. E. A. Aleshin, "I. V. Michurin i nauka," *Puti sel'skogo khoziaistva,* 1927, No. 6–7, pp. 118–125. Cf. the analogous comment on Michurin in Zhegalov, pp. 476–480. For Michurin's hostile attitude toward Zhegalov, as expressed in his notebook, see Michurin, *Sochineniia* (2nd ed.), IV, 422, 517.
60. On Shull and Burbank, see again W. L. Howard, *Luther Burbank.* For the record of Aleshin's brief stay at Michurinsk, see his biography in Lipshits, I. Cf. the attacks on Aleshin in A. Sakhaltuev, "O razvitii i znachenii rabot Michurina," *PZM,* 1932, No. 9–10, p. 237, and in *Sotsialisticheskoe plodo-ovoshchnoe khoziaistvo,* 1932, No. 6, pp. 41–5.
61. Thus P. N. Iakovlev reminisces, in *I. V. Michurin v vospominaniiakh,* p. 106. Cf. Vavilov's comment, *SRSKh,* 1936, No. 12, p. 44. Cf. also the heavily censored letters from Michurin to Iakovlev, in Michurin, *Sochineniia* (2nd ed.), IV, 592 ff., and the fictional version of the correspondence in Lebedev, pp. 304–5. Lebedev obviously saw the letters or at least heard about them from one or both of the correspondents.
62. Otto Renner, "Vererbung bei Artbastarden," *Zeitschrift für induktive Abstammungs- und Vererbungslehre,* 33:317–347 (1924). B. A. Keller, the well-known botanist, was the first to cite Renner on Michurin's behalf, and Michurin's press agent was quick to point with pride. See A. N. Bakharev, "I. V. Michurin," *Kraevedenie,* 1926, No. 2, p. 266. Bakharev was still pointing with pride in 1933. See his *Selektsionno-geneticheskaia stantsiia,* p. 31.
63. I. S. Gorshkov, "Iz rabot I. V. Michurina," *Sad i ogorod,* 1927, No. 12, p. 6.
64. Quoted by Gorshkov in *I. V. Michurin v vospominaniiakh,* p. 23. One of the two scientists was A. I. Luss, who was repeatedly attacked by the Michurinists in the 1930s. He died in the defense of Leningrad

against the Germans. The other scientist was Lev Iosifovich Reibort, who seems to have fallen victim to the terror in the 1930s.

65. I. S. Gorshkov, "K iubileiu I. V. Michurina," *Agronom,* 1925, No. 10, p. 21.

66. Gorshkov, pp. 24–6.

67. See, for example, D. D. Romashov, "O metodakh raboty I. V. Michurina," *Zhurnal obshchei biologii,* 1940, No. 2, pp. 177–204; most recently, Alikhanian; and Dubinin.

68. Gorshkov, p. 22.

69. *Trudy vsesoiuznogo s'ezda po genetike,* I, pp. 13, 114, 125.

70. *Ibid.,* pp. 121–2.

71. *Ibid.,* III, pp. 189–99.

72. *Ibid.,* I, p. 129.

73. Feiginson, *Chego dobilis',* p. 84. Cf. Feiginson and Terent'ev, *Komsomol'tsy labkory,* p. 7.

74. See works cited in note 73. Cf. M. M. Bushuev, "Opytnoe sel'skokhoziaistvennoe delo po raionam SSSR i zadachi ego v rekonstruktsii krest'ianskogo khoziaistva," *Planovoe khoziaistvo,* 1928, No. 9, pp. 275–7.

75. T. I. Shpil'man (or Spielmann), "O novykh putiakh povysheniia urozhainosti," *Pravda,* Jan. 22, 1926.

76. Chizhevskii, *Fizicheskie faktory.* See review in *PZM,* 1924, No. 8–9, pp. 314–15.

77. Chizhevskii's big boost did not come until 1931. See *SZ,* 1931, Oct. 11 and 29.

78. See below, p. 80.

79. N. A. Maksimov, "Stimuliatsiia semennogo materiala, kak sredstvo povysheniia urozhaia," *TPBGS,* XIV (1925), No. 5, pp. 115–131. See also B. Ia. Kurbatov and S. A. Glikman, "Materialy k voprosu o stimuliatsii semian," *ibid.,* XXIII (1930), No. 2, pp. 155–298.

80. Feiginson, *Chego dobilis',* pp. 97–100.

81. D. D. Artsybashev, "Otchet po rabotam Tul'skoi akklimatizatsionnoi stantsii za 1923–24 g.," *TPBGS,* XIV (1925), No. 4, pp. 31 ff.

82. G. K. Meister, "Perspektivy selektsii ozimoi pshenitsy," *ibid., prilozhenie* 34 (1929), pp. 299–303. Cf. Meister, "Problema mezhvidovoi gibridizatsii v osveshchenii sovremennogo eksperimental'nogo metoda," *Zhurnal opytnoi agronomii iugo-vostoka,* 1927, No. 1, pp. 3–86.

83. Quoted in Feiginson, *Chego dobilis',* pp. 98–99.

84. Dunin, *Stimulirovanie,* p. 3.

85. Akademiia Nauk SSSR, Institut Etnografii, *Trudy, novaia seriia,* XLI, p. 128.

86. See, e.g., Dunin, *Stimulirovanie,* pp. 10 ff. and Feiginson and Terent'ev *Komsomol'tsy labkory.*

87. Vit. Fedorovich, "Polia zimoi," *Pravda,* August 7, 1927. I wish to thank Dr. Zh. A. Medvedev for calling this article to my attention.

88. *Ibid.*

89. *Ibid.*

90. Lysenko, *Vliianie termicheskogo faktora.*

91. N. A. Maksimov, "Fiziologicheskie faktory opredeliaiushchie dlinu vegetatsionnogo perioda," *TPBGS,* XX (1929), pp. 169–212. See also A. L. Shatskii, "K voprosu o summe temperatur kak sel'skokhoziaistvenno-klimaticheskom summe," *Trudy po sel'skokhoziaistvennoi meteorologii,* XXI (1930), no. 6, pp. 259–267.

92. Lysenko and Dolgushin, "K voprosu o sushchnosti ozimi," *Trudy Vsesoiuznogo s'ezda,* III, 189–99; and Lysenko, "V chem sushchnost' gipotezy 'ozimosti' rastenii," *Sel'skokhoziaistvennaia gazeta,* Dec. 7, 1929.

93. Maksimov, "Fiziologicheskie sposoby regulirovaniia dliny vegetatsionnogo perioda," *Trudy vsesoiuznogo s'ezda,* III, pp. 3–20, an abridged version of Maksimov's speech, does not contain these actual words, which are quoted from Maksimov's article in *Sel'skokhoziaistvennaia gazeta,* 1929, Nov. 19.

94. "Otkrytie agronoma Lysenko," *Ekonomicheskaia zhizn',* Aug. 4, 1929.

95. *TPBGS, prilozhenie* 34 (1929), pp. 338–9 and 380.

96. See, for example, *ibid.*, pp. 294–5 and 406–7.

97. Ia. A. Iakovlev, "Za udvoenie urozhaia!" *Ekonomicheskaia zhizn',* July 3, 1928. He implicitly endorses the accusation more in sorrow than in anger.

98. *TPBGS, prilozhenie* 34 (1929), p. 274.

99. Vavilov gives this figure in his *Genetika na sluzhbe,* p. 35.

100. See *Sel'skokhoziaistvennaia gazeta,* Oct. 10, Nov. 13, 19, and Dec. 7, 1929.

101. *Ibid.*, Nov. 19 for Maksimov's comment.

102. See above, pp. 36–7.

4. RAISING STALIN'S HAND

1. See above, p. 35.

2. *KPSS v rezoliutsiiakh,* II, p. 696. For an economist justifying this boast without falling into falsehood, see Ia. P. Nikulikhin, "Marks o ratsional'nom zemledelii," *Na agrarnom fronte,* 1933, No. 1, p. 127.

3. *Comparisons,* I, 211. Cf. *Sel'skoe khoziaistvo SSSR* (1960), p. 196, which implies a decline of 5 percent by the mid-1930s. The American economists use the average yield for the period 1925–1929 as their base. The Soviet handbook uses the period 1928–1932, thus running together two years before and two years after mass collectivization had begun. For one of the last reminders that the original target had been a 35 percent increase, see Cherdantsev, p. 91.

4. See below, Chapter 5, "Terror."

5. Cited in *Agrotekhpropaganda,* 1933, No. 1, p. 42.

6. See the convenient summary by Williams' disciple, A. Ia. Bush, in *Akademik V. R. Vil'iams,* especially pp. 129–130.

7. See Bogdenko, pp. 22–33. Bogdenko omits the names of Tulaikov and Williams, but his references make the facts clear.

8. Stalin, XI, pp. 190–2. For the decree of the Central Committee, see *KPSS v rezoliutsiiakh,* II, pp. 397–8.

9. The figures are in Bogdenko, pp. 187 *et passim.*

10. Stalin, XII, pp. 129–130. Repeated on several occasions, this view of *rentabel' nost'* was turned into a dictionary definition. See Ushakov, IV.

11. Bogdenko, p. 247.

12. See, e.g., N. M. Tulaikov, "Agrotekhnika i sevooboroty v bor'be s zasukhoi," *SZ,* 1931, 23 Oct.

13. See I. I. Prezent, "Protiv vredneishei 'filosofii' agronomii," *PZM,* 1934, No. 3, pp. 198–202, for a diatribe against specialists who dared to suggest that socialist farmers might upset the balance of nature.

14. Williams could not attend because of illness. The chief speakers

for his cause were Bush and Bushinskii. See especially A. Ia. Bush, in SZ, 1931, 24 Oct. The record of the conference is contained in Vsesoiuznaia konferentsiia po bor'be s zasukhoi, *Biulleten'*, Nos. 1–7, and *Tezisy i materialy*, Nos. 1–3. For Williams' written communication, see *ibid.*, No. 1.

15. See the resolution in *Ibid., Proekt postanovlenii*, p. 4, point 5. For a summary of the conference decisions, see SZ, 1931, 2 Nov.

16. Molotov's speech is in SZ, 1931, 4 Nov.

17. Tulaikov, "Bor'ba s zasukhoi," in Bondarenko, p. 149.

18. See *Izvestiia*, 1933, 23 March and 3 June. For Tulaikov's original suggestion of shallow plowing, see his "K voprosu ob osnovnoi vspashke pochvy," SZ, 1932, 4 May. The open-minded reader will note that Tulaikov offered this suggestion very tentatively.

19. Of course, no one in the 1930s recalled Kalinin's suggestion of simplified agronomy. For the record, see above, p. 29.

20. For Samarin's condemnation, see M. Strukov, "Protiv 'uproshcheniia' agrotekhniki," *Na agrarnom fronte*, 1933, No. 2, pp. 119–121. Tulaikov published repeated apologies, beginning with "Protiv vrednoi teorii preimushchestv melkoi pakhoty," *Izvestiia*, 1933, 3 June. See the brief account in his *Izbrannye*, I, 26–27.

21. Stalin, XIII, 329–332. Cf. Robert G. Jensen, "The Soviet Concept of Agricultural Regionalization and Its Development," in Karcz, *Soviet and East European Agriculture*, pp. 77–103.

22. The first official signs of strong favor for Williams' scheme of crop rotation and land use were a decree of 14 July 1934, "Concerning the Experimental Verification of the *travopol'noi* System of Agriculture," and a decree of 11 Nov. 1934 ordering a celebration of Williams' forthcoming anniversary. Both were issued by the Council of People's Commissars. For the former, see *Sobranie postanovlenii pravitel'stva SSSR*, 1962, No. 3, p. 47. For the latter, see *Akademik V. R. Vil'iams*, p. 5. For examples of the use of Tulaikov as a scapegoat, see V. N. Stoletov, "O vrednykh teoriiakh v agrotekhnike," *SRSKh*, 1933, No. 4–5, pp. 3–25, and, by the same author, "Protiv chuzhdykh teorii v agronomii," *Pravda*, 1937, April 11. The latter article was almost a call for Tulaikov's arrest, which did in fact occur soon after. For more on the problem of crop rotation, see Chapter 9, "Land."

23. See Joravsky, *Soviet Marxism and Natural Science*.

24. Vsesoiuznaia konferentsiia po planirovaniiu genetiko-selektsionnykh issledovanii, *Materialy*, p. 5.

25. *Ibid.*

26. Imperial Bureau of Plant Genetics, *Bulletin*, No. 13 (Nov., 1933).

27. *Izvestiia*, 1930, 21 July.

28. See Vsesoiuznyi issledovatel'skii institut plodovogo i iagodnogo khoziaistva, *Trudy*, vypusk 1 (1931), p. 3.

29. *Sotsialisticheskoe plodo-ovoshchnoe khoziaistvo*, 1932, No. 3, p. 10.

30. V. G. Shmerling, "'Chelovek na griadke,'" *Oktiabr'*, 1930, No. 5–6, p. 224.

31. *Ibid.*, pp. 224–5.

32. *Ibid.*, No. 7, p. 186.

33. *Ibid.* Shmerling expanded this two-part article into a book. See Bibliography.

34. See the record of the conference, published as the first number of the Institute's *Proceedings*, as cited in note 28.

35. Pashkevich, ed., *K standartizatsiiu.*

36. The decree was published in several places. See, e.g., *Sadovodstvo i ogorodnichestvo,* 1931, No. 7–8, p. 6.

37. See above, p. 52.

38. See V. A. Odintsov, "O rabote Nauchno-issledovatel'skogo instituta plodovodstva im. Michurina," *Sotsialisticheskoe plodo-ovoshchnoe khoziaistvo,* 1933, No. 11, pp. 8–13.

39. See the biographical information in Luss, pp. 5–6.

40. *Sotsialisticheskoe plodo-ovoshchnoe khoziaistvo,* 1932, No. 3, p. 14.

41. V. L. Simirenko, "K itogam Vsesoiuznogo soveshchaniia po standartizatsii assortimentov plodovykh i iagodnykh rastenii," *ibid.,* pp. 9–14.

42. Odintsov, pp. 7–8.

43. Bakharev, *Selektsionno-geneticheskaia stantsiia,* p. 31. Bakharev repeated this line in his *I. V. Michurin* (1934), pp. 25, 38. Cf. also I. S. Gorshkov, "O nauchnykh dostizheniiakh I. V. Michurina," in *Trudy sessii VASKhNILa* (1933), pp. 196–211.

44. See, e.g., P. N. Iakovlev, "Rekonstruktor flory," *Priroda,* 1934, No. 9, p. 52. See also Lebedev. This journalist's biography of Michurin is obviously based on a good deal of conversation with the old man and on examination of his papers. By the time it appeared the Lysenkoites had assumed control of Michurinism, and fictional names were given to several key figures. It is clear that "Academician Veshchilov" was actually Vavilov. See pp. 295–7 for Lebedev's version of the popular story of Vavilov in America getting the Golden Delicious apple for Michurin. As Michurin told the story to reporters, American nurserymen refused to sell any cuttings to Vavilov, so he sent apples to Michurin. Michurin got the last laugh on the selfish Americans by making the appleseed breed true. American agricultural specialists were quite annoyed by this obvious invention. They had been trying hard for good relations with their Soviet colleagues. See File 10213 of the USDA, National Archives, Washington, D.C.

45. *Izvestiia,* 1933, 12 March.

46. Ia. P. Nikulikhin, "Burzhuaznye teorii i vreditel'stvo v osvoenii novoi tekhniki zemledeliia," *Na agrarnom fronte,* 1933, No. 2, pp. 29 ff.

47. Ia. I. Potapenko, "Nauka o plodovodstve," *Sotsialisticheskoe plodo-ovoshchnoe khoziaistvo,* 1933, No. 11, p. 8.

48. *Ibid.,* p. 9.

49. To be precise, he was the scientific director of the affiliated All-Union Fruit and Berry Institute in the village of Mleev. The director of the Kiev Institute was T. I. Tyl'nyi, who disappeared from the public record, but was not explicitly called a "wrecker," as Simirenko was.

50. A. I. Luss and M. A. Rozanova were utterly uncompromising critics of Michurinism in the field of fruit breeding. Neither was arrested. A. P. Koval' and N. V. Kovalev were also leading fruit specialists who angered the Michurinists; both seem to have been arrested, but after Michurin's death. P. N. Shteinberg and V. V. Pashkevich were leading fruit specialists who criticized Michurin in the 1920s and early 1930s. Subsequently Pashkevich paid diplomatic tribute to Michurin. E. I. Aleshin was an eminent pomologist who argued in the 1920s that Michurin's data might have significance for science. (See above, p. 51.) In the early 1930s he was appointed head of the department of breeding at Michurin's Institute. The result was a bitter quarrel, as in the earlier, analogous case of Luss, who also tried to do a scientific check of Michurin's results. Aleshin left

the Institute in 1933, and died a natural death in Leningrad in 1941.

51. See I. V. Belokhonov, "Do kontsa likvidirovat' posledstviia vreditel'stva i uproshchenchestva v plodovodstve," *Za michurinskoe plodovodstvo*, 1938, No. 1, p. 10, for news of Odintsov's arrest, along with names of other "wreckers" among fruit specialists. By this time the Michurinist cause had been absorbed into the Lysenkoite movement.

52. See below, Chapter 5, "Terror."

53. V. A. Lebedev, "Sad chudes," *Tridtsat' dnei*, 1934, No. 10, p. 63.

54. Quoted by P. N. Iakovlev, "Michurin i nauka," *SRSKh*, 1934, No. 11, p. 22, from *Nasha pravda* (Michurinsk), 1934, 20 Sept. The quotation serves as an introduction to this article by a major discipline of Michurin's, but the implicit Lysenkoism is not made explicit. P. N. Iakovlev, who was then a graduate student under Vavilov's supervision, became a crusader against genetics after Michurin's death. Cf. I. S. Gorshkov and P. N. Iakovlev, "Vozmozhnosti otkryvaemye dostizheniiami I. V. Michurina," *ibid.*, p. 17, for the usual tribute to Vavilov as a friend of Michurin's.

55. See G. D. Karpechenko, "Teoriia otdalennoi gibridizatsii," in Vavilov, *Teoreticheskie osnovy*, I, pp. 293–354, which implicitly establishes the extremely modest nature of Michurin's contribution. Michurin himself had expressed his annoyance at Karpechenko in 1932, shortly after the young man's prize-winning cross of radish and cabbage — achieved by induced ploypioidy — was celebrated in the popular press. Michurin, *Sochineniia* (2nd ed.), IV, 243–4.

56. On 21 Feb. 1933 Lysenko sent a letter to Michurin's chief of publicity, "with an evaluation of I. V. Michurin's doctrine of heredity in plants. . . ." That cryptic description is given in Bakharev, *Michurin* (1952), item 907. Obviously neither Bakharev nor Michurin replied, or else they replied in an unsuitable manner, for this brief entry in a catalogue of Michurin's archive is the only news of the episode that the Lysenkoites have given. Lysenko had no better luck on 21 April 1933, when he sent Michurin a copy of his *Bulletin* inscribed "To dear teacher Ivan Vladimirovich, from a student unknown to you, T. Lysenko." See the pathetic treatment of this episode in Prezent, *V sodruzhestve s prirodoi*, pp. 101–3. In 1934 Lysenko began to make public hints of a special affinity between his views and Michurin's, but he was restrained in such hints until Michurin's death set him free.

57. The genuine scientist was D. F. Petrov. According to Baranov and Lebedev, Michurin put him in charge of the department of cytogenetics on the recommendation of Vavilov. *Botanicheskii zhurnal*, 1955, No. 5, p. 754. For evidence of Vavilov and Michurin fostering the story of their friendship, see references cited in notes 44, 53, 54, 58, and 60.

58. N. I. Vavilov, "Prazdnik sovetskogo sadovodstva," *Novyi mir*, 1934, No. 11, pp. 140–143. The careful reader, who is aware of the facts, will find that Vavilov did not actually tell untruths about Michurin's service to science. But he certainly gave a misleading impression.

59. The motion is reproduced in Vasil'chenko, *I. V. Michurin* (1963), pp. 104–5. Vavilov was one of twelve sponsors. They were responding to a suggestion from Michurin's colleagues, whose letter appears in *ibid.*, pp. 103–4.

60. N. I. Vavilov, "Puti sovetskoi rastenievodcheskoi nauki," *SRSKh*, 1936, No. 12, pp. 43–44. For other efforts to prove that Michurin was a friend of genetics, see references in note 67 of Chapter 3.

61. For an inadvertent proof of this appraisal of Michurin's attitude toward genetics, see Prezent's effort to demonstrate that Michurin carried on a ceaseless battle with Mendelism. The careful reader of Prezent, *I. V. Michurin,* pp. 177–196, will note first, that the quotations from Michurin reveal his failure to understand genetics, and second, that the quotations are mostly drawn from letters and notes that Michurin never bothered to publish.

62. Quoted in Revenkova, p. 250.

63. For a representative sampling of such declarations, see Vavilov, *Izbrannye,* V. Volume I contains a fairly complete bibliography of his publications, which reached a peak in the early 1930s.

64. A. K. Kol', "Prikladnaia botanika ili leninskoe obnovlenie zemli," *Ekonomicheskaia zhizn',* 1931, 29 Jan.

65. N. I. Vavilov, "Rabota VIRa v oblasti introduktsii novykh rastenii," *ibid.,* 1931, 13 March.

66. Shteingardt, "Agrotekhnicheskie voprosy podgotovki k vesennemu sevu," *Bol'shevik,* 1932, No. 23–24, p. 62.

67. *Pravda,* 1931, March 15.

68. Stalin, XIII, 38–9.

69. T. Postolovskaia, "Osobennosti klassovoi bor'by na Ukraine v period mezhdu XVI i XVII parts'ezdami," *Pod markso-leninskim znamenem,* 1934, No. 1, p. 61. In the Ukraine the situation may have been worse than in most other parts of the Soviet Union.

70. In the early 1930s a great deal was published on the subject. For the official attitude as the craze reached its peak, see Shteingardt, as cited in note 66, pp. 57–59. For a convenient summary with a brief bibliography, see *Sel'skokhoziaistvennaia entsiklopediia* (1st ed.), IV, 75–77, and 79.

71. See the decree of Sovnarkom, "O seve iarovykh pod zimu," *Sobranie uzakonenii . . . RSFSR,* 1932, vypusk 18, No. 98, pp. 323–4. Cf. Kukol.

72. In 1926 Professor P. N. Shteinberg cited this proverb in a long list of peasant stupidities. See *Vestnik znaniia,* 1926, No. 1, p. 13.

73. Cited by Shteingardt, *Bol'shevik,* 1932, No. 23–24, p. 59.

74. Ia. A. Iakovlev, "O knige Anisimova, N. I., 'Agrotekhnika,'" *SRSKh,* 1935, No. 5, p. 124. The ban on superearly sowing was part of the annual decree "O gosudarstvennom plane vesennego seva," *Sobranie zakonov . . . SSSR,* 1936, No. 8, stat'ia 67, p. 121.

75. See, e.g., A. S. Kruzhilin, "Protiv 'teorii' sverkhrannego i podzimnego poseva," *Sotsialisticheskoe zernovoe khoziaistvo,* 1936, No. 2, pp. 147–149.

76. Chizhevskii was given space in *Pravda,* 1934, May 6, to tell of his achievements. *Pravda,* 1936, 3 June, found counterrevolution in his articles, and Kol'man, chief of science and technology for the Moscow Party Organization, called Chizhevskii "this fascist bastard [*svoloch'*], this crook." See Ernst Kol'man, "Chernosotennyi bred fashizma i nasha mediko-biologicheskaia nauka," PZM, 1936, No. 11, pp. 71–2.

Spielmann's biontization was endorsed by decrees of the Commissariat of Agriculture in 1930, and the Workers' and Peasants' Inspection in 1931; a laboratory was organized for him at the Lenin Academy of Agricultural Sciences. See VASKhNIL, Laboratoriia . . . , *Biontizatsiia semian.* In 1935 he published his magnum opus, *Teoriia i praktika biontizatsii,* which was blasted with merciless justice in *Vestnik sel'skokhoziaistvennoi literatury,* 1936, No. 1, pp. 46–7.

77. Prezent, *Teoriia Darvina,* p. 19.

78. V. G. Shmerling, "Nachalo," *Nashi dostizheniia,* 1934, No. 4, pp. 123–126. The hero's name was L. Ia. Novoselov.

79. Meister tried this cross in 1914. His subordinate, A. P. Shekhurdin, began to try it in 1922, and then passed the job to Verushkin. See Shekhurdin, *Izbrannye,* I, 18, for the editor's weasel-worded tribute to Tsitsin, who is said to have achieved what Shekhurdin could not. Cf. A. P. Shekhurdin and S. M. Verushkin, "Hybrids between Wheat and Couch Grass," *Journal of Heredity,* 24:328–335 (Sept. 1933). For Tsitsin's effort to prove himself superior to these predecessors, see his *Problema,* pp. 94–99.

80. See Tsitsin, *ibid.,* for an account of these events that simply omits mention of V. R. Berg. See V. Rudnyi, "Pshenitsa porodnilas' s pyreem," *Nashi dostizheniia,* 1934, No. 7–8, pp. 103–112, for a journalist's story, based on an interview with Tsitsin, which condemns Berg as a "wrecker." For Berg's expression of cautious hope that crosses of wheat and couch grass might prove practical, see his "Pshenichno-pyreinyi gibrid," *SZ,* 1932, 10 Jan.

81. Tsitsin, *Problema,* pp. 98–9. Cf. references in note 80.

82. See N. V. Tsitsin, "Problema ozimykh i mnogoletnikh pshenits budet reshena!" *Pravda,* 1934, 9 July.

83. *Pravda,* 1935, 30 Dec. Quoted again in an editorial, *Pravda,* 1936, 3 Jan., and repeated many times later on, especially by Tsitsin himself. Cf. *Pravda,* 1938, 23 Aug., for further colloquy between Tsitsin and Stalin.

84. See, e.g., S. M. Verushkin, "Vazhneishie napravleniia v rabote s pshenichno-pyreinymi gibridami," *PZM,* 1936, No. 7, pp. 144–158.

85. See D. V. Goriunov, "Ozimye pshenichno-pyreinye gibridy v proizvodstve," in Tsitsin, ed., *Otdalennaia* (1958), pp. 232–282, for the most extensive discussion of the practical results of Tsitsin's distant crosses. Goriunov claims that 140,000 hectares were sown to 3 certified varieties in 1955. Cf. Tsitsin, *Bol'shoi kolos* (1960), pp. 14–15, for a confession that these varieties are little different from ordinary wheat. Note too his omission of figures on acreage, except for a new hybrid, then undergoing production tests on 18,000 hectares, pp. 15–17. And note, on p. 22, his admission that perennial wheat is still an unsolved problem.

86. *Izvestiia,* 1935, 15 Feb.

87. T. D. Lysenko, "Iarovizatsiia v khoziaistvennykh usloviiakh," *SZ,* 1930, 2 July.

88. *Ibid.* In subsequent comments on the experiments of 1930 Lysenko was far less candid. He even compiled selected data to prove that they had been a great success. See his "Rezul'taty opytov 1930 g. s iarovizirovannymi posevami," *Biulleten' iarovizatsii,* 1932, No. 1, pp. 57–61. Cf. also his *Agrobiologiia* (1949), p. 22, for a statement that "hundreds of collective-farm experimenters were involved in 1930," with actual results reported for four hectares in four farms. To call such a man a liar is to misread his mind and character.

89. See *Biulleten' iarovizatsii,* 1932, No. 1, pp. 71–2.

90. Only brief references to this conflict have appeared in print. See, e.g., Tikhon Kholodnyi's biography of Lysenko in *Pravda,* 1938, 7 Dec., and in *Novyi mir,* 1938, No. 7, pp. 193–5.

91. The resolutions are collected in *Biulleten' iarovizatsii,* 1932, No. 1, pp. 73–80.

92. Reported in *SZ*, 1931, 29 Oct. The text of Iakovlev's speech may be found in Vsesoiuznaia konferentsiia po bor'be s zasukhoi, *Biulleten'*, No. 8 and No. 4, p. 16.

93. *Ibid.*

94. *Biulleten' iarovizatsii*, 1932, No. 1, p. 80.

95. N. A. Maksimov and M. A. Krotkina, "Issledovaniia nad posledeistviem ponizhennoi temperatury," *TPBGS*, XXIII (1930), vypusk 2, pp. 427–78. For an analysis of the scientific issues, see below, Chapter 7, "Plant Physiology."

96. Ia. A. Iakovlev, p. 161. The speech was originally given in 1931.

97. Winterization and superearly sowing exposed seed to the possibility of total destruction. If vernalization was limited to a quick moistening, as it seems to have been in many cases, the seed was not greatly endangered.

98. Between November 1932 and February 1934 Lysenko had only one article in a major newspaper ("Ovladet' tekhnikoi iarovizatsii," *SZ*, 1933, 27 March). Cf. A. A. Savchenko-Bel'skii, "Massovyi opyt vooruzhit' teoriei," *SZ*, 1933, 24 January, a piece on experimental campaigns by one of Lysenko's chief journalistic promoters. He put greatest stress on superearly sowing, which he pictured as vernalizing grain naturally. Cf. Ia. A. Iakovlev, pp. 164–8, a 1933 republication of his remarks at the 1931 Conference on Drought Control, omitting the very strong endorsement of vernalization quoted above. 1933 is also noteworthy for the considerable number of published criticisms of Lysenko's theorizing. See, e.g., V. N. Liubimenko, "K teorii iskusstvennogo regulirovaniia dliny vegetatsionnogo perioda," *Sovetskaia botanika*, 1933, No. 6, pp. 3–30; E. V. Lebedintseva, "Znachenie dliny dnia dlia vykolashivaniia ozimykh," *TPBGS*, Series III, 1933, No. 3 (5), pp. 141–154; Chailakhian, articles in *Sovetskaia botanika*, 1933, Nos. 5, 6, and in *Doklady AN SSSR*, 1933, No. 5 (n.s.).

99. For the 1933 goal of superearly sowing, see above, note 73. For the 1931 success of winterization, see note 71. For the 1933 acreage under vernalized seed, see Lysenko, "Iarovizatsiia," *Sel'skokhoziaistvennaia entsiklopediia* (1st ed.), I (1934), column 1046.

100. See below, Chapters 6 and 9.

101. See the brief account by Ol'shanskii, in VASKhNIL, *Trudy*, vypusk 43, pp. 34–6.

102. Lysenko and Baskova, *Iarovizatsiia.*

103. Two pamphlets, published in 1935, were as far as Lysenko ever got in carrying out his repeated promises to complete his theory of stages. For a convenient republication, see his *Agrobiologiia* (1949), pp. 3–117. For analysis of the scientific issues, see below, Chapter 7, "Plant Physiology."

104. See, most notably, P. N. Konstantinov *et al.*, "Neskol'ko slov o rabotakh Odesskogo instituta selektsii i genetiki," *SRSKh*, 1936, No. 11, pp. 121–130.

105. Compare successive editions of N. A. Maksimov, *Kratkii kurs*, from the 1st in 1927 through the 4th in 1932. In the 4th edition see the purely Lysenkoite passage on vernalization, pp. 272–3. In later editions more Lysenkoite material was attached.

106. A. S. Sereiskii, "Problema individual'nogo razvitiia v osveshchenii sovremennoi fiziologii rastenii," *Za marksistsko-leninskoe estestvoznanie*, 1932, No. 5–6, p. 28. Sereiskii was originally a student of the distinguished plant physiologist, N. G. Kholodnyi.

107. "O selektsii i semenovodstve," *Pravda,* 1931, 3 August.
108. *Ibid.*
109. See, e.g., V. E. Pisarev, N. V. Kovalev, and N. A. Maksimov, "Za industrializatsiiu raboty sel'skokhoziaistvennykh nauchno-issledovatel'skikh institutov," *Izvestiia,* 1931, 30 Sept.; and Pisarev's articles in *Semenovodstvo,* 1931, Nos. 21, 23.
110. I. Sip, "Montazh," *Nashi dostizheniia,* 1934, No. 7–8, p. 119. Cf. V. G. Shmerling, "Laboratoriia v million gektarov," *ibid.,* 1934, No. 4, p. 122. Lysenko's first public statement was somewhat less threatening to genetics. See his "Fiziologiia razvitiia rastenii v selektsionnom dele," *Semenovodstvo,* 1934, No. 2, pp. 20–31.
111. See the items cited in note 110, especially the article by Sip. Toward the end of 1934 Lysenko and Prezent sent to press their scientific statement of these themes: *Selektsiia i teoriia stadiinogo razvitiia rastenii.* It is conveniently reprinted in Lysenko's *Agrobiologiia,* all editions, various languages.
112. *Iarovizatsiia,* 1935, No. 1.
113. The episode, which took place at the end of 1932 or beginning of 1933, is briefly recalled in *Riadom s Vavilovym,* p. 119.
114. VASKhNIL, *Trudy,* vypusk 43, p. 101. Cf. *Izvestiia,* 1934, 24 May and 20 July.
115. See, e.g., N. I. Vavilov, "Sovetskoe nauchnoe rastenievodstvo za period sotsialisticheskoi rekonstruktsii," in Bondarenko, ed., *Sel'skokhoziaistvennaia nauka* (1934), pp. 1–20. This may well be a version of the report that Vavilov gave to SNK in May, 1934.
116. V. Smirnov, "Na sessii sel'skokhoziaistvennoi akademii nauk," *Partrabotnik,* 1932, No. 21–22, p. 40.
117. "O rabote VASKhNIL," *Sobranie zakonov. . . . SSSR,* 1934, No. 37, stat'ia 296.
118. See the report in *Front nauki i tekhniki,* 1935, No. 7, pp. 76–9. Note that Ia. A. Iakovlev himself came to the first meeting of the reorganized Academy to criticize and exhort.
119. A. I. Muralov, "Ne otstavat' ot zhizni," *Izvestiia,* 1935, 29 June.
120. The former director was the eminent breeder of wheat, A. A. Sapegin. Cf. references in note 90.
121. See above, p. 84.
122. See Postyshev's speeches to leaders of "hut labs," in *Izvestiia,* 1934, 10 June, and 1935, 15 March. Lysenko repeatedly thanked Postyshev for help. See especially *Soveshchanie peredovikov urozhainosti,* p. 181, for public gratitude expressed in Postyshev's presence. It should be noted, however, that Postyshev's praise for vernalization in the speech of June, 1934 was quite perfunctory. On band wagons questions of sincerity are usually pointless.
123. T. D. Lysenko, "Iarovizatsiia kartofelia na iuge," *SZ,* 1935, 26 March.
124. See below, Chapter 9, "Potatoes."
125. T. D. Lysenko, "Obnovlenie semian," *Izvestiia,* 1935, 15 July.
126. *SZ,* 1935, 16 Sept.
127. For Prezent's respect for Mendelian genetics, see his introduction to Filipchenko, *Eksperimental'naia zoologiia* (1932). For Marxist-Leninist attitudes toward genetics, see below, Chapter 8, "Philosophy."

128. Lysenko and Prezent, *Selektsiia i teoriia* (1935), pp. 21, 25.

129. N. I. Vavilov, "Blizhaishie zadachi sovetskoi agronomicheskoi nauki," *Front nauki i tekhniki*, 1936, No. 2, p. 63. For a depressing review of Vavilov's tributes to Lysenko, several of them drawn from archives, see M. Popovskii, "Tysiacha dnei Akademika Vavilova," *Prostor*, 1966, Nos. 7, 8. As early as February, 1934, according to Popovskii, Vavilov went so far as to sponsor Lysenko for candidate membership in the Academy of Sciences. While Popovskii's article is valuable for such revelations, it is spoiled by a picture of Vavilov as a dupe, who really believed in the scientific value of Lysenko's work. See the criticism of Popovskii by Zh. Medvedev, *Novyi mir*, 1967, No. 4, pp. 226–234. Cf. D. Joravsky, "The Vavilov Brothers," *Slavic Review*, 24:381–394 (Sept., 1965).

130. See *Iarovizatsiia*, 1935, No. 1, for various articles on these themes.

131. *Izvestiia*, 1935, 30 October. Cf. his tribute to Lysenko in *ibid.*, 9 October.

132. *Pravda*, 1935, 9 Dec. He did not name Lysenko, but the implication was clear.

133. *Soveshchanie peredovikov urozhainosti*, p. 186.

134. *Pravda*, 1935, Nov. 22; Stalin, XIV (I), p. 94.

135. *Soveshchanie peredovikov urozhainosti*, p. 157.

136. The astute reader may object that Stalin's famous remarks of December 1929 and November 1935 did not actually specify that practical success *in agriculture* is the ultimate criterion of theoretical truth *in biological science*. None of Lysenko's scientific opponents dared to be so astute.

137. In the middle of 1935 some Lysenkoites were forced into the Lenin Academy of Agricultural Sciences, and some people already there suffered conversion. Even so, only seven or possibly eight of the forty-seven academicians were Lysenkoites in 1935: Lysenko, S. S. Perov, B. A. Keller, E. F. Liskun, I. G. Eikhfel'd, I. V. Iakushkin, Ia. A. Iakovlev, and possibly V. R. Vil'iams (possibly, because there was subsequently a little friction between him and the Lysenkoites). Eleven of the academicians were firm opponents of Lysenkoism, and four, including the new president, were compromisers. The other twenty-four were in unaffected fields and took no public stand. Aside from the Lenin Academy and other institutions of agricultural research and education, scientific institutions such as the Academy of Sciences and the universities were, in 1935, almost solidly anti-Lysenkoite. See below, pp. 120–1, for the instructive experience of Leningrad and Moscow Universities.

138. Bukharin, who was the most important official on the scientific front in 1935, showed his coldness most notably by his policy as editor of the journal *Sotsialisticheskaia rekonstruktsiia i nauka* (usually called *Sorena* for short). See also his non-Lysenkoite tribute to Michurin, *Izvestiia*, 1935, 8 June. N. P. Gorbunov, as Lenin's chief administrator for questions of science and technology, had sent the telegram that was supposed to prove the discovery of Michurin by Lenin himself. Gorbunov himself would not repeat that legend. In the late 1920s and early 1930s he was the chief Party official involved in the organization of the Lenin Academy. When the Lenin Academy was cut back and forced to accept Lysenkoites in 1934–35, Gorbunov was transferred to the Academy of Sciences. He is still remembered as a special friend of Vavilov's. See *Riadom s Vavilovym*,

p. 213, where A. S. Enukidze, another official on the scientific front, is remembered as a friend of Vavilov's. A. A. Zhdanov is also remembered this way. See below, pp. 137 *et seq.*

139. See below, pp. 229 *et seq.*

140. The mathematician was N. N. Luzin. See *Pravda*, 1936, 3–15 July, for the opening of the campaign. For a convenient summary of the charges against Luzin and the conclusions that scientists were supposed to draw from his case, see V. Molodshii, "Ob uchenom vrage v sovetskoi maske," *PZM*, 1936, No. 9. (Although Luzin was endlessly denounced as an "enemy," he was not executed. In fact, he was not even jailed.)

141. G. Frizen, "Genetika i fashizm," *PZM*, 1935, No. 3, pp. 86–95. In accordance with Soviet editorial mores, the journal endorsed Frizen's views by publishing his article without the tag "For discussion." For full analysis of the philosophers' reaction to Lysenkoism, see below, Chapter 7.

142. Quoted from the archives of the Lenin Academy of Agricultural Sciences, by M. Popovskii, *Prostor*, 1966, No. 7, p. 18. If I interpret Popovskii correctly, the date was February 1934. Subsequently Lysenko became more extravagant in his voluntarist and obscurantist declarations. See especially his exchange with Kerkis in 1939, quoted below, p. 110.

5. STALINIST SELF-DEFEAT, 1936–1950

1. See *Pravda*, Dec. 9, 1935, for the origin of this much quoted phrase, in a report of a reception for officials of the Lenin Academy.

2. *Soveshchanie peredovikov zhivotnovodstva*, p. 56.

3. *Ibid.*, p. 266.

4. B. Tal', "O zadachakh zhurnala 'PZM,'" *PZM*, 1936, No. 2–3, p. 7.

5. *Sobranie zakonov*, 1936, No. 8, *st.* 67, p. 139.

6. "Upriazhneniia reaktsionnykh botanikov," *Pravda*, Feb. 27, 1936.

7. "Zhurnal peredovoi agronomicheskoi mysli," *ibid.*

8. A. A. Maksimov put the case very simply: "The measures recommended by Comrade Lysenko have won recognition for their usefulness — that is what has broken the resistance of his diehard opponents and forced them to be quiet." See A. A. Maksimov, "Filosofiia i estestvoznanie za piat' let," *PZM*, 1936, No. 1, pp. 64–5.

9. Within the Lenin Academy of Agricultural Sciences Lysenko received support from B. A. Keller, an eminent botanist, S. S. Perov, an incompetent biochemist, E. F. Liskun, an animal breeder, I. G. Eikhfel'd, a plant breeder, and I. V. Iakushkin, a crop specialist. V. R. Vil'iams or Williams, the agrobiologist of soil science, wavered for a while, and then came out strongly for Lysenko. Against these six — eight, counting Lysenko and Ia. A. Iakovlev — there were eleven anti-Lysenkoite academicians, and four who tried to effect a compromise. The reader should bear in mind that we are speaking of the Lenin Academy in 1936, *after* the forced reorganization of 1935 had raised men like Lysenko, Perov, and Eikhfel'd to the rank of academician, on a level with men like Vavilov and Meister. For the staff of VASKhNIL at various times, see its *Kalendar'-spravochnik*, 1936 *et seq.*

10. This happened early in 1936. See *Iarovizatsiia*, 1937, No. 1, pp. 162–3.

11. See, e.g., Vavilov's remarks at a meeting in the Academy of Sciences in February, 1936, reported in *VAN*, 1936, No. 4–5, p. 102. For an example of scientific hypotheses offered to the Lysenkoites in support of

their contentions, see D. F. Petrov's paper in VASKhNIL, *Selektsionnaia rabota.* This record of a conference in June, 1936 contains some of the most characteristic exchanges in the "discussion." The featured polemics were published in *SRSKh,* 1936, Nos. 7–12.

12. See Joravsky, especially Chapter 16.

13. *Ibid.,* especially Chapter 17.

14. A decree of TsIK SSSR on May 7, 1936 seems to have marked the beginning of the end for the Timiriazev Institute. See *VAN,* 1936, No. 6, pp. 75–6. For further news, see *ibid.,* No. 8–9, pp. 98–99; No. 11–12, p. 129; 1937, No. 1, pp. 140–141; and No. 4–5, p. 134. For descriptions of the Institute at its height, see O. Krasovskaia, "Biologicheskii institut im. K. A. Timiriazeva," *FNIT,* 1934, No. 4, pp. 82–86; and B. P. Tokin, "V Biologicheskim Institute im. K. A. Timiriazeva," *PZM,* 1934, No. 4, pp. 184–189. For the major publications of the Institute, see Kommunisticheskaia Akademiia, Biologicheskii institut, *Probleme der theoretischen Biologie* (1935), and *Pamiati K. A. Timiriazeva* (1936).

15. See, most notably, M. M. Zavadovskii, "Protiv zagibov v napadkakh na genetiku," *SRSKh,* 1936, No. 8, pp. 84–96. It must not be thought that all or even most of the biologists associated with the Communist Academy became vigorous opponents of Lysenkoism. Some — mentioned in the text above — did so. Some were removed by the terror (Levit, Agol, Levin, Uspenskii, and Bauer). Some tried to effect a compromise between science and Lysenkoism (B. M. Zavadovskii, Krenke, and P. P. Bondarenko). Some, including the last director of the Timiriazev Institute, B. P. Tokin, evaded the controversy. Only Lepeshinskaia, the would-be revolutionizer of cytology, and P. I. Valeskaln, a would-be philosopher of biology, became partisans of Lysenkoism. (I am ignoring Navashin's brief defection to Lysenkoism in the early 1950s.)

16. P. N. Konstantinov, P. I. Lisitsyn, and Doncho Kostov, "Neskol'ko slov o rabotakh Odesskogo instituta," *SRSKh,* 1936, No. 11, pp. 121–130.

17. *Spornye voprosy,* pp. 162–3.

18. See the majority of the speeches in *Dostizheniia sovetskoi selektsii.* This is the record of the "noncontroversial" part of the conference of December 1936. In fact a minority of speakers used the opportunity to endorse Lysenkoism, while another minority expressed feeble demurrers, and the majority were silently evasive. For another characteristic meeting, at which a handful of Lysenkoites attacked a majority of silent and evasive agricultural specialists, see *Kul'tura pshenitsy.*

19. A notable example of this type was the plant breeder V. Ia. Iur'ev, who managed, throughout his long life, to keep the good will of Lysenkoites and anti-Lysenkoites. The more famous breeder, G. K. Meister, might have accomplished the same feat, if the terror had not taken him. But I am inclined to doubt that he could have kept the good will of the Lysenkoites, for he got heavily involved in the theoretical issues of genetics, as Iur'ev did not. In this respect the experience of Grishko and Delone is very instructive. See below, p. 108.

20. Serebrovskii recounts the episode in *Spornye voprosy,* p. 451.

21. K. Ia. Bauman, "Polozhenie i zadachi sovetskoi nauki," *Pravda,* Sept. 6, 1936. Reprinted in *PZM,* 1936, No. 9, and *VAN,* 1936, No. 10.

22. See especially the speeches of A. I. Muralov, President of VASKhNIL, and of G. K. Meister, who gave the official summary of the conference, both in *Spornye voprosy.*

23. In his opening speech, *ibid.*, pp. 11–38, Vavilov continued his former effort at compromise. In his closing remarks, pp. 462–473, he came down quietly but firmly on the anti-Lysenkoite side.

24. See, e.g., *Izvestiia*, May 26, 1934. Cf. also the editorial note appended to H. J. Muller, "Sovremennoe polozhenie mutatsionnoi teorii," *Priroda*, 1936, No. 6, pp. 40–49. Shortly after the conference of Dec. 1936, Muller left the Soviet Union, as part of a medical mission to the embattled Spanish republic. He never returned.

25. For his explicit endorsement of Prezent, see his concluding remarks, *Spornye voprosy*, pp. 452–461.

26. *Ibid.*, pp. 308–9.

27. *Ibid.*, p. 454.

28. *Ibid.*, p. 455.

29. Stalin, XIV (I), pp. 275–6.

30. "Derzat' v nauke," *Pravda*, Nov. 21, 1938. Note that Tsitsin is also named as a young scientist who has shown what Stalin meant by his toast.

31. The arrested presidents were A. I. Muralov and G. K. Meister. By June 1937 Muralov seems to have been in serious trouble. See *Iarovizatsiia*, 1937, No. 4 (13), p. 4. In July he was arrested. See "Pod krylyshkom vreditelei," SZ, Sept. 28, 1937. Meister was arrested and Lysenko succeeded him early in 1938. See Medvedev, pp. 258–259.

32. *VAN*, 1938, No. 5, p. 72. Note, on p. 101, evidence that the Presidium of the Academy had already tried to head off criticism for its hostility to Lysenkoism. The Institute of Genetics had been ordered to establish a Laboratory of Evolutionary Ecology under the direction of B. A. Keller, the only academician who was on Lysenko's side.

33. *VAN*, 1938, No. 6, pp. 75–77.

34. *Ibid.* Cf. also pp. 39–40.

35. *Ibid.*, No. 7–8, pp. 119, 123.

36. *Ibid.*, pp. 119–128.

37. See below, pp. 258–9.

38. *Pravda*, January 11, 1939. The letter also attacked L. S. Berg, the distinguished zoologist and geographer, who refused to disavow the book in which he had expounded a vitalist theory of evolution, *Nomogenesis*.

39. Reported by Prezent, in *PZM*, 1939, No. 5, p. 146.

40. The bare facts are given in Akademiia Nauk SSSR, *220 let*, pp. 205–206. A tendentious report by Kol'tsov's successor is in *VAN*, 1948, No. 9, pp. 88–89.

41. Note in particular his role in the 1929 purge of the Academy of Sciences. See Ul'ianovskaia, p. 163.

42. Komarov, *Proiskhozhdenie kul'turnykh rastenii* (2nd ed., 1938), *passim*. The few brief tributes to Lysenko inserted in this edition did not change the essential argument. Komarov also showed his hand as editor of *Izvestiia Akademii Nauk, seriia biologicheskaia*. See especially the discussion he organized in the 1937 issues.

For the most part Komarov ignored Lysenko. In *Pravda*, March 14, 1941 he published a brief, perfunctory tribute to him.

43. The man in question was G. K. Khrushchov. See his explanations in *VAN*, 1948, No. 9, pp. 85–91. I call him a complete opportunist, in part because of his effort to throw the blame on other people, but mainly because of his betrayal of his profession. See his endorsements of Lepeshin-

skaia's anticytology, in *Kul'tura i zhizn'*, May 31, 1950, and in *Soveshchanie po probleme zhivogo veshchestva*, pp. 84–9.

44. The director was A. A. Rikhter. See *Pravda*, July 26, 1938 and *VAN*, 1938, No. 7–8, pp. 55–61. His successor was the biochemist A. N. Bakh, who was especially dear to the Bolshevik chiefs for his willing cooperation in political matters. Rikhter continued to work at the Institute, and so did M. Kh. Chailakhian, who was one of the plant physiologists most hated by the Lysenkoites.

45. S. S. Chetverikov, "O nekotorykh momentakh evoliutsionnogo protsessa s tochki zreniia sovremennoi genetiki," *Zhurnal eksperimental'noi biologii*, Seriia A, 1927, vypusk 1, pp. 3–54. For a translation, with an appreciation by I. Michael Lerner, see American Philosophical Society, *Proceedings*, vol. 105, No. 2. Chetverikov's arrest in the early 1930s had nothing to do with Lysenkoism; it was apparently the result of his bourgeois origins. See *Za proletarskie kadry*, Feb. 25, 1931. He was released by the mid-1930s and allowed to work at the University of Gorky. By that time Lysenkoism was making itself felt; very few Soviet scientists continued the kind of research that Chetverikov had projected. England and the United States became the major centers of population genetics.

46. See, e.g., *Moskovskii universitet*, May 19, 1938. Cf. the complaint of the dean of the biological faculty at Leningrad University, in *PZM*, 1939, No. 11, p. 102. A Lysenkoite *dotsent* disputed the dean's bleak picture and called him two-faced (p. 213), which he was, in an effort to appease the Lysenkoites. For a first-hand picture of the clash at Leningrad U., see *Leningradskii universitet*, 1939, Feb. 28, March 31, June 22, *et passim*. In the issue of April 7, 1937 Karpechenko, chairman of the *Kafedra* of genetics, complained that the "moral atmosphere" at the University was making it impossible to teach genetics.

47. See, e.g., Paramonov, *Kurs darvinizma*, and I. M. Poliakov, *Kurs darvinizma*. Cf. the discussion of the course in *Sovetskaia nauka*, 1938, No. 1.

48. *PZM*, 1939, No. 11, p. 145. Cf. p. 159 for Lysenko himself giving a similar appraisal.

49. *VAN*, 1948, No. 9, p. 53.

50. Quoted from the archives by M. Popovskii, "1000 dnei Akademika Vavilova," *Prostor*, 1966, No. 8, p. 114.

51. Johann Gansovich Eikhfel'd was one of the first of Vavilov's protégés to endorse Lysenkoism. See *Dostizheniia sovetskoi selektsii*, pp. 91–5. Even earlier, while he was still flattering Vavilov, he had already learned the Lysenkoite style of public relations. See his "Ukroshchenie Subarktiki," *Izvestiia*, July 11, 1934.

52. Delone and Grishko, *Kurs genetiki*.

53. SZ, June 14, 1939. Cf. other attacks in *Biologiia v shkole*, 1939, No. 1, pp. 91–2; *Iarovizatsiia*, 1938, No. 4–5, pp. 221–236; *Sovetskaia nauka*, 1939, No. 1, pp. 171–6; and *Vestnik s.-kh. literatury*, 1939, No. 1, pp. 31–38.

54. *SZ*, 1939, Feb. 1, March 3, June 14, and Sept. 8.

55. Reported by Serebrovskii, in PZM, 1939, No. 11, p. 96.

56. Andreev endorsed Lysenko in his report on agriculture to the Eighteenth Party Congress. Quoted in *Iarovizatsiia*, 1939, No. 1 (2), pp. 4–5. At a conference on plant breeding and seed production, Commissar of

Agriculture Benediktov declared that his commissariat "supports Academician Lysenko in his practical work and his theoretical views, and obliges the breeding stations of the USSR to apply his methods in seed production and breeding work." SZ, March 5, 1939, and quoted in *Iarovizatsiia,* 1939, No. 2 (23), p. 121.

57. See, e.g., Popovskii, *Prostor,* 1966, No. 8, p. 115.

58. It is a mistake to suppose that Prezent spoke for Soviet philosophers in the 1930s. See below, Chapter 7.

59. *PZM,* 1939, No. 11, pp. 134–5.

60. M. B. Mitin, "Za peredovuiu sovetskuiu genetichesguiu nauku," *PZM,* 1939, No. 10, pp. 174–5.

61. *Ibid.,* No. 11, p. 94.

62. Iur'ev *et al., Obshchaia selektsiia* (1940).

63. *Zhurnal obshchei biologii,* edited by Schamlhausen, Z. I. Berman, and Sobol'. Note the invocation of Mitin's authority in the opening editorial, 1940, No. 1, pp. 3–4. Each issue featured articles that implicitly criticized Lysenkoism. For explicit criticism, see V. V. Khvostova, "Problema genotipicheskogo vliianiia pri privivkakh i transplantatsiiakh," *ibid.,* No. 3, pp. 468–9.

64. *SZ,* March 27, 1940. The article in question was T. K. Enin, "Gibridizatsiia gorokhov v dele vyvedeniia novykh sortov," *Selektsiia i semenovodstvo,* 1939, No. 2–3, pp. 32–35. The angry Lysenkoites granted that the journal printed many more articles for their side, but they were still dissatisfied.

65. The transformation of biological education in the secondary schools is clearly revealed in successive issues of *Biologiia v shkole,* 1937–1939. For the result, see Mel'nikov *et al., Osnovy darvinizma* (1941). The reader will note that Mel'nikov supported genetics in 1937. In his 1941 textbook chromosomes and genes have simply disappeared.

66. *SZ,* 1939, June 15, July 9, and Oct. 10.

67. See Lysenko, *Nekotorye voprosy,* pp. 26–7. The campaign occurred in the spring of 1941. Cf. SZ, 1940, Oct. 23, and Nov. 13 and 20. Note the friction between Lysenko and Tsitsin.

68. See Lysenko's account in *PZM,* 1939, No. 11, pp. 149–150.

69. *Ibid.,* p. 146.

70. The three were Ia. A. Iakovlev, M. A. Chernov, and A. I. Muralov, who were respectively: Party chief of agriculture, Union Commissar of Agriculture, and President of VASKhNIL. See *Iarovizatsiia,* 1935, No. 1, p. 3. Cf. *Pravda,* Feb. 15, 1935, for an especially warm tribute to Iakovlev. If one adds Lysenko's acknowledgment of the help of Postyshev (*FNIT,* 1936, No. 2, pp. 60–61), the score becomes four out of four.

71. Of course biologists had been arrested before. But no one perceived a connection with Lysenko's crusade. For example, when the geneticist Chetverikov was arrested, Lysenko had not yet entered the field of genetics. Lysenko ultimately profited from this arrest — see above, note 45 — but accidentally. Similarly with the arrest in 1930 of I. I. Ivanov, the pioneer in artificial insemination of livestock. His place was taken by a young careerist, V. K. Milovanov, who became a major supporter of Lysenkoism when it moved from plant physiology to breeding and genetics.

72. For the first public sign that Levit and his institute were in serious trouble, see "Protiv antinauchnykh vrazhdebnykh 'teorii,'" *Komsomol'skaia*

pravda, Nov. 15, 1936. At this time Levit was still at large. See also Ernst Kol'man, "Chernosotennyi bred fashizma i nasha mediko-biologicheskaia nauka," *PZM,* 1936, No. 11, pp. 64–72; and Karlik's attack on Levit's publications, *ibid.,* No. 12, pp. 178–186.

73. See, e.g., *Kniga i proletarskaia revoliutsiia,* 1937, No. 2, p. 118; *PZM,* 1937, No. 10, p. 76; and *Za proletarskie kadry,* June 2, 1937. The last item includes hostility to Lysenkoism among the charges against Levin.

74. *Izvestiia,* Dec. 22, 1936. The *New York Times* dispatches are conveniently anthologized in Zirkle, *Death of a Science,* pp. 1–5.

75. The public record is fairly informative concerning the earliest or Cheka peak. See E. J. Scott, "The Cheka," *St. Antony's Papers,* 1:1–23 (1956); and Carr, *Bolshevik Revolution,* I, pp. 158–170, 179–181, 212. Cf. Slusser and Wolin, for documents. For the later peaks the only creditable studies based on archival research, as far as I know, are T. Postolovskaia, "Osobennosti klassovoi bor'by na Ukraine v period mezhdu XVI i XVII parts'ezdami," *Pod markso-leninskim znamenem,* 1934, No. 1, pp. 39–68; and Fainsod, *Smolensk under Soviet Rule, passim.* Weissberg, *The Accused,* is a good personal account by a physicist who was jailed in the late 1930s. For general studies, see Andics, Conquest, and Lewytskij.

76. See the reports in the newspaper of Moscow State University, *Za proletarskie kadry,* 1937, Jan. 9, March 28, April 11, May 15. The attack on physics was renewed before Hessen's arrest, most notably at the March 1936 session of the Academy of Sciences. See *VAN,* 1936, No. 4–5, pp. 62–75; *Izvestiia AN, seriia fizicheskaia,* 1936, No. 1–2; and *Uspekhi fiz. nauk,* 1936, No. 7, pp. 837–976. The chief polemics were published in *PZM* during 1937, and insinuations of subversion occasionally appeared in the obscurantists' attacks on the physicists. See especially Kol'man's letter, *PZM,* 1937, No. 11–12, pp. 232–3. Kol'man, science chief of the Moscow Party organization, was perhaps the only person on the obscurantists' side who genuinely understood the issues. In this letter he spurned the outstretched hand of A. F. Ioffe, the dean of Soviet physicists, calling attention to the fact that "enemy" Hessen was Ioffe's student.

77. For the list of repressed physicists, see Appendix A. I consider it very likely that some obscurantists did suffer repression. For example, G. A. Kharazov vanished with disturbing suddenness following his "rebuttals" of relativity. (See Joravsky, p. 280.) And T. N. Gornshtein was suspiciously silent from the mid-1930s to the late 1950s. But perhaps she could not be called an obscurantist, merely a philosopher who began to discuss physics before she understood it. (See *ibid.,* pp. 290, 383.) In any case, the evidence is not strong enough to include these people in the list of the repressed — which stands to reason: the obscurantists, on the whole, were much less distinguished people than the defenders of physics, and therefore left much less evidence of themselves in the public record.

78. See Appendix A.

79. The size of the two groups was estimated on the basis of *Nauka i nauchnye rabotniki SSSR.* The number of repressed was arbitrarily guessed to be three times the number on my list. This may well be an exaggeration. A recent Soviet writer, for example, counts eighteen members of VIR arrested by 1940, out of a staff of "a thousand." (Popovskii, *Prostor,* 1966, No. 8, pp. 98–9.) It must be remembered that VIR, the chief insti-

tution of plant science, was particularly exposed to scrutiny by the terrorists. Cf. also H. J. Muller et al., "Vazhneishie rezul'taty raboty Instituta genetiki AN SSSR," *Izvestiia Akademii Nauk, seriia biol.,* 1937, No. 5, pp. 1469–92, for a list of the staff at the Institute of Genetics, the chief institution in that field. Of the thirty-three it is certain that Director Vavilov subsequently became a victim of repression; two others — Lepin and Panshin — probably became victims. See note 81 for the fate of the rest.

80. The two most well-known turncoats were Kh. F. Kushner and N. I. Nuzhdin. Less famous were: R. L. Dozortseva, K. V. Kosikov, and N. S. Butarin. It may be said in defense of the last three that they did not really understand the science they turned against. No such defense can be made of Kushner and Nuzhdin.

81. The total number of geneticists is estimated from the data in *Nauka i nauchnye rabotniki.* It is a depressing experience to trace the careers of the thirty people who were on the staff of the Institute of Genetics in 1937. (See note 79. I am omitting the three Americans from this count.) Twenty-one were still — or once again — at work in the post-Stalin period, as evidenced by publications and entries in professional directories. Only four of them (turned Lysenkoite) were still at the Institute of Genetics. The rest were scattered in a variety of biological and agricultural institutions doing everything but genetics.

82. See items cited in note 72.

83. See their speeches in *Spornye voprosy.* K. Ia. Bauman was the chief of the Central Committee's Division of Science. He did not speak for the record, but H. J. Muller recalled the instructions that he gave in private. Muller told me with obvious pleasure how he had defied Bauman's order to leave human genetics out of his speech.

84. *Spornye voprosy,* p. 144, the official record of the conference, abridges Muller's argument almost to the point of unintelligibility. He was kind enough to let me see the original typescript.

85. *Ibid.,* pp. 182–3. The speaker was B. M. Zavadovskii.

86. *Ibid.,* p. 341.

87. *Ibid.* See speeches by Ermakov, Bosse, Perov, and Prezent.

88. *Ibid.,* p. 423.

89. The victims were V. P. Efroimson and D. D. Romashov. See Medvedev, pp. 128–9, where a third victim is mentioned, a student.

90. See, e.g., Iu. Chernichenko, "Russkaia pshenitsa," *Novyi mir,* 1965, No. 11, p. 191; and Popovskii, *Prostor,* 1966, Nos. 7, 8. Khrushchev told the Central Committee that "some outstanding scientists who did not agree with the *travopol'naia* system of V. R. Williams were declared enemies of the people." *Plenum . . . 5–9 marta 1962,* p. 41. The psychological implication — they were condemned because they did not agree — is very strong.

91. Prezent, *Klassovaia bor'ba,* an attack on B. E. Raikov. *PZM,* 1934, No. 3, 198–202, an attack on Demchinskii and Ianata. *Iarovizatsiia,* 1937, No. 3, p. 63, an attack on Uranovskii and Busygin, "enemy" administrators of Leningrad University held responsible for termination of Prezent's course in Darwinism. The fourth is a borderline case: In SZ, April 12, 1937, Prezent and A. A. Nurinov (who probably perished as an "enemy" himself), attacked Kol'tsov in language that suggested an "enemy" charge without actually using the word. "Who has been the teacher to

whom, the fascists to Kol'tsov or Kol'tsov to the fascists, does not change matters." Kol'tsov was not arrested.

Medvedev, p. 258, argues that Prezent's denunciation of Levit in the spring of 1937 "began the persecution of this scientist." Medvedev seems to be unaware of Kol'man's attack on Levit in November 1936, and of Perov's statement, at the Conference of December, 1936, that Levit was already exposed as an "enemy."

92. In 1948, after the condemnation of genetics was formal and total, this polemical device became fairly common. See, for an extreme example, N. V. Turbin, "Torzhestvo michurinskoi biologii," *Vestnik Leningradskogo Universiteta,* 1948, No. 10, *passim.* Turbin was extreme in his public mention of men who had perished as "enemies," in this case Vavilov and Karpechenko. Public mention of "enemies'" names was usually taboo, except for the immediate aftermath of their arrest, during "the liquidation of the consequences of wrecking." Even at such meetings, most speakers used circumlocutions to refer to the arrested person. "Nonperson" is a poor term; it points to the emotionless end of a process that was intensely emotional. The "unnameable" or "unspeakable" would be a better term, especially in view of the fact that the prominent ones were remembered though (or because?) most people feared to speak their names.

93. See *PZM,* 1939, No. 11, p. 96, for the secretary's disapproving report of Shlykov's speech. See *ibid.,* No. 10, p. 158 for Mitin's disapproval.

94. Reported in Popovskii, *Prostor,* 1966, No. 8, p. 112.

95. The other man was A. K. Kol'. He was the first man to make a public attack on Vavilov. See *Ekonomicheskaia zhizn',* Jan. 29, 1931. The evidence that he and Shlykov were arrested is circumstantial but very strong.

96. *Spornye voprosy,* p. 322. Perov did a similar trick at the August Session of 1948. See *O polozhenii,* pp. 121–2.

97. *Za proletarskie kadry,* June 2, 1937. *Moskovskii universitet,* May 19, 1938.

98. That was D. A. Sabinin. See first citation in note 97.

99. *Moskovskii universitet,* June 29, 1937. M. A. Gremiatskii was the anthropologist. I have omitted the newspaper's reference to another "enemy," Turetskii, for I have no other information on him.

100. Anna Markovna Bykhovskaia. See her autobiography in *FNIT,* 1936, No. 3, p. 71. Cf. *Za proletarskie kadry,* June 17, and Oct. 17, 1937. The June article attacks her for being soft on "enemies"; the October article announces that she herself was an "enemy."

101. G. G. Bosse. See *ibid.,* June 17, 1937, for news of his expulsion. For evidence of his reinstatement, see his autobiography in Lipshits, I (1947).

102. S. I. Alikhanian. See his article in *Moskovskii universitet,* May 9, 1938, attacking the existing leadership. See his article in *ibid.,* Sept. 16, hailing the new textbook by Grishko and Delone, calling upon the biology faculty to put new life into its courses, and upon students to return to the study of genetics. Alikhanian was chairman of the *kafedra* of genetics. S. D. Iudintsev emerged as the dean of the biology faculty. See *ibid.,* March 21, 1938 for an editorial apologizing to him for charges made in a previous issue, and for the news that he has been readmitted into the

Party. In the June 13 issue, as "acting dean," he defended a professor who had been attacked in a previous issue for deriding Lysenko. In 1948, still dean of the biology faculty, he refused to speak at the August Session, even though *Pravda* called upon him to do so.

103. See *Leningradskii universitet,* January 15, 1938 for an attack on a number of biologists, including E. Sh. Airapet'iants, who had been expelled from the komsomol for consorting with "enemies," and M. E. Lobashev, who had been put under surveillance for the same reason. The attacks on Karpechenko and Levitskii were comparatively mild in their insinuations of subversion. See, e.g., *ibid.*, Feb. 7 and May 21, 1937. The reader should also bear in mind the fact that the newspaper carried attacks on other biologists, but they do not figure in this analysis because I have no other information on them.

104. *Ibid.*, March 31, 1939.

105. *Ibid.*, Feb. 28, 1939. For the expulsion see note 103.

106. *Ibid.* Cf. Feb. 13, 1940, for the continuing conflict between Lysenkoites and Airapet'iants; and Sept. 2, 1940, for the news that he had been awarded a "Stalin Stipend."

107. For Lobashev's resistance until 1948 see Prezent's complaint in *O polozhenii,* p. 508. For Lobashev's subsequent kowtowing, see his *Ocherki.* For his return to intransigence, see his *Genetika.*

108. N. I. Muralov was removed from the post to figure in one of the show trials. His successor, S. [G.?] Kolesnev, was in trouble from the start of his brief tenure. See *SZ,* April 8, 1937. If I have his middle initial right, he survived the terror.

109. *O polozhenii,* p. 472.

110. It has been necessary to omit slandered people who were otherwise entirely undistinguished. See, e.g., those referred to in note 103. A few eminent specialists also had to be omitted from the count. For example, I. L. Nikitin, a cotton expert who was one of Vavilov's assistants at the Lenin Academy, was publicly charged with protecting "enemies." SZ, Sept. 28, 1937. I cannot, however, learn whether he was arrested. If such cases, both the eminent and the obscure, were added on, the list of publicly slandered biologists and agricultural specialists would be approximately doubled. A final caution: the reader must bear in mind that I examined only a few local newspapers.

111. The fifteen cases are: N. K. Kol'tsov, N. I. Vavilov, A. S. Serebrovskii, D. A. Sabinin, M. A. Gremiatskii, G. G. Bosse, S. D. Iudintsev, E. Sh. Airapet'iants, M. E. Lobashev, G. D. Karpechenko, G. A. Levitskii, P. N. Konstantinov, D. N. Prianishnikov, N. M. Tulaikov, and M. M. Zavadovskii. The four arrested were: Vavilov, Karpechenko, Levitskii, and Tulaikov. Lysenkoites were given the positions of Vavilov and Tulaikov. Much of the evidence for these assertions is cited in the preceding footnotes. The rest will be supplied in response to specific requests.

112. The only geneticists arrested after 1948 (Efroimson and Romashov) were not denounced in public. One of the most vicious speeches of 1948 was Glushchenko's, in the Academy of Sciences, virtually pinning the name of "enemy" on Rapoport and Dubinin. *VAN,* 1948, No. 9, pp. 59–66. See p. 101 for Kushner's forecast of their fate: they were transferred to "practical" work, Dubinin in a forestry station, Rapoport in a pharmaceutical institute.

113. Secret, even anonymous letters of denunciation are a stock theme

of anti-Lysenkoite novels. See Troepol'skii, *Kandidat nauk,* and Mishnev, *Uchenaia stepen'*. These authors, like many Soviet scientists with whom I have spoken, tend to take it for granted that letters of denunciation usually had dire results. It seems to me that they were less serious than a published denunciation, if only because publication required the co-operation of several people, who thus publicly staked their reputations on the denunciation. See Medvedev, pp. 129–131, for a letter of denunciation found in the archives. Note that none of the individuals denounced in this letter was arrested.

114. See Appendix A. They were Shlykov, Kol', Ermakov, Odintsov, Zarkevich, Nurinov.

115. I have in mind such people as Serebrovskii, M. M. Zavadovskii, Lisitsyn, Konstantinov, Chailakhian, M. S. Navashin, Kol'tsov, Kostov, Rozanova, Dubinin, Zhebrak, Schmalhausen, Rapoport, Zhukovskii. These, and many other brave people, spoke and wrote against Lysenkoism when they had reason to fear that they were risking arrest by doing so. The fact that most of them were not arrested does not detract from their enormous courage. Neither does the fact that at least one — Navashin — subsequently caved in and endorsed Lysenkoism.

116. See Brzezinski, *Permanent Purge.*

117. *Bor'ba s bolezniami* (1953), pp. 3–4. As the date indicates, the quoted phrase is not contemporaneous with the height of the terror for specialists, when the trend toward such nativism was only beginning. See the analysis in the text, pp. 125–6.

118. *Soveshchanie peredovikov urozhainosti* (1936), pp. 245–52. Reprinted in *FNIT*, 1936, No. 2, pp. 64–66. The reader will not appreciate the shocking boldness of Prianishnikov's comments unless he reads the self-abasement of the other non-Lysenkoite specialists at this conference.

119. *Pravda,* Nov. 16, 1937. Cf. *SZ,* 1937, Nov. 14–20. See also *SZ,* April 8, 1937, for a report of Prianishnikov's outspoken criticism of agrobiology.

120. Beria's wife was a student of Prianishnikov at the time. See Medvedev, p. 73. A slightly different account of Prianishnikov's intervention on Vavilov's behalf is in Chernichenko, *Novyi mir,* 1965, No. 11, p. 191.

121. *Selektsionnaia rabota,* p. 83.

122. For a prime example, see Lysenko and Preznet, "O 'logiiakh,' 'agogiiakh' i deistvitel'noi nauke," *Pravda,* June 26, 1936. Cf. I. A. Sizov, "Intsukht i ego primenenie v selektsii," *SRSKh,* 1936, No. 12, p. 95, for a characteristic foreshadowing of the anticosmopolitan theme: "Our geneticists, and with them the breeders, uncritically and hastily accept what is offered from without, and adopt this as a standard."

123. *Sotsialisticheskoe zernovoe khoziaistvo,* 1938, No. 1, p. 14. See earlier issues for the turmoil over agrobiology at the Saratov Institute. The one man who had publicly attacked Lysenkoism was not arrested. That was A. P. Shekhurdin. See *Spornye voprosy,* pp. 156–9, for his vigorous criticism in 1936. Cf. p. 286 for Lysenkoite counterattack and defense of Shekhurdin by Ia. A. Iakovlev. Shekhurdin sat silently through the 1948 August Session where he was cited by a defender of genetics as a man who had bred great varieties with the aid of genetics. The Lysenkoites appealed to him to say it was not so, but he kept his mouth shut. See *O polozhenii,* pp. 292, 321, and 519. In 1937 he had seen his most illustrious colleagues — Tulaikov, Meister, David, Plachek — snatched by

the terror. (Tulaikov and Meister had made a great effort to compromise with the Lysenkoites. Plachek had made a slight effort, while defending inbreeding.)

124. See Appendix A for Ermakov, Odintsov, Kol', Shlykov, and Nurinov.

125. Of course, the arrest of such people was not likely to leave a trace in the public record. Neither was the arrest of young scientists in training. My sample is restricted to people who were sufficiently distinguished — by publications, positions, and public comment — to leave notice of themselves in the public record. Whether this results in serious distortion only the archives can show.

126. SZ, Nov. 23, 1937. Italics added.

127. See Joravsky, Chapters 16, 18, and 19.

128. In January 1933 the Central Committee resolved that "wrecking and sabotage . . . should, in the final analysis, play the same beneficial role in the matter of organizing new Bolshevik cadres for the collective and state farms as wrecking and the 'Shakhty Trial' played in industry. . . . There is no reason to doubt that wrecking and sabotage . . . will serve as the same kind of turning point in the business of developing revolutionary vigilance among our village and district Communists [i.e., in the business of watching over senior specialists], and the business of selecting new, Bolshevik cadres for the collective and state farms." *KPSS v rezoliutsiiakh,* II, pp. 736–7.

129. M. V. Sulkovskii, "Vreditel'stvo ekonomicheskoe," *Sel'skokhoziaistvennaia entsiklopediia* (1st ed.), I (1932), p. 772.

130. "Ot kollegii OGPU," *Izvestiia,* March 12, 1933.

131. "Nikakoi poshchady vragam naroda!" *Pravda,* March 15, 1933. The pledge of each *udarnik* to watch ten other peasants was resolved upon at the congress of *kolkhozniki-udarniki.*

132. SZ, 1938, January 9. Cf. March 21 for Lobanov, Commissar of Agriculture of the RSFSR, charging "enemies" with responsibility for the shortage of fruit and vegetables. Cf. Ia. A. Iakovlev, "O merakh po uluchsheniiu semian zernovykh kul'tur," *Iarovizatsiia,* 1937, No. 4 (13), pp. 3–5, for analogous charges. Iakovlev was himself on the verge of arrest.

133. The reader who wishes to see the difference at a glance should compare the article cited in note 129 with the items cited in note 135.

134. See, e.g., N. N. Terent'ev, "Iskorenit' posledstviia vreditel'stva v institute zhivotnovodstva," *Problemy zhivotnovodstva,* 1938, No. 1, pp. 74–81; and SZ, 1938, Jan. 9, 11, *et passim.*

135. See, e.g., *Meditsinskii rabotnik,* 1938, April 1 and Aug. 17; *Za proletarskie kadry,* 1937, Jan. 9, April 11, 22, *et passim; SZ,* 1937, April 8, Sept. 28, Nov. 22, 1939, Jan. 25, *et passim; VAN,* 1937, No. 4–5, pp. 7–14; *FNIT,* 1937, No. 7, pp. 138–140; *Timiriazevka,* Jan. 7, 1938, *et passim.* See also many citations in preceding notes.

136. For a convenient collection of Lysenko's wartime articles, see his *Raboty v dni.* Pp. 182–219 are devoted to the potato. In the next two years he published many more articles on the subject. For the Civil War recipe, see N. V. Zhilin, "O posadke kartofelia glazkami," *Russkoe sadovodstvo i ogorodnichestvo,* 1919, No. 4–5, pp. 29–31.

137. Lysenko, *Raboty v dni,* pp. 116–130. Cf. also his *Nekotorye voprosy agrotekhniki.*

138. Lysenko, *Raboty v dni,* pp. 170–181.

139. See Lepeshinskaia, p. 3, for Lysenko's endorsement, signed October 27, 1945. In November 1945, he gave a series of lectures on "Natural Selection and Intraspecific Competition." They were first printed in SZ, 1946, Jan. 5, 6, 8, 10, 12, and in *Selektsiia i semenovodstvo*, 1946, No. 1–2. For a convenient reprint, see Lysenko, *Agrobiologiia* (1949), pp. 525–59.

140. Lysenko, "Ruchnoi gnezdovoi posev kok-sagyza," *Pravda*, April 17, 1943 marks the beginning of this recipe. For the climax, see his *Agrobiologiia* (1949), pp. 657–75.

141. It is quite possible that Lysenko merely attached his cluster method of planting to a campaign originating among the disciples of Williams. For one of the earliest proposals for the campaign, see A. I. Krylov, "Travopol'naia sistema zemledeliia v chernozemnykh stepiakh SSSR," in Vsesoiuznoe soveshchanie nauchno-issledovatel'skikh uchrezhdenii, *Materialy*, pp. 122–142. This meeting was held in October, 1945. Williams' disciples were there in force; Lysenko's were only feebly represented; and the keynote speech by Basiuk praised *travopol'e*, while ignoring Lysenko's recipes. See pp. 5–20. For a history of the campaign by a disciple of Williams, see V. S. Dmitriev, especially pp. 90–2. Cf. G. N. Vysotskii, "Shelterbelts in the Steppes of Russia," *Journal of Forestry*, 33:781–788 (Sept. 1935), for the history before the ascendance of agrobiology.

142. See, most notably, P. M. Zhukovskii, "Darvinizm v krivom zerkale," *Selektsiia i semenovodstvo*, 1946, No. 1–2, pp. 71–79.

143. See Chapter 7, "Genetics."

144. The abandonment of natural selection began with the lectures of November 1945. (See note 139.) In 1946 the first volume of Stalin's *Sochineniia* appeared, containing a previously unknown endorsement of Lamarckism (pp. 301, 303, 309). Nevertheless, the Lysenkoites were slow in taking advantage. Not until the August Session did they forthrightly take a stand for Lamarckism, leaning on Stalin for support. See *O polozhenii*, pp. 14, 185–6, *et passim*.

145. See *Sobranie postanovlenii pravitel'stva SSSR*, 1962, No. 3, *stat'ia* 24, pp. 47–59, for the repeal of previous decrees in support of *travopol'e*. 1945 was a peak year with 21 decrees, exceeded only by 1949 with 31 decrees. See again the record of the conference held in October 1945, as cited in note 141. Cf. also *Sovetskaia agronomiia* during 1946. It was strongly in favor of *travopol'e*, but annoyed the Lysenkoites by its cool attitude toward them.

146. P. N. Konstantinov, "Protiv uproshchenchestva v agronomii," SZ, Sept. 22, 1945. Lysenko's article had appeared in the Sept. 20 issue.

147. See SZ, 1945, Sept.–Dec. *passim*.

148. *Selektsiia i semenovodstvo* and *Sovetskaia agronomiia* were the two journals. The former carried Zhukovskii's article, as cited in note 142. Cf. also the suggestive evidence in note 141.

149. *KPSS v rezoliutsiiakh*, II, 1046.

150. This guess is inspired for the most part by the evidence of compromise in the resolutions, but also by subsequent references, especially those of Khrushchev. See, e.g., Khrushchev, VI, 177–9.

151. *KPSS v rezoliutsiiakh*, II, 1052.

152. *Ibid.*, 1053.

153. See Lysenko, *Agrobiologiia* (1949), pp. 592–3. As originally pub-

lished in SZ, March 21, 1947, this speech did not endorse the use of inbred lines.

154. P. M. Zhukovskii, "O doktorskikh dissertatsiiakh i otvetstvennosti opponentov," *Vestnik vysshei shkoly,* 1945, No. 4. I have been obliged to rely on reports of this article, which I was unable to obtain.

155. See *ibid.,* 1946, No. 11–12, and 1947, Nos. 1–3.

156. *Ibid.,* 1947, No. 1, pp. 29–30. Note his criticism of the elective course in Darwinism offered at the Timiriazev Academy by Paramonov, who had initiated the discussion.

157. *Ibid.,* No. 3, pp. 44–45.

158. D. A. Kislovskii, *ibid.,* 1946, No. 11–12, pp. 16–17. He was a stockbreeder who attached himself to the Lysenkoite cause in the mid-1930s.

159. The facts were subsequently revealed in reproaches and confessions during and after the August Session. See *O polozhenii,* p. 100, and see especially *VAN,* 1948, No. 9, pp. 26, 89–91, *et passim.*

160. *Ibid.,* pp. 41–2. Perhaps the parasitologist, E. N. Pavlovskii, should be added to this unique list, for he also supported Lysenko.

161. Quoted in *ibid.,* p. 64.

162. A. A. Zhdanov, "Vystuplenie," *Voprosy filosofii,* 1947, No. 1, p. 271. For a convenient English translation, see his *Essays,* pp. 73–4. For the most reliable indication of Zhdanov's policies, see the newspaper, *Kul'tura i zhizn',* which began to appear in June 1946, as the anticosmopolitan campaign was getting under way. Through the first half of 1948 it simply ignored the "discussion" in biology.

163. The chief evidence is his letter to Stalin, published in *Pravda,* August 7, 1948. Indirect confirmatory evidence is the behavior of the chief philosophers of science. See below, Chapter 8.

164. Letter of Julian Huxley to H. J. Muller, Dec. 31, 1945. Huxley reports the soundings of Eric Ashby, Australian cultural chargé in Moscow and a botanist by profession. Serebrovskii, Zhebrak, and Dubinin responded with delight, Orbeli with caution. In the personal papers of H. J. Muller.

165. The complete record, covering the years 1935–1939, is in the personal papers of L. C. Dunn.

166. A. R. Zhebrak, "Soviet Biology," *Science,* 102:357–358 (Oct. 1945). See Zirkle, *Death of a Science,* pp. 51–62, for a convenient reprinting, along with the articles by Dunn and Sax that occasioned Zhebrak's. In part his article was also the result of conversations with American scientists during his visit to San Francisco to attend the founding of the United Nations. See again the personal papers of L. C. Dunn. See also N. P. Dubinin, "The Work of Soviet Biologists: Theoretical Genetics," *Science,* 105:109–112 (Jan. 1947).

167. How daring it was can be gathered from Leonov's novel, *Russkii les,* Chapter 12, part 5, in which, toward the end of the war, the villainous scientist tries to inveigle the virtuous scientist into a compromising meeting with an Australian scientist. (Cf. note 164.) The virtuous hero refuses.

168. I. D. Laptev, "Antipatrioticheskie postupki," *Pravda,* Sept. 2, 1947. *Pravda* started the "discussion," but *Literaturnaia gazeta* was the main continuer. See the issues of Oct. 1, 11, 18; Nov. 29; Dec. 10, 27.

169. Quoted in Grashchenkov, p. 35.

170. See again note 163. One of the first signs that the ideological es-

tablishment was beginning to move toward total support of Lysenko was the article of M. B. Mitin, "Za rastsvet sovetskoi agrobiologicheskoi nauki," *Literaturnaia gazeta,* Dec. 27, 1947. But it is noteworthy that Mitin does not explicitly condemn genetics root and branch.

171. Reported in *O polozhenii,* pp. 161–2. See also the report in SZ, Nov. 12, 1947, and Jan. 9, 1948.

172. I. S. Galkin, "Za boevuiu nauchnuiu kritiku; protiv nizkopoklonstva v nauke," *Vestnik vysshei shkoly,* 1947, No. 12, pp. 14–15.

173. See A. A. Avakian, "Stadiinye protesessy i tak nazyvaemye gormony tsveteniia," *Agrobiologiia,* 1948, No. 1, pp. 47–77, for the Lysenkoite condemnation of such work, which precipitated the meeting. For reports of the meeting, see *VAN,* 1948, No. 9, pp. 94 and 134. Cf. N. G. Kholodyni, "V zaschitu ucheniia o gormonakh rastenii," *BZ,* 1954, No. 3, pp. 403–414.

174. *Tezisy dokladov konferentsii po problemam darvinizma.* Cf. the brief report in *Estestvoznanie v shkole,* 1948, No. 3, pp. 85–6.

175. N. Khlopin et al., "Ob odnoi neudachnoi kontseptsii," *Meditsinskii rabotnik,* July 7, 1948. See *ibid.,* Nov. 24, for Lepeshinskaia's reply.

176. It was published in *Pravda,* August 7, 1948.

177. See SZ, July 28, 1948 for publication of the decree and the front-page article, "Pod zashchitoi lesnykh polos." There had been a trickle of articles on afforestation in preceding issues; from this point on they became a flood.

178. *O polozhenii,* p. 13.

179. See *VAN,* 1948, No. 9, pp. 48–51, 157, *et passim,* for denunciations of L. A. Orbeli at the follow-up meeting in the Academy of Sciences. See his speeches, pp. 27–37, 164–170. Given the context, he conducted himself with great dignity and independence. For example, he endorsed the condemnation of genetics but he would not condemn his colleague Davidenkov.

180. *O polozhenii,* p. 472.

181. *Ibid.,* p. 134. This public record of his remarks has been obviously abridged.

182. *Ibid.,* pp. 523–8.

183. *Pravda,* Aug. 7, 1948.

184. See below, pp. 153–4. See also Iu. Zhdanov, *Lenin i estestvoznanie,* pp. 35–6, for his routine repetition of the current orthodoxy even in 1959, when he was no longer head of the Central Committee's Division of Science and could have expressed his own thoughts, if he had had any.

185. That was the philosophical analysis of chemistry. See his articles, as listed in *Bibliographie der sowjetischen Philosophie.*

186. *Pravda,* Aug. 7, 1948.

187. *O polozhenii,* p. 221.

188. *Ibid.,* p. 512.

189. I. A. Benediktov, speaking to the Academy of Sciences. See *VAN,* 1948, No. 9, p. 74.

190. See above, pp. 86 *et passim.* For the pre-Lysenkoite edition of his textbook, see *A Textbook of Plant Physiology* (N.Y., 1930). For the growth of Lysenkoite passages, see editions three (1931) to eight (1948) of Maksimov, *Kratkii kurs fiziologii rastenii.* The last time he tried printed criticism of Lysenko was in his *The Theoretical Significance of Vernalization,* written for the Imperial Bureau of Plant Genetics, and published

as Bulletin No. 16 (1934). His criticism was very restrained, but the Lysenkoites replied with a very angry blast, in *Iarovizatsiia,* 1935, No. 1, pp. 127–8. In April, 1936 Tulaikov was obliged to call a "self-critical" meeting at the Saratov Institute, to which Maksimov had been sent, and Maksimov felt obliged to attack his own ideas at this meeting. See *Sotsialisticheskoe zernovoe khoziaistvo,* 1936, No. 3, pp. 115–116. Note, however, the ultimate ambiguity of Maksimov's remarks.

191. *VAN,* 1948, No. 9, pp. 92–6.

192. *Ibid.*

193. A. L. Kursanov. See his list of publications in Lipshits, IV. Cf. his bland disruption of Lysenkoite plant physiology in 1957, when he organized the conference that produced *Itogi i perspektivy issledovanii razvitiia rastenii.* Note especially p. 210, for a Lysenkoite citing endorsements of Lysenkoism by Maksimov and by Kursanov, and cf. pp. 217–19, for Kursanov's evasive comment.

194. Compare Iur'ev, et al., *Obshchaia selektsiia* (1940) with the second edition of 1950. Compare Grishko and Delone, *Kurs genetiki* (1938) with Turbin, *Genetika s osnovami selektsii* (1950). Of course some elements of genuine science persisted in the new textbooks. See, e.g., Shennikov, *Ekologiia rastenii* (1950), which rejected such terms as "intraspecific competition" but managed to convey the concept. See the attack by O. N. Chizhikov, "O nekotorykh voprosakh teorii razvitiia rastitel'nosti," *Agrobiologiia,* 1952, No. 1, pp. 157–172; and the reply of A. P. Shennikov, "O nekotorykh sposobakh kritiki teorii razvitiia rastitel'nosti," *BZ,* 1952, No. 4, pp. 507–518. For the transformation of the biology curriculum on the secondary level, see M. I. Mel'nikov, "O perestroike prepodavaniia biologii v srednei shokole," *Estestvoznanie v shkole,* 1948, No. 5, pp. 44–54. Note that the previous editors have been fired; see pp. 10–15 for a catalogue of their sins. Cf. Mel'nikov, ed., *Iz opyta* (1950), and compare successive editions of Mel'nikov, *Osnovy darvinizma.*

195. S. V. Kaftanov, *Za bezrazdel'noe gospodstvo.* For his former favor to genetics, see above p. 134. See *Vestnik vysshei shkoly,* 1948, Nos. 10 and 11, for reports of the meeting at which he vehemently endorsed the new policy. Cf. also *VAN,* 1948, No. 9, pp. 48–59.

196. The painting is conveniently reproduced in Harcave, p. 660.

197. See *Pravda,* Oct. 24, 1948, for the basic decree. See *SZ,* 1948–51, *passim,* for many articles on the Great Plan.

198. As Leonov worked on his *Russkii les* confidence was replaced by doubt. The book, published in 1953, comes to the beginning of the Great Plan and stops.

199. See *VAN,* 1948, No. 9, pp. 44–48. Cf. Sukachev, *Stalinskii plan.* Previously Sukachev was most well known precisely for his study of the depressing effects of close planting. See, e.g., Gray, p. 364. For biographical articles and a list of Sukachev's works that express his genuine thoughts, see *Akademiku V N. Sukachevu k 75-letiiu.*

200. V. Ia. Koldanov, "Nekotorye itogi i vyvody po polezashchitnomu lesorazvedeniiu za istekshie piat' let," *Lesnoe khoziaistvo,* 1954, No. 3, pp. 10–18. Cf. the report of a conference of forest specialists, *ibid.,* 1955, No. 3, pp. 37–51.

201. See below, pp. 155 ff. Even at the height of Lysenko's power V. N. Sukachev managed to publish a few things that diverged from the

official creed. See, e.g., his "O nekotorykh osnovynkh voprosakh fitotsenologii," in *Problemy botaniki,* I (1950), pp. 449–464.

202. Iu. Zhdanov, "O kritike i samokritike v nauchnoi rabote," *Bol'shevik,* 1951, No. 21, pp. 38–39. The reader should bear in mind that Zhdanov published this unqualified endorsement of Lysenko and Lepeshinskaia at the very time that Sukachev and other biologists were taking the first tentative steps toward public criticism. See also *ibid.,* 1952, Nos. 2 and 4 for further endorsements of Lysenko's views on natural selection and species formation, while Zhdanov was still head of the Division of Science. In his *Vozdeistvie,* sent to press in July 1952, he endorsed Lysenko and condemned genetics, but he was silent on the issues of cytology and natural selection.

203. For denial of viruses, see L. V. Rozhalin and O. D. Belova, "K voprosu o gotike ili veretenovidnosti kartofelia," *Agrobiologiia,* 1948, No. 6. For hormones, see Avakian, *ibid.,* No. 1. For a Lysenkoite virologist, K. S. Sukhov, attempting to have his penny and his cake, see AN SSSR, Inst. Gen., *Protiv,* pp. 124–153.

204. Lysenko, *Agrobiologiia* (1948), pp. 368–9; (1952), p. 515. There is a very large body of Lysenkoite writings on this subject. For example, see Glushchenko, *Vegetativaia gibridizatsiia rastenii.*

205. For a thorough survey of the Lysenkoite literature on this subject, see V. V. Skripchinskii, "Prevrashchenie ozimykh zlakov v iarovye i iarovykh v ozimye," *BZ,* 1955, No. 1, pp. 64–90.

206. T. D. Lysenko, "Vid," *BSE* (2nd ed.), VIII (1950), 18.

207. *Ibid.* Cf. Bosh'ian for the transformation of viruses into bacteria and vice versa.

208. T. D. Lysenko, "K bespredel'nomu rostu urozhaev," *SZ,* Jan. 1, 1949.

6. SELF-CONQUEST, 1950–1965

1. See Harold Berman, Chapter 2.

2. M. A. Ol'shanskii, "Protiv dezinformatsii i klevety," *SZ,* 1964, 29 Aug. The bio-chemist is Zh. A. Medvedev.

3. See *Osobo opasnye gosudarstvennye prestupleniia,* for a recent effort by Soviet judicial authorities to define the law in this area.

4. *KPSS v rezoliutsiiakh,* II, 1061 and 1074.

5. See *Comparisons,* I, 211. Cf. *Sel'skoe khoziaistvo SSSR* (1960), p. 196; and *Narodnoe khoziaistvo SSSR v 1960 godu,* p. 377. The last named handbook pictures average grain yields for the period 1949–1953 as lower than the average for 1940, which was approximately equal to that of 1913, the best prerevolutionary year.

6. For Williams' argument on this subject, see his *Sobranie sochinenii,* VI, 405–408. For a convenient review of Williams' writings on this matter, with Lysenkoite criticism, see the editorial comment, *ibid.,* VI, 5 *et seq.*

7. Khrushchev, I, 138.

8. See "*rentabel'nost',*" in Ushakov. Cf. young Zhdanov's distinction between the criterion of practice, which was valid, and "*deliacheskii podkhod k praktike, pogonia za kopeikoi,*" which was contemptible. *Pravda,* 1948, Aug. 7.

9. See Khrushchev, VI, *passim.*

10. T. D. Lysenko, "Ob agronomicheskom uchenii V. R. Vil'iamsa," *Pravda*, 1950, 15 July. For a characteristic reaction, even years afterward, see S. P. Anikeev, "Sel'skokhoziaistvennoe proizvodstvo i nauka o pochve," *VF*, 1954, no. 1, pp. 135 ff.

11. See, e.g., Stalin, XVI [3], 167–8, for his annoyance at a correspondent who addressed such an appeal to him.

12. *Ibid.*, p. 144.

13. See especially the conclusion of the second article, *ibid.*, p. 157, for a call to Soviet linguists to "take first place in world linguistics." The argument that diverse languages will persist into the indefinite future is one of Stalin's main themes in these articles, yet some authors have virtually overlooked it, in their eagerness to stress the Russian element in Stalin's national policy. See, e.g., Barghoorn, pp. 244–5, and 253–5. For a substantial analysis of the school of linguistics that Stalin was disestablishing, see Thomas. For a convenient anthology, see *The Soviet Linguistics Controversy.*

14. In Poland natural science was immediately and officially declared to be separate from the ideological superstructure. See Jordan, pp. 475 ff. For Soviet comment on this subject see above Chapter 1.

15. Stalin, XVI [3], 145.

16. Stalin's first article on linguistics appeared in *Pravda*, 1950, June 20. Lysenko's criticism of *travopol'e* appeared on July 15.

17. See *VF*, 1951, No. 2, and following issues, for the resumption of the defense, which had lapsed in 1949. For reviews of Soviet discussions of physics, see Wetter, *passim;* Müller-Markus; David Comey, "The Soviet Controversies over Relativity Theory," in *The State of Soviet Science*, pp. 186–199; and Loren Graham, *et al.*, "Quantum Mechanics and Dialectical Materialism," *Slavic Review*, 25:381–420 (Sept. 1966).

18. A. I. Rudenko, *Opredelenie faz.* Tributes to Lysenko and his stage theory are scattered throughout the book, but Rudenko manages to evade the theory by distinguishing between *stages* in the development of plants, which he leaves to Lysenko's theory, and *phases*, which he discusses in his book. The author's preface is signed January 1948. The book was set in type in February 1950, and "signed for publication" on June 26, 1950.

19. See *BZ*, 1951, No. 2, pp. 113–114, for an editorial that uses Stalin's words on the freedom of science to back up a demand for revival of discussion of plant phylogeny. In the same issue articles by Kozo-Polianskii and M. M. Il'in opened the discussion. See further issues for further articles.

20. V. N. Sukachev, ed., *Nauchnaia sessiia.* This conference was convened by the Biological Division of the Academy of Sciences, but the subject of discussion and the people involved strongly suggest connections with the agricultural establishment. Note especially the leading role of Sukachev, who was simultaneously organizing the anti-Lysenkoite discussion in the *Botanical Journal*, of which he was editor-in-chief, and was also beginning, as one of the leading specialists on afforestation, to turn the agricultural establishment against Lysenko's cluster method of planting trees. At the conference of October 1951, some speakers made perfunctory obeisance to Lysenko, and one, Iu. V. Rakitin, devoted a long paper to attacking standard plant physiologists like Kholodnyi and Chailakhian. But on the whole, at this conference on biology and agriculture, Lysenko's theories and recipes were ignored.

21. Johann Gansovich Eikhfel'd was the thoroughly Lysenkoite director, replaced by Zhukovskii in 1951. Eikhfel'd had been director since 1940, when Vavilov, who had founded the Institute, was arrested. See biographies in *Biograficheskii slovar'*.

22. Iu. Zhdanov, "O kritike i samokritike v nauchnoi rabote," *Bol'shevik,* 1951, No. 21, pp. 38–9.

23. For an early public reference to the inventory of tree plantings, see M. A. Ol'shanskii, "O teoreticheskikh oshibkakh i nepravil'nykh prakticheskikh predlozheniiakh akad. Sukacheva," *SZ,* 1952, Jan. 8. This is an attack on Sukachev, who was in charge of the *Kompleksnaia nauchnaia ekspeditsiia [Akademii Nauk SSSR] po voprosam polezashchitnogo lesorazvedeniia,* which seems to have supervised the inventory. See its *Trudy,* I–II (1951–1953), for much detail, mostly on other issues than Lysenko's cluster method of planting. Ol'shanskii was enraged by the silence.

24. For a selective report, see V. Ia. Koldanov, "Nekotorye itogi i vyvody po polezashchitnomu lesorazvedeniiu za istekshie piat' let," *Lesnoe khoziaistvo,* 1954, No. 3, pp. 10–18. Cf. 1955, No. 3, pp. 37–51 for a report of a conference that discussed Lysenko's method. Koldanov subsequently estimated a loss of one billion rubles resulting from the method. See his "Gnezdovye posevy drevesnykh porod i srastanie ikh kornevykh sistem," *BZ,* 1958, No. 5, p. 715. For a brief summary by Sukachev, see *VAN,* 1965, No. 3, pp. 96–8. The disastrous results of the inventory were cryptically conceded by Ol'shanskii, in SZ, 1952, Jan. 8. In typical Lysenkoite fashion he blamed the failure, not on the cluster method of planting, but on the sloppiness of those using the method.

25. *SZ,* 1952, April 2.

26. Lysenko, "Rezul'taty opytnykh i proizvodstvennykh posevov lesnykh polos gnezdovym sposobom," *SZ,* 1952, April 3.

27. See *Bol'shevik,* 1952, No. 2, pp. 56–65, and No. 4, pp. 74–80, for unqualified endorsements of Lysenko's theory of species formation. From then on *Bol'shevik* had nothing to say about biology and agricultural science, until June 1953, when A. N. Nesmeianov and A. V. Topchiev gave a brief, ambiguous, evasive account of the situation. See their "Zadachi estestvoznaniia i tekhnicheskikh nauk v piatoi piatiletke," *Kommunist,* 1953, No. 9, pp. 59–61.

28. Malenkov, p. 58.

29. *Ibid.,* pp. 95–6.

30. See above, pp. 138–9, for Zhdanov's 1948 stipulation. See V. N. Sukachev, "K voprosu o razvitii rastitel'nosti," *BZ,* 1952, No. 4, pp. 496–507, for one of the first strong replies to a virulent Lysenkoite attack. Even so, note that Sukachev, who was at that time the leader of the anti-Lysenkoite movement, replied to lesser Lysenkoites and refrained from explicit attack on Lysenko. At one point (p. 501) Sukachev even declared himself to be a supporter of Williams' ideas, though he had been the first scientist to give an extensive and devastating critique of Williams' ideas. (See above, p. 25.) Explicit criticism of Lysenko himself began in *BZ,* 1952, No. 6.

31. See, e.g., O. N. Chizhikov, "O nekotorykh voprosakh teorii razvitiia rastitel'nosti," *Agrobiologiia,* 1952, No. 1, pp. 157–172; V. S. Dmitriev, "I. V. Stalin i razvitie sovetskoi agronomicheskoi nauki," *Selektsiia i*

semenovodstvo, 1953, No. 4, pp. 3–13; and F. A. Dvoriankin, "Za darvinizm v teorii vidoobrazovaniia," *ibid.,* 1953, No. 2, pp. 68–76.

32. Lysenko, "Novoe v nauke o biologicheskom vide," *Pravda,* 1950, Nov. 3; cf. his "Vid," *BSE* (2nd ed.), VIII, 14–19.

33. N. V. Turbin, "Darvinizm i novoe uchenie o vide," *BZ,* 1952, No. 6, pp. 798–818. Cf. his "Torzhestvo michurinskoi biologii," *Vestnik leningradskogo universiteta,* 1948, No. 10, pp. 21–51; and his *Genetika* (1950), one of the most widely used Lysenkoite textbooks.

34. For a Lysenkoite demanding the suppression of anti-Lysenkoite publications, see V. S. Dmitriev, "O nekotorykh neobychnykh diskussiakh," *Selektsiia i semenovodstvo,* 1953, No. 2, p. 68. By the time a non-Lysenkoite replied to this polemic, the citation of Stalin was getting to be embarrassing; decrees of the Central Committee and remarks of Lenin served the same purpose, in this case, defense of free discussion. See P. A. Baranov, "O vidoobrazovanii," *BZ,* 1953, No. 5, pp. 693–4.

35. For a convenient collection of Lysenko's articles on fertilizer, see his *Pochvennoe pitanie* (1st ed., 1955) and (3rd ed., 1962). Unfortunately, there is no analogous collection of his articles on butterfat. For representative reports, see S. L. Ioannisian, "Nekotorye voprosy plemennoi raboty v molochnom skotovodstve," *Agrobiologiia,* 1954, No. 1, pp. 4–17; Lysenko, in *Pravda,* 1957, July 17; and Lysenko, "O povyshenii zhirnosti moloka u korov," *Agroliologiia,* 1958, No. 6, pp. 7–22. Cf. the critical review by N. A. Kravchenko, "O nauchnoi rabote po zhivotnovodstvu v 'Gorkakh leninskikh,'" *Zhivotnovodstvo,* 1966, No. 1, pp. 74–89.

36. A. V. Sokolov, "Nedobrokachestvennoe obobshchenie opytov s granulirovannym udobreniiam," *SZ,* 1953, Feb. 3. Neither Lysenko nor his work is mentioned, but the author's attack on the notion that the mixture of organic and mineral fertilizer improves their effectiveness points unmistakably to Lysenko. For Lysenko's angry response, see *Zemledelie,* 1953, No. 1, p. 100.

37. On the transformation of wheat into rye, see E. V. Bobko, "K voprosu o metodike izucheniia obrazovaniia novykh vidov," *BZ,* 1953, No. 3, pp. 401–406. For the hazelnut-hornbeam fraud, see A. A. Rukhkian, "Ob opisannom S. K. Karapetianom sluchae porozhdeniia leshchiny grabom," *ibid.,* No. 6, pp. 885–891. Cf. p. 891 for Lysenko's feeble letter in defense of the fraud, which had been reported in his journal.

38. *Ibid.,* 1955, No. 2, pp. 206–16. At the end of this editorial there is a long bibliography of the discussion of species formation. For another, and presumably fuller bibliography, see *Uchenye zapiski Tomskogo gosudarstvennogo universiteta,* 1956, No. 27, pp. 161–175, which was not available to me.

39. *VF,* the main journal of philosophy, evaded the discussion until the end of 1954, when it came down on Lysenko's side. See No. 6, pp. 47–59 and 116–132, for articles by Kaganov and Platonov.

40. Lysenko made the cuckoo suggestion at a conference in February, 1953. Reported by P. A. Baranov, "O vidoobrazovanii," *BZ,* 1953, No. 5, p. 675.

41. Khrushchev, I, 68. This was included in his famous report to the CC Plenum of September, 1953.

42. See Turbin's biography, with brief bibliography, in *Biograficheskii slovar'*. For examples of further work by the institute he headed in Minsk, see Turbin, ed., *Geterozis,* and Turbin, ed., *Genetika* (1962). For the

upheaval at Leningrad University in 1948, see *Vestnik leningradskogo universiteta,* 1948, No. 10, pp. 109–113. Until the Department of Genetics was restored, Lobashev seems to have worked in an Institute of Physiology.

43. Stoletov made a name for himself by attacking Tulaikov. See V. N. Stoletov, "O vrednykh teoriiakh v agrotekhnike," *SRSKh,* 1933, No. 4–5, pp. 3–25, and especially Stoletov, "Protiv chuzhdykh teorii v agronomii," *Pravda,* 1937, April 11, which virtually called Tulaikov a "wrecker." For some of Stoletov's fulsome tributes to Lysenko, see "Vydaiushchiisia uchenyi," *Izvestiia,* 1940, Dec. 19, and "Put' agrobiologa," in AN SSSR, Inst. Genetiki, *Trudy,* XVI (1948), pp. 5–18. For Stoletov's "scientific" views, see his "Agrobiologiia," *BSE* (2nd ed.), I (1949), 354–361. That he did not change his views in the Khrushchev era is clear in many of his publications, including his chef d'oeuvre, *Vnutrividovye prevrashcheniia.*

44. The journal, or nonperiodical serial, was *Issledovaniia po genetike,* which began to appear in 1961. The textbook was Lobashev, *Genetika.* The last previous textbook had been Delone and Grishko. In 1963 the Belorussian Ministry of Higher Education announced the forthcoming publication of a second edition of N. N. Medvedev's textbook or laboratory guide, which first appeared in the 1930s. This text did not appear when promised. It did appear after the final downfall of Lysenkoism.

45. For an example of Lysenkoite fuming, see A. S. Musiiko et al., "O teoreticheskoi nesostoiatel'nosti i prakticheskoi besplodnosti formal'noi genetiki," *Vestnik sel'skokhoziaistvennoi nauki,* 1964, No. 8, pp. 149–154.

46. Oparin was the only really distinguished biologist who gave really strong support to Lysenkoism, a special distinction that should be known in the scientific community. The most succinct evidence is Oparin's speech to a meeting of the Academy of Sciences in 1948. See *VAN,* 1948, No. 9, pp. 38–44. Subsequently Oparin became Secretary of the Academy's Biology Division, replacing the physiologist Orbeli, who had tried to defend scientific integrity. For one of the most scandalous episodes in Oparin's career, see his chairmanship of *Soveshchanie po probleme zhivogo veshchestva,* the 1950 meeting that enthroned Lepeshinskaia's anticytology. For a brief biography of V. A. Engel'gardt, see *Biograficheskii slovar'.* For Lysenkoite complaints against him, see Lysenko and Nuzhdin, *Za materializm,* p. 25. Cf. Lysenko's comment at *Plenum TsK* (1958), p. 236.

47. In March 1955, the Presidium of the Academy adopted a resolution to improve the publication, but *Izvestiia Akademii Nauk SSSR, seriia biol.* did not change until the second half of 1956. Compare 1956, No. 4 with No. 5. Note, in particular the change in the editorial board in No. 5, and the editorial, pp. 3–10.

48. This was the Dmitriev affair. See pp. 171–2.

49. See Stoletov, *Vnutrividovye prevrashcheniia,* for the book that grew out of the dissertation. For a little public evidence of the continual trouble over Lysenkoite dissertations, see *O polozhenii,* p. 392, and Stoletov's comments on 484–5.

50. N. P. Dubinin, "Ptitsy lesov nizhnei chasti doliny reki Ural," in AN SSSR, Inst. Lesa, *Trudy,* XVIII (1953) and XXXII (1956).

51. From 1955 to 1960 Dubinin seems to have commuted between the Laboratory in Moscow and the Institute in Novosibirsk. Only fragments of this complex story have appeared in the Soviet press. See, in particu-

lar, Oleg Pisarzhevskii, "Pust' uchenye sporiat," *Literaturnaia gazeta,* 1964, Nov. 17; and the reports in *ibid.,* Nov. 24. Cf. the interview with Dubinin in *Ekonomicheskaia gazeta,* 1966, April, No. 17, p. 11. See *Pravda,* 1958, Dec. 14, for an attack on Dubinin. It is noteworthy that G. M. Frank, director of the Institute of Biological Physics, was a Party member, as were other people — for example, the secretary of the Party cell in Novosibirsk — who stood with Dubinin against the highest officials. For Khrushchev's public intervention, see *Pravda,* 1959, July 2.

52. For a convenient survey of laboratories working at genetics, molecular biology, and other slightly disreputable subjects, see *VAN,* 1965, No. 3, pp. 54–61. The author of the survey, N. M. Sisakian, was another renegade from Lysenkoism. See, e.g., his speech at the *Soveshchanie po probleme zhivogo veshchestva,* and his essay in AN SSSR, Inst. Genetiki, *Protiv reaktsionnogo mendelizma-morganisma.* Tsitsin's defection from Lysenkoism began already in the 1940s. See *O polozhenii,* pp. 173, 299, 368; cf. Tsitsin's letter of explanation in *VAN,* 1948, No. 9, p. 85, and a Lysenkoite rejoinder, p. 131. Apparently there was a personal element in Tsitsin's split with Lysenko; see Khrushchev's angry outburst in *Plenum TsK* (1958), pp. 233–4. The director of the Institute of Physical Chemistry was N. N. Semenov, one of the strongest partisans of genetics.

53. The truckling geneticist was S. M. Gershenzon. See his comments at the 1939 conference (*PZM,* 1939, No. 11, p. 118), and at a 1960 conference (Akademiia Nauk Ukrainskoi SSR, *Filosofskie voprosy,* pp. 362–6). The satisfied response of a Lysenkoite at the 1960 conference (p. 397) should have made Gershenzon blush.

54. Hans Stubbe, "Über die vegetative Hybridisierung von Pflanzen," *Die Kulturpflanze,* II (1954), as cited in D. V. Lebedev, "Novye dannye k voprosu o vegetativnoi gibridizatsii," *BZ,* 1955, No. 4, pp. 603–4. Cf. the article by Böhme, summarized in No. 3, pp. 434–7. In Czechoslovakia Lysenkoism seems to have withered while Stalin was alive. See S. I. Alikhanian, "Geneticheskie issledovaniia v Chekhoslovakii," *BMOIP, otdel biol.,* 1961, No. 4, pp. 140–143. At the October, 1958, Vsesoiuznoe soveshchanie po filosofskim voprosam estestvoznaniia, the Lysenkoites were able to produce a Bulgarian to speak for them, but note his denunciation of a Polish symposium, *Biology and Politics,* which drew "anti-Soviet" lessons from the Lysenko affair (*Filosofskie problemy,* pp. 525–6), and note that the Rumanian philosopher who spoke a little later was anti-Lysenkoite (*ibid.,* pp. 548–554). See especially *Agrobiologiia,* for a steady trickle of unconvincing efforts to prove that there was significant support of Lysenkoism in foreign countries.

55. *Plenum TsK* (1964), pp. 352–3. The speaker was the stockbreeder M. M. Lebedev, who also attacked Lobashev's textbook in *Zhivotnovodstvo,* 1964, No. 2. Before the year was over he deserted Lysenkoism.

56. See below, Chapter 8, "Philosophy."

57. See *Pravda,* 1962, June 12, and *VAN,* 1962, No. 8.

58. V. A. Kirillin was the genuine scientist. See his biography in *Biograficheskii slovar'.* S. P. Trapeznikov took his place in 1965. I wish to thank Jerry Hough for information about the political career of Trapeznikov, whose pseudoscholarship is evidenced in his published works. See below, pp. 310–11.

59. *Pravda,* 1958, Dec. 14. Cf. *O polozhenii,* p. 160, for Lysenkoite indignation in 1948 at the *razviaznost' i prenebrezhenie,* the disrespect

and scorn, that scientists had all along displayed toward the practical triumphs of agrobiology.

60. *Plenum TsK* (1958), p. 233.

61. For a convenient compilation of the original draft and the final version, see Triska, p. 118. Note that a sentence endorsing Michurinism was added to the final version, and also a clause calling for "various methods" of studying and controlling heredity.

62. *Izvestiia,* 1963, Jan. 26.

63. "The Michurinist trend in biological science . . . derives from the principle that the conditions of life determine the development of the organic world." *Ibid.* The same words appear in the Party Program. Lest the reader imagine that my translation has obscured what was clear in the original, here it is: "Michurinskoe napravlenie v biologicheskoi nauke . . . iskhodit iz togo, chto usloviia zhizni iavliaiutsia vedushchimi v razvitii organicheskogo mira."

64. T. D. Lysenko, "Teoreticheskie osnovy napravlennogo izmeneniia nasledstvennosti s.-kh. rastenii," *Pravda* and *Izvestiia,* 1963, Jan. 29.

65. See, e.g., Khrushchev, II, 408–9.

66. L. F. Il'ichev, "Metodologicheskie problemy estestvennykh i obshchestvennykh nauk," in Akademiia Nauk SSSR, *Metodologicheskie problemy nauki,* p. 100.

67. See again Il'ichev's speech, and cf. M. A. Ol'shanskii, "Biologicheskaia nauka i sel'skokhoziaistvennoe proizvodstvo," *Kommunist,* 1963, No. 4, pp. 14–26.

68. The journal *Biologiia v shkole* provides steady evidence. In the classrooms there seems to have been a small number of teachers with sufficient knowledge and courage to teach genuine biology. See, e.g., the story told by B. Kh. Sokolovskaia, *ibid.,* 1965, No. 2, pp. 31–6.

69. See below, pp. 289–90.

70. For the most daring attack on the Lysenkoite record in corn breeding, see P. A. Baranov *et al.,* "Problema gibridnoi kukuruzy," *BZ,* 1955, No. 4, pp. 481–507.

71. See, e.g., Akademiia Nauk SSSR, Institut Genetiki, *Nasledstvennost' i izmenchivost'.* Many of the papers read at this jubilee meeting — organized by the Lysenkoites to celebrate the fortieth anniversary of the Bolshevik state — contained only perfunctory reference to "Michurinism," or even none at all.

72. Kushner, *Nasledstvennost'.* See also his unapologetic Russian version of Ivar Johansson, *Handbuch der Tierzüchtung.*

73. See three editions of Iur'ev. Cf. also two editions of Iakushkin.

74. Quoted in Iur'ev, pp. 294, 291, and 241 in successive editions. The original may be found in Lenin, *Sochineniia,* III, 398.

75. A. Maliugin, "Protiv matematicheskogo fetishizma v agronomicheskom nauke," *FNIT,* 1936, No. 12, pp. 35–41.

76. See, e.g., the chemist N. N. Semenov, "O sootnoshenii khimii i biologii," *VF,* 1959, No. 10, pp. 95–102; the chemist I. L. Knuniants, "Taina zhiznennoi sily," *Izvestiia,* 1962, June 22, and "Shifr zhizni," *Komsomolskaia pravda,* 1962, March 17; the physicist P. L. Kapitsa, "Teoriia, eksperiment, praktika," *Ekonomicheskaia gazeta,* 1962, March 26; the geneticist V. S. Kirpichnikov and the biochemist Zh. A. Medvedev, "Perspektivy," *Neva,* 1963, No. 3; the journalist M. A. Popovskii, "Selektsionery," *Novyi mir,* 1961, No. 8, pp. 197–212.

77. Mishnev, p. 10. Cf. p. 31 for another especially forthright statement.

78. Troepol'skii, pp. 205–6.

79. *Ibid.*, p. 123.

80. O. N. Pisarzhevskii, "Pust' uchenye sporiat," *Literaturnaia gazeta*, 1964, Nov. 17.

81. Undoubtedly there were many who spoke the truth that they could not print. Yet it is noteworthy that Ol'shanskii, when attacking critics of agrobiology, mentioned only one, the forest specialist F. D. [*sic;* should be L.] Shchepot'ev, who derided Lysenko's cluster method of planting at a conference in 1963. If we can believe Ol'shanskii, the conference condemned Shchepot'ev's "slander." See SZ, 1964, Aug. 29.

82. Khrushchev, I, 68.

83. V. Koldanov, *BZ*, 1958, No. 5.

84. S. S. Khokhlov, "'Novoe v nauke o biologicheskom vide' i praktika sel'skogo khoziaistva," *BZ*, 1954, No. 3, p. 377.

85. Khrushchev, II, 45. Cf. p. 40, the opening of this speech, in which Khrushchev confessed that "we," the Party leaders, were also responsible for pressing *travopol'e* upon farmers.

86. *Ibid.*, p. 56. Cf. also VI, 322–4, for a blast at a group of Baltic scientists who dared to defend grass.

87. *Ibid.*, VI, 284.

88. *Ibid.* For Khrushchev's praise of Tulaikov's resistance to official pressure, see p. 383 ff. The implication that Tulaikov paid with his life occurs on p. 385. I have taken the liberty of running together remarks that Khrushchev made in different speeches.

89. *Osobo opasnye gosudarstvennye prestupleniia*, p. 126.

90. Khrushchev, VI, 283.

91. For a review of post-Stalin reforms, see Karcz, ed., *Soviet and East European Agriculture*. Sidney Ploss, in his *Conflict and Decision-Making*, makes a commendable effort to discover the connections between agricultural problems and the political process. It seems to me that he fails, largely because he repeatedly misunderstands the agricultural problems, and because he makes unwarranted assumptions about the political process.

92. For a rare publication of Stalin's private expression of this attitude toward the peasantry, see the letter quoted by Khrushchev, in *Pravda*, 1963, March 10. Otherwise one must rely on the words put into the mouths of Stalinist officials by liberal writers. See the review of "The Peasants in Soviet Literature Since Stalin," in Nove.

93. Gertrude Himmelfarb argues that Malthus changed his thought, as well as his mode of expression, in the later editions of the *Essay*. I am not convinced. See, e.g., the 6th ed., II, 479, where Malthus opposes any contraceptive methods "both on account of their immorality and their tendency to remove a necessary stimulus to industry."

94. See the discussion of peasant motivation by liberal Soviet farm officials and specialists, in Troepol'skii, pp. 137–140. Note the similarity to American liberal discussions of the Negro problem: the peasants show little initiative and enterprise because they have been beaten down by exploitation and oppression. See also A. V. Chaianov, *The Theory of Peasant Economy*, where he argues that peasants pay little attention to the value of their own labor. He makes this seem an extension of family

attitudes — a mother does not calculate her wages for cooking meals and sitting up with sick children — and he is probably right. (Compare Solzhenitsyn's heroine, Matryona, who could not turn down any request for her labor.) But I cannot help seeing also in peasant work habits an internalization of the masters' traditional refusal to calculate the price of labor as of other inputs.

95. Between 1928 and 1932 total sown acreage increased 21.4 million hectares, with a decline following in 1933. Between 1953 and 1956 the increase was 37.5 million hectares. Even the rate of increase was greater in the second case. See *Comparisons,* I, 228–229.

96. See Khrushchev's proposal to the Politburo, in Khrushchev, I, 85–100. The appended memo by Lysenko was presumably the same thing that was published in *Izvestiia,* 1954, February 20.

97. A recent biography by Mariagin makes no mention of Lysenko, and pictures A. N. Sokolovskii, a disciple of Williams', as the principal agricultural adviser of Postyshev. Of course, Lysenko was in disgrace when this biography appeared. In any case the published sources of the 1930s suggest that Lysenko was Postyshev's favorite. See, in particular, Postyshev's tribute to Lysenko (*Izvestiia,* 1935, March 15) and Lysenko's tribute to Postyshev (*Soveshchanie peredovikov urozhainosti,* p. 181), which was given at a major public meeting with Postyshev sitting on the platform. For Khrushchev's fond recollection of the Ukrainian beginning of his friendship with Lysenko, see *Plenum TsK* (1962), p. 444. Lysenko boasted of Khrushchev's support already in a 1944 publication: Lysenko; *Nekotorye voprosy,* p. 26.

98. This was the main emphasis of Khrushchev's remarks about Dmitriev. See Khrushchev, I, 157 and 271–2.

99. See S. Stankov's letter in *Pravda,* 1954, March 26. Compare editorials in *Kommunist,* 1954, No. 5, and No. 15.

100. See *Pravda,* 1954, Feb. 5; *Izvestiia,* 1954, Feb. 20; *Pravda,* 1954, April 15.

101. In his famous report to the September 1953 Plenum Khrushchev endorsed Lysenko's summer planting of potatoes (Khrushchev, I, 37). In February 1954, only a few days before Khrushchev criticized Lysenko for protecting Dmitriev, he praised him for good advice on soil treatment (*ibid.,* p. 207). For more praise in following years, see *ibid.,* II, 28, 119–120, 151, 375, 405–10 (note here Khrushchev's rejection of the charge that Lysenko is intolerant of opposition), 514; V, 54, 56, 78, 81–3, 325–7; VI, 57–8, 73, 240–1, 283, 397; VII, 67, 140–2, 147, 452–3, 466–7. See also *Plenum TsK* (1958), pp. 233–4, 468; *Plenum TsK . . . dek. 1959* (M., 1960), p. 398; *Plenum TsK* (1961), pp. 340, 537–8; *Plenum TsK . . . dek. 1963* (M., 1964), pp. 45, 410.

102. Khrushchev, VII, 141.

103. For Polianskii's public endorsements of Lysenko, see *Plenum TsK . . . dek. 1959* (M., 1960), p. 12 (see also Lysenko's thanks, p. 330); *Izvestiia,* 1962, Nov. 4. A careful search would probably turn up more cases. N. V. Podgornyi was another member of the Politburo who gave Lysenko explicit public blessing. Suslov's indirect indications of support are described above on pp. 161 ff.

104. Khrushchev, II, 151.

105. *Ibid.,* p. 409.

106. *Ibid.*

107. *Plenum TsK* (1958), pp. 233–5, 240, 421–33, 468.
108. *Plenum TsK . . . dek. 1959* (M., 1960), pp. 319–332, 398. Note also the speech of the Lysenkoite F. G. Kirichenko, pp. 332–8, fawning gratefully on Khrushchev and Podgornyi.
109. At the January 1961 meeting of the Central Committee Khrushchev made a sarcastic comment that can be interpreted this way. See *Plenum TsK* (1961), p. 340.
110. M. A. Ol'shanskii was one of the earliest Lysenkoites, an original editor of *Biulleten' iarovizatsii* in the early 1930s, the cotton specialist of the Odessa Institute, who promised a breed that would grow in the Ukraine. Though (or because) he never fulfilled the promise, he rose to be vice-president of the Lenin Academy toward the end of Stalin's reign, and became All-Union Minister of Agriculture in 1961. In 1962 he returned to the Lenin Academy as president, and was still at that post when Lysenko's fall forced him into retirement at the age of 57.
111. Khrushchev, VI, 240–241.
112. *Plenum TsK* (1964), p. 41. Note his perfunctory reference to Lysenko on p. 33, as one of a list of meritorious scientists.
113. *Ibid.*, pp. 341, 343.
114. *Ibid.*, pp. 251–3.
115. *Ibid.*, pp. 318–22. Note the soothing syrup at the end of his speech.
116. *Ibid.*, pp. 411–13. For other Lysenkoite speeches, see pp. 305–9, 322–27, 349–53. The Lysenkoites had their characteristic advantage in number of speakers and the right to speak their minds forcefully, but the opposition's comparative restraint now expressed a greater advantage: bureaucratic resistance to uncongenial policies.
117. *Ibid.*, pp. 424–5.
118. See *Pravda*, 1964, Oct. 16, 17. Cf. also *Plenum TsK* (1965), for the follow-up meeting that made the charges public in a decently vague and unruffled form.
119. See below, Chapter 7.
120. See Appendix B for a representative list of such officials.
121. For a brief biography of Lobanov, see *BSE* (2nd ed.), LI (1958), and *Ezhegodnik* (1959). For his post-Khrushchev desertion of Lysenkoism, see *VAN*, 1965, p. 119. Having sat silently through a conference that blasted Lysenko's reputation, Lobanov was asked if he would like to say something. He simply answered "No." See also his historical sketch, "Sel'skokhoziaistvennye nauki," in *SSSR* (M, 1967), pp. 319–21, and his "Sel'skokhoziaistvennaia nauka — proizvodstvu," *Kommunist*, 1967, No. 3, pp. 37–48. For a handy insight into his Lysenkoite period, see his speech in *O polozhenii*, pp. 458–69, and his *Sel'skokhoziaistvennaia nauka.*
122. Leonov, *Russkii les*, p. 363. Another novelist called such people *fliugeroidy*, which literally means weathercocks. Troepol'skii, p. 257.
123. See a brief biography of Benediktov in *BSE* (2nd ed.), *Ezhegodnik* (1962). For Khrushchev's rebuke, see Khrushchev, II, 409.
124. For a brief biography of Matskevich, see *BSE* (2nd ed.), *Ezhegodnik* (1966). For Khrushchev's rebukes on the subject of Lysenkoism, see Khrushchev, II, 409, and *Plenum TsK* (1961), p. 340. The speech reported in *SZ*, 1956, June 22, shows Matskevich to be a very cautious reformer. On the other hand, his frequent praise for eminent Lysenkoites — other than Lysenko himself — and his unreasoning passion for corn

in the mid-1950s (see, e.g., his "Kukuruza — kul'tura neogranichennykh vozmozhnostei," *Zhivotnovodstvo,* 1955, No. 2) reveal more than a touch of the bureaucratic intellectual and the agrobiologist.

125. *Plenum TsK . . . dek. 1963* (M., 1964), p. 215.

126. For Volovchenko's brief biography, see *BSE* (2nd ed.), *Ezhegodnik* (1963). Khrushchev seems to have discovered him in the course of his constant trips of agricultural inspection. See Khrushchev, V, 15–16. Perhaps the conclusive evidence of a bureaucratic shift that the Minister was powerless to oppose is the experience of Ol'shanskii as Minister. See above, p. 174.

127. For one other example, angrily publicized by Khrushchev, see Khrushchev, VI, 322–4. In this case a group of Lithuanian specialists had signed a petition arguing, quite sensibly, that their region should be exempted from the drive against *travopol'e,* since grass is an efficient fodder crop in the Baltic area. In the immediate aftermath of Khrushchev's fall a little group of Baltic officials and specialists called for "Travy bez travopol'ia," *Izvestiia,* 1964, October 25; that is, grass without Williams' crankish doctrine. Note, however, the endorsement of dung-earth composts in this article, further evidence of self-perpetuating ignorance mingled with common sense.

128. It is impossible to give here the evidence supporting this generalization. The cases are too numerous, and many are open to conflicting interpretation. See, e.g., V. Dudintsev's case study in *Komsomolskaia pravda,* 1964, Oct. 23, and 1966, July 8. He presents the plant breeder N. A. Lebedeva as an intransigent defender of science, while her supervisor, S. M. Bukasov, is made to seem a complete opportunist. I consider them both pliable defenders of their discipline. I do not doubt the facts as stated by Dudintsev — Bukasov did not stand up for Lebedeva when Lysenkoites pushed her out of her job — but I note other facts that Dudintsev overlooked. See, e.g., Bukasov and Lebedeva, *Michurinskie metody,* in which the Lysenkoite passages seem to have been written by Lebedeva. It is also noteworthy that Bukasov, the dean of Soviet potato specialists, steadily eliminated Lysenkoite nonsense from his publications in the 1950s.

129. See Lysenko's account of the ministry's resistance, his memo writing, and the grudging issuance of a decree. *Plenum TsK . . . ianv. 1961* (M., 1961), pp. 338–41.

130. In 1961, 1214 Dutch bulls were imported; in 1962 the number rose to 1501. *Zhivotnovodstvo,* 1965, No. 2, p. 20. In February 1964, the director of Lysenko's farm claimed that "over 1000" bulls in Soviet farms were descendants of Lysenko's bulls. *Plenum TsK . . . fev. 1964* (M., 1964), p. 323. Lysenko gave more modest figures, p. 343. In any case, a chronic complaint of the Lysenkoites was that farms which bought Lysenkoite stock subsequently butchered the expensive animals to avoid pollution of their herds.

131. See I. G. Zorin, "Itogi raboty po povysheniiu zhirnomolochnosti korov na Ukraine," *Zhivotnovodstvo,* 1966, No. 3, pp. 11–18; and N. A. Kravchenko, "O nauchnoi rabote po zhivotonovodstvu v 'Gorkakh leninskikh'," *ibid.,* No. 1, pp. 74–89.

132. See *VAN,* 1965, No. 11, *passim.*

133. See, e.g., Vsesoiuznoe soveshchanie rabotnikov, *Materialy,* p. 17, for Matskevich, in a speech dedicated to reform, stubbornly refusing to men-

tion Lysenko, yet praising such an arch-Lysenkoite as F. G. Kirichenko. This habit survived the fall of Lysenko and Matskevich's return to the ministerial office. See Matskevich, "Nashe sel'skoe khoziaistvo vchera, segodnia, i zavtra," *Nauka i zhizn'*, 1965, No. 7, pp. 7–8. Of the six scientists he singled out for praise four had a record of strong support for Lysenkoism, one of compromise with it, and one no connection. Obviously Matskevich was trying to reassure Lysenkoites, but his effort seems excessive, until one realizes that he probably felt genuine sympathy and kinship with them.

134. See again Matskevich's speech at the 1956 meeting, cited in the preceding note. Note on p. 12 his bland comments on intervarietal hybrids of corn, which the Lysenkoites had insisted upon in preference to hybrids of inbred lines. Of all Soviet agricultural officials I pick Matskevich as an example because he showed the *least* tendency to engage in this kind of evasion. For particularly egregious examples, see the editorial in *Pravda*, 1958, Dec. 14, and *Agrobiologiia*, 1963, No. 4, pp. 589–606. The last item, an effort to prove that Lysenko's tree planting campaign was successful, is briefly recapitulated in Ól'shanskii, as cited in note 67, p. 25. Cf. also Trapeznikov, *Istoricheskii opyt*, p. 337, for praise of vernalization by the man who would become the Central Committee's chief of science in 1965.

135. Khrushchev, II, 408.

136. See, e.g., *Komsomolskaia pravda*, 1964, Oct. 23, Nov. 17, 29, Dec. 2; *Literaturnaia gazeta*, 1964, Nov. 17, 24; *Pravda*, Nov. 22. Perhaps the most effective piece was by Anatolii Agranovskii, "Nauka na veru nichego ne prinimaet," *Literaturnaia gazeta*, 1965, Jan. 23. It precipitated the investigation of Lysenko's experimental farm, which brought the final destruction of his reputation. See *VAN*, 1965, No 11, pp. 30–31, 79.

137. V. N. Stoletov, "Protiv chuzhdykh teorii v agronomii," *Pravda* 1937, April 11.

138. V. N. Stoletov, "Usloviia zhizni i razvitie organizmov," *VF*, 1964, No. 10, pp. 45–58. The article is an extended exercise in evasion. Note also the Lysenkoite complaint, in February 1964, that Stoletov was too tolerant of genetics in institutions of higher learning. See above, p. 161.

139. For a particularly astonishing performance, see his keynote speech at the founding convention of a society of geneticists. V. N. Stoletov, "O zadachakh obshchestva genetikov i selektsionerov," *Genetika*, 1966, No. 10, pp. 4–12.

140. See the transformation in the editorial policy, and the editorial staff, of *Biologiia v shkole*, 1965, Nos. 2 through 6. For the new curriculum see Polianskii.

141. See especially G. V. Platonov, "Dogmy starye i dogmy novye," *Oktiabr'*, 1965, No. 8, pp. 149–162; and the subsequent "discussion," "Za partiinuiu printsipial'nost' v nauke," *ibid.*, 1966, No. 2, pp. 144–172.

142. *Biologiia v shkole*, 1966, No. 3, p. 47. The editor himself, Iu. I. Polianskii, wrote these hard hitting words.

143. See, e.g., *Selektsiia i semenovodstvo*, 1965, No. 2, pp. 49–51; No. 3, pp. 36–42; 1966, No. 2, pp. 48–52; No. 6, pp. 19–21; 1967, No. 1, pp. 71–2; No. 6, pp. 65–70. See also *Oktiabr'*, 1966, No. 12, pp. 135–141, for an article that continues Williams' approach to soil science, without mentioning Williams' name.

144. *Literaturnaia gazeta*, 1964, Nov. 17.

145. Many conferences, spreading the word of science on many special topics, were convened in the aftermath of Lysenko's fall. For a complete index, see *Sel'skokhoziaistvennaia literatura SSSR,* 1965 *et seq.* For a convenient survey, see the transformation that occurred in such journals as *Zemledelie, Selektsiia i semenovodstvo,* and *Biologicheskie nauki.* The attentive reader will note a lingering influence of agrobiology.

146. For a bibliography of these articles, see *Biologiia v shkole,* 1965, No. 4, pp. 86–90. Their frequency dropped sharply in the course of 1965. The most comprehensive and incisive was N. N. Semenov, "Nauka ne terpit sub'ektivizma," *Nauka i zhiizn',* 1965, No. 4, pp. 38–43, 132.

147. *VAN,* 1965, No. 2, pp. 5–10.

148. See especially the speech of Astaurov, *ibid.,* p. 103, and the resolution, p. 113.

149. *VAN,* 1965, No. 11.

150. There are a few exceptions to the enforcement of silence on diehards. The conservative literary magazine *Oktiabr'* has published an occasional retrospective defense of agrobiology. See, for example, in 1967, No. 7, pp. 212–15, Davitashvili's review of a novel that dealt with the clash of biology and Lysenkoism. Cf. also G. V. Platonov and M. A. Lisavenko, "Michurin i sovremennaia biologiia," *Vestnik sel'skokhoziaistvennoi nauki,* 1967, No. 4, pp. 118–123.

151. Medvedev, *Rise and Fall.*

152. *VAN,* 1965, No. 11, pp. 80–1.

153. *Ibid.,* p. 125.

7. ACADEMIC ISSUES: SCIENCE

1. See Bernal, *Science in History* (1954), pp. 665 ff. Compare the third ed. (1965), p. 702, where Bernal backs away from his earlier argument, without admitting any error, still clinging to the notion that Lysenko made significant contributions to plant physiology. For general histories of science that give an impression of Lysenko as an important contributor to plant physiology, see Pledge, p. 249, and Taton, p. 465. R. O. Whyte, "History of Research in Vernalization," in Murneek, gives an account that is mostly accurate, though it fails to bring out fully the quackery of Lysenko. The failure is compounded by a picture of Lysenko facing the title page of this scientific symposium, with a caption that naïvely repeats the myth of Lysenko as a master of practical agriculture. The caption derives from Ashby, *Scientist in Ruusia,* which gives a fairly accurate picture of Lysenkoism, except for the uncritical report of his practical successes.

2. See, e.g., Taton, p. 446, for one of many authors who derive Lysenkoism from Lamarckism. See also Gillispie, pp. 346–351.

3. See above, Chapter 4.

4. M. I. Kalinin, "Nauka i liudi truda," *Novaia Petrovka,* 1923, No. 3–4, pp. 5, 7.

5. See the recent survey of occupational preferences, reported by George Fischer, pp. 26–7.

6. Ushakov, III.

7. See Joravsky, p. 255, for one example of this remark and the Stalinist objection to it.

8. See *ibid.,* pp. 204, 264, *et passim.*

9. For one of many unwitting revelations of these attitudes, see Aka-

demiia obshchestvennykh nauk, pp. 137 *et passim,* where *praktik* is used to mean a person holding a professional job though lacking the proper diploma, and the dwindling of this category is reported with pride. Note also, p. 171, that two-thirds of the rural schoolteachers were still *praktiki* in this special sense in the mid-1960s. That sort of statistic suggests that the sense of status difference between diplomaed professional and *praktik* is greater in the Soviet Union than in other advanced countries.

10. I. E. Glushchenko, "Uchenyi iz naroda, T. D. Lysenko," *Biologiia v shkole,* 1937, No. 5, p. 30. Lysenko's parents told a slightly different story about his education. See their letter in *Pravda,* 1936, Jan. 3.

11. *Pravda,* 1927, Aug. 7.

12. The correspondent was imitating the style of L. S. Sosnovskii, chief of agitprop in 1921, and founder of the rural newspaper *Bednota.* See his autobiography in *Deiateli,* XLI, pt. 3, pp. 90–107, in which he boasts that he originated "militant *feuilletons* in a Party organ" on technical aspects of agriculture. After a period of political eclipse because of his Trotskyism, Sosnovskii returned to journalistic agriculture in the 1930s, with "militant *feuilletons*" in support of Michurin and Lysenko. See e.g., *Izvestiia,* 1935, June 8, Oct. 9, and Oct. 30. Subsequently he fell victim to the terror, and remains an unacknowledged pioneer of agrobiology.

13. *Sovetskoe studenchestvo,* 1936, No. 2–3, p. 22. The letter originally appeared in *Pravda,* 1936, Jan. 3.

14. M. B. Mitin, "Za peredovuiu sovetskuiu geneticheskuiu nauku," *PZM,* 1939, No. 10, pp. 174–5.

15. *VAN,* 1948, No. 9, p. 123.

16. For a neat account of this contretemps, see N. A. Maksimov and M. A. Krotkina, "Issledovaniia nad posledeistviem ponizhennoi temperatury na dlinu vegetatsionnogo perioda," *TPBGS,* XXIII (1929–30), No. 2, pp. 465–9. For Lysenko's exposition of the formula concerning "the amount of heat," see his *Vliianie.* See also N. A. Maksimov, "Fiziologicheskie faktory, opredeliaiushchie dlinu vegetatsionnogo perioda," *TPBGS,* XX (1929), pp. 169–212.

17. For Lysenko's articles on vernalization, see the bibliography at the end of his *Agrobiologiia,* especially the sixth ed., 1952, and in *Materialy k biobibliografii uchenykh SSSR. Seriia biologicheskikh nauk. Agrobiologiia. Vypusk 1* (M., 1953).

18. N. A. Maksimov, "Ob uskorenii razvitiia rastenii," *Sorena,* 1932, No. 1, p. 51.

19. The abridged version of Maksimov's speech in *Trudy* [*I*] *vsesoiuznogo s'ezda,* III, pp. 3–20, does not contain these actual words, which are quoted from Maksimov's article in *Sel'skokhoziaistvennaia gazeta,* 1929, Nov. 19.

20. A. G. Shlikhter, "O poseve ozimykh kul'tur vesnoi," *Pravda,* 1929, Oct. 8.

21. *Sel'skokhoziaistvennaia gazeta,* 1929, Nov. 13. For Lenin's interest in Lisitsyn, see Lenin, *Polnoe sobranie,* XLIV, 395, and LIV, 300–1, and 662–3. See also *Istoricheskii arkhiv,* 1962, No. 1, pp. 60–1, and P. I. Lisitsyn, "Pionery semenovodstva," *Nashi dostizheniia* 1929, No. 4..

22. A. Popovskii, *Iskusstvo,* p. 38.

23. There are many reviews of these and later experiments. See, e.g., Whyte, as cited in note 1. See also the extensive bibliography in P. Chouard, "Vernalization and Its Relation to Dormancy," *Annual Re-*

view of Plant Physiology, 11:191–238 (1960). Zirkle, *Death of a Science*, p. 27, turned up instances even earlier than the fairly well-known Ohio report of 1857. For details on the Russian experiments of the nineteenth century, see Maksimov and Krotkina, as cited in note 16, pp. 432–4. For another, very informative Russian survey, see V. N. Liubimenko, "K teorii iskusstvennogo regulirovaniia dliny vegetatsionnogo perioda," *Sovetskaia botanika*, 1933, No. 6, pp. 3–30.

24. Klebs' pioneering book, *Willkürliche Entwickelungsänderungen*, was quickly translated into Russian by K. A. Timiriazev. It is included in Timiriazev, VI. Note especially pp. 332–5, for Klebs' comments on the induction of flowering by chilling. Gassner's 1918 paper is cited and evaluated in all the reviews listed in note 23.

25. Maksimov in Russia and McKinney in the United States established this already in the 1920s. See again reviews cited in note 23. See *Annual Review of Plant Physiology*, XI, 218, for a chart of the various relationships of photoperiodic and chilling requirements of many different plants. The popular work by Fogg, *The Growth of Plants*, is misleading on this point. The statement that "unvernalized plants are indifferent to day length . . ." (p. 272) and what follows sound like Lysenko's incorrect assertions.

26. Polianskii, p. 253.

27. For an encyclopedic survey of the present state of knowledge about plant development, see Ruhland, XV and XVI. See especially A. Lang, "Physiology of Flower Initiation," *ibid.*, XV, pp. 1380–1536.

28. I. M. Tolmachev, "K voprosu o fiziologicheskoi prirode steble-obrazovaniia u ozimei i sakharnoi sveklovitsy," in *Trudy* [*I*] *vsesoiuznogo s'ezda*, III (1929), pp. 539–553. See the reaction of Maksimov and Krotkina, as cited in note 16. Cf. also M. Kh. Chailakhian, "K probleme iarovizatsii rastenni," *Sovetskaia botanika*, 1933, No. 5, pp. 111–133, and No. 6, pp. 30–45.

29. For the decree of the Commissariat of Agriculture granting Lysenko a journal of his own in connection with his mass trial of vernalization, see *Biulleten' iarovizatsii*, 1932, No. 1, pp. 71–2. For Lysenko's favorite places of publication see the bibliographies cited in note 17.

30. In the serials that were published by Lysenko's institutes — *Biulleten' iarovizatsii*, *Iarovizatsiia*, *Agrobiologiia*, and *Trudy Instituta Genetiki* — not the slightest criticism of his views was ever published, except for extremely rare polemical exchanges between himself and an outsider, as in *Iarovizatsiia*, 1939, No. 1, which reprinted his debate with Vavilov on the teaching of genetics. In virtually every other case, the polemics in his periodicals were purely Lysenkoite, without rebuttals or rejoinders.

31. At the August Session of 1948 Lysenko brushed off complaints of repression with the remark that he had been too tolerant in the past. See *O polozhenii*, p. 23. For a characteristically violent Lysenkoite response to the revival of criticism in 1952–53, see V. S. Dmitriev, "O nekotorykh neobychnykh diskussiakh," *Selektsiia i semenovodstvo*, 1953, No. 2, and F. Dvoriankin, "Za darvinizm v teorri vidoobrazovaniia," *ibid.* Not all Lysenkoite polemics were as explicit as Dmitriev in calling for the repression of the opposition, but all took it for granted that the decisions of the August Session must be considered inviolable.

32. Quoted in M. Popovskii, "1000 dnei Akademika Vavilova," *Prostor*, 1966, No. 7, p. 18.

33. Favorov and Kotov, p. 34. This flat declaration is unusual. Ordinarily the Lysenkoites followed their master's example of evading talk about photoperiod in favor of the fuzzy concept of "the light stage." See his *Agrobiologiia,* 1952 ed., with a convenient subject index. In the English edition the light stage, *svetovaia stadiia,* is translated as the "photo phase."

34. A. I. Potapov, et al., "Lzhenauchnye metody Akademika Rikhtera," *Pravda,* 1938, July 26. For a review of Lysenko's evasive hostility to plant hormones, see N. G. Kholodnyi, "V zashchitu ucheniia o gormonakh rastenii," *BZ,* 1954, No. 3, pp. 406–8. From the mid-1950s on, the Lysenkoites tried to deny that they had ever been opposed to such research, and cited in their defense *Problemy biokhimii v michurinskoi biologii* (Moscow, 1949). Actually this slim collection of essays in pseudochemistry was designed to prove the inheritance of acquired characters.

35. Lysenko and Baskova, *Iarovizatsiia,* pp. 4–5.

36. Lysenko seems to have confused potatoes with fruits, whose development is indeed connected with flowering. For a thorough review of the factors that influence tuber formation, see L. E. Gregory, "Physiology of Tuberization," in Ruhland, XV, 1328–54. In view of the evidence that fruit formation in the potato plant may depress the yield of tubers, some specialists recommend the breeding of varieties which will not flower at all. See *Selektsiia i semenovodstvo,* 1946, No. 1–2, p. 80.

37. Feiginson, *Chego dobilis',* p. 41.

38. See Chounard, as cited in note 23. Lysenko was especially angry at I. M. Vasil'ev for his careful experiments with vernalization. See the bibliography at the end of Vasil'ev's biography in Lipshits, II.

39. Chouard, as cited in note 23, p. 215, wrestles to make sense of Lysenko's stage theory, and resigns in defeat, finding it "a faith that does not fall within the scope of scientific experimental analysis."

40. Lysenko, "Iarovizatsiia pozhnivnogo prosa — eto bor'ba za dobavochnyi urozhai," *SZ,* 1934, June 26.

41. See, e.g., Razumov, *Sreda i osobennosti razvitiia rastenii.* For the same author coming close to an admission that the stage theory is an abortion, see his "Upravlenie individual'nym razvitiem rastenii," in Akademiia Nauk, Institut Genetiki, *Genetika — sel'skomu khoziaistvu,* p. 30. Razumov was comparatively astute in his defense of the stage theory. For something closer to vintage Lysenkoism, see Avakian, *Biologiia razvitiia.* See also Kuperman and Rzhanova, *Biologiia razvitiia rastenii.*

42. See, e.g., V. I. Razumov, "Vzaimovliianie dliny dnia i iarovizatsii rastenii v ontogeneze ozimykh pshenits," in *Erevanskii simpozium* pp. 138–152. Compare Razumov, "Dvadtsat' piat' let teorii stadiinogo razvitiia," in *Itogi i perspektivy,* pp. 34–49, for his outlook while Lysenko had power. Contrast also Kuperman's speech, *ibid.,* pp. 50–65, with the articles she has published since Lysenko's fall.

43. President Keldysh, while telling the Academy of Sciences in February 1965, that Lysenko would no longer be immune to criticism, warned against wholesale rejection of his stage theory, which, "in the opinion of several eminent scientists, has scientific significance." *VAN,* 1965, No. 3, p. 9. Cf. also the official *Razvitie biologii v SSSR* (M., 1967), pp. 135–6, for a sketchy history of the stage theory, which states that the significance of the theory was "exaggerated," but fails to make clear what that significance was.

44. See citations in note 19. The newspaper article of Nov. 1929 contains the words quoted, but otherwise marks the beginning of Maksimov's departure from truth-telling.
45. N. A. Maksimov, *The Theoretical Significance of Vernalization.*
46. See the review by A. I. Vorob'ev in *Iarovizatsiia,* 1935, No. 1, pp. 127–8. Cf. A. A. Avakian, "O tak nazyvaemoi 'iarovizatsii' rastenii svetom," *ibid.,* pp. 65–107.
47. For a brief account of the meeting, see *Sotsialisticheskoe zernovoe khoziaistvo,* 1936, No. 3, pp. 115–16.
48. Compare Maksimov, *Kratkii kurs fiziologii rastenii* (2nd ed., 1929), p. 325, where he describes Gassner's work without mentioning Lysenko, to the same passage in (3rd ed., 1931), where he has inserted a brief tribute to the practical value of Lysenko's work. In (4th ed., 1932), pp. 272–3, there is a completely uncritical repetition of Lysenko's views on vernalization. Subsequent editions, culminating in (8th ed., 1948), increased such passages. See also the posthumous (9th ed., 1958).
49. *VAN,* 1948, No. 9, pp. 92–6. Cf. pp. 133–5 for Avakian responding with undiminished hostility.
50. N. A. Maksimov and P. A. Genkel', "Teoriia stadiinogo razvitiia i ee znachenie dlia fiziologii rastenii," *Zhurnal obshchei biologii,* 1949, No. 1, pp. 3–12. The tribute to Maksimov, *Pamiati Akademiku N. A. Maksimov,* is cruelly silent about the true significance of these enforced lapses from the truth as Maksimov knew it.
51. See T. D. Lysenko and D. A. Dolgushin, "Uskorenie razvitiia kartofelia v polevykh usloviiakh sotsialisticheskogo khoziaistva," *Biulleten' iarovizatsii,* 1932, No. 2–3, pp. 35–46.
52. For a Lysenkoite celebration of Razumov, see *BZ,* 1962, No. 12, pp. 1855–7. His eclipse since the end of 1964 has not been total. See the paper that he published in 1966, as cited in note 42.
53. For Vlasiuk's sudden and vigorous turn against plant hormones in 1948, see Kholodnyi, *BZ,* 1954, No. 3, p. 407. It must not be imagined that Vlasiuk had neglected to praise agrobiology before 1948. See his biography with bibliography in Lipshits, II. For a further list of his administrative positions, see *Biograficheskii slovar',* I, p. 176.
54. See above, p. 168.
55. P. A. Vlasiuk and P. Z. Lisoval, "Vliianie organicheskikh i mineral'nykh udobrenii," *Agrobiologiia,* 1965, No. 1, pp. 10–17.
56. See his articles, as listed in *Sel'skokhoziaistvennaia literatura SSSR,* 1967–68, *passim.*
57. For the insinuation of Sabinin's disloyalty in a report of his opposition to Lysenkoism, see the newspaper of Moscow University, *Za proletarskie kadry,* 1937, June 2. For an angry response to his ridiculing of Lysenkoism in the lecture hall, see *O polozhenii,* p. 310. For Sabinin's tiny bit of tribute to Lysenkoism, in the midst of criticism, see *VAN,* 1938, No. 7–8, pp. 55–61 (the report of a commission that included Sabinin), and *Literaturnaia gazeta,* 1947, Nov. 29 (a letter signed by Sabinin and others). See also Sabinin, *Fiziologiia razvitiia,* pp. 105–6 and 153.
58. See Medvedev, p. 128. Chailakhian's tribute to Sabinin in *Fiziologiia rastenii,* 1965, No. 5, pp. 761–774, withholds the grimmest details.
59. Sabinin, *Fiziologiia razvitiia,* pp. 105–6 and 153.
60. From the early 1930s, when Maksimov cited Chailakhian's work

as a refutation of Lysenko's intuitions, the Lysenkoites expressed a continuous hostility to Chailakhian. But he would not quit the kind of research that undermined their pet beliefs, and he gave only rare and ambiguous tributes to them. See, e.g., the letter he wrote with A. A. Rikhter, in *SZ,* 1934, Dec. 26. Cf. the Lysenkoite attack on both men, in *Pravda,* 1938, July 26; and A. A. Avakian, "Stadiinye protsessy i tak nazyvaemye gormony tsveteniia," *Agrobiologiia,* 1948, No. 1, pp. 47–77. For a very thorough review of the issues in the prolonged conflict, see M. Kh. Chailakhian, "Osnovnye zakonomernosti ontogeneza vysshikh rastenii," in *Itogi i perspektivy,* pp. 5–33.

61. See Kholodnyi, *BZ,* 1954, No. 3, pp. 408–9, where he argues that Chailakhian's views on plant hormones are indeed tainted with the decadent bourgeois notions that are mistakenly charged against his own views. Chailakhian did not stoop to this kind of argument.

62. This is an intuitive generalization of much biographical and institutional information. The reader who wishes to get some notion of the people and institutions involved might begin with N. A. Maksimov's historical survey, "Fiziologiia rastenii," in *Ocherki po istorii russkoi botaniki,* pp. 211–273.

63. See A. Lang, "Physiology of Development — A New Era?" in Ruhland, XV.

64. Even in those years, non-Lysenkoite work on the physiology of plant development was possible, if the author paid proper tribute to the famous stage theory. For example, A. I. Rudenko praised Lysenko's "stages," and then devoted his book to a non-Lysenkoite analysis of "phases" in the life of plants. For a publication on "growth substances," see I. N. Konovalov, "Izmenenie morfologii plodov shipovnika pod deistviem rostovykh veshchestv," *Trudy Instituta fiziologii rastenii,* 1950, No. 1, pp. 241–246.

65. D. Hilbert, "Naturerkennen und Logik," *Die Naturwissenschaften,* 1930, p. 960.

66. For a convenient reprint, see Peters, p. 2.

67. See Olby, *Origins, passim;* and E. B. Gasking, "Why Was Mendel's Work Ignored?" *Journal of the History of Ideas,* 20:60–84 (Jan., 1959); and William Coleman, "Cell, Nucleus, and Inheritance," *Proceedings of Am. Phil. Soc.,* 109:124–158.

68. There is a very large Soviet literature on Timiriazev, most of it spoiled by tendentious interpretation. He lived long enough to win Lenin's admiration by endorsing the Bolshevik Revolution. The Lysenkoites turned admiration into veneration, mercilessly twisting his thought to make it resemble their own. The best way to discover Timiriazev's actual views is to read his own works, which still capture and hold the reader's interest. For his views on heredity, Lamarckism, Mendelianism, and such, consult the subject index of Timiriazev, X.

69. For intelligent summaries of this outlook, see A. P. Vladimirskii, *Peredaiutsia li,* and, by the same author, "Sovremennoe sostoianie voprosa o nasledovanii priobretennykh priznakov," in *Nauka XX veka,* II. See also Komarov, *Lamark,* Chapter VIII, and, by the same author, "Lamark i ego nauchnoe znachenie," in Lamark [i.e., Lamarck], *Filosofiia zoologii,* I, xi–xcvi.

70. There are many useful studies of the history of genetics and evolutionary doctrines. To cite a few here would be to invite the erroneous

thought that they are responsible for the one-sided interpretation that I have offered above. I have emphasized — perhaps exaggerated — those aspects of the history of genetics that are relevant to the Lysenko affair. Those who wish to see a balanced history should begin with Dunn, *A Short History of Genetics,* and proceed with Dunn, ed., *Genetics in the Twentieth Century.* Those who wish to correct the usual neglect of Russian writers should read Gaisinovich.

71. Komarov, *Izbrannye,* XII, 251, for a ritualistic tribute inserted in the 1938 edition of his *Origin of Cultivated Plants.* Cf. also the successive editions of his *Proiskhozhdenie rastenii,* from the first in 1933 through the seventh in 1943. For Komarov's continuing, historical interest in Lamarck see the works cited in note 69. Evidence of Komarov's support for the geneticists abound, *inter alia* in the hostility of leading Lysenkoites. See I. I. Prezent, "Uchenie Lenina o krizise estestvoznaniia i krizis burzhuaznoi biologicheskoi nauki," in Ral'tsevich, p. 248; and Shlykov, pp. 110–113, *et passim.* It is therefore impossible to take seriously the extravagant tribute to Lysenko which is attributed to Komarov by Stoletov in SZ, 1945, June 12.

72. *Spornye voprosy* p. 67. For other denials of Lamarckism see pp. 57 and 327.

73. Favorable references to Lamarckism are scattered through the Lysenkoite speeches at the August Session of 1948. See *O polozhenii,* pp. 11, 14, and especially 185, 304–5, and 507. For subsequent Lysenkoite celebrations of Lamarck, see *Istoriia estestvoznaniia,* III, 361. See also Prezent, *Biolog-materialist,* and Poliakov, *Zh.-B. Lamark,* especially the final chapter.

74. T. D. Lysenko, "Fiziologiia razvitiia rastenii v selektsionnom dele," *Semenovodstvo,* 1934, No. 2, p. 25.

75. *Ibid.,* p. 28. It hardly needs to be said that Lysenko's fundamental assumption — the earlier the variety the better — and his use of the stage theory to analyze earliness — short vernalization stage plus short light stage — were mistaken in the first case and senseless in the second, quite apart from his further assumptions about the hereditary pattern of earliness and lateness.

76. One of the most admirable of these silent conformists was the zoologist and geographer, L. S. Berg. In his *Nomogenez* he rejected the Darwinian analysis of evolution as the enthronement of chance, the subversion of natural law. For his frank endorsement of teleology he was attacked by Soviet Marxists as well as by geneticists. His responsive silence was therefore defiant rather than meek; he would not give a self-criticism even when a letter in *Pravda,* 1939, Jan. 11, connected him with fascist ideology. Toward the end of his life he was officially lionized for his patriotic writings on the history of Russian science. A lesser man might have added to his official status by endorsing Lysenkoism, which resembled Berg's nomogenesis as, say, Billy Graham's views resemble Paul Tillich's. But Berg maintained his silence on genetics and evolution. Similar courageous silence was maintained by a number of distinguished old biologists who held more or less Lamarckist views. Indeed, B. A. Keller was the only notable person of this type who joined the Lysenkoites.

77. See *Agrobiologiia* and other Lysenkoite journals in 1953, as they responded to the first, very limited criticism of the master's theory of species formation.

78. Lysenko, *Agrobiologiia* (1949), p. 486. Cf. the 1952 ed., p. 456, for a slightly different version, and the English translation of 1954, pp. 415, 476, 482, *et passim.*

79. *Spornye voprosy,* p. 185. The speaker was P. N. Iakovlev, one of Michurin's disciples who became a leading Lysenkoite.

80. *Iarovizatsiia,* 1937, No. 4 (13), p. 25.

81. There is a very large Lysenkoite literature on this subject. For a critical review, see Kraevoi.

82. The people were T. K. Enin and N. I. Ermolaeva. Their articles are cited in A. N. Kolmogorov, "On a New Confirmation of Mendel's Laws," *Doklady Akademii Nauk SSSR,* 1940, XXVII, No. 1, pp. 37–41.

83. *Ibid.*

84. T. D. Lysenko, "In Response to the Article by A. N. Kolmogorov," *ibid.,* 1940, XXVIII, No. 9, pp. 832–3. Note that E. Kol'man's reply to Kolmogorov, pp. 834–8, is less extravagant than Lysenko's but makes essentially the same point.

85. Lysenko, "Po povodu stat'i akademika N. I. Vavilova," *SZ,* 1939, Feb. 1. Reprinted in *Iarovizatsiia,* 1939, No. 1.

86. Lepeshinskaia, *Proiskhozhdenie kletok,* with an introduction by Lysenko. For her earliest major articles on this subject, see *Biologicheskii zhurnal,* 1932, No. 2, and 1934, No. 2, and *PZM,* 1935, No. 2.

87. See, e.g., O. B. Lepeshinskaia, "Otvet M. S. Navashinu," *PZM,* 1937, No. 2, pp. 139–140.

88. See, e.g., his remarks on cytology in *Spornye voprosy,* pp. 70–1.

89. The highpoint was reached at the *Soveshchanie po probleme zhivogo veshchestva.* See especially Lysenko's speech, pp. 109–112.

90. For the first article of this new faith, see Lysenko's "Estestvennyi otbor i vnutrividovaia konkurentsiia," originally given as a lecture in November 1945, and widely reprinted, e.g., in his *Agrobiologiia* (1946 and subsequent editions).

91. See above, pp. 138–9. Cf. also *O polozhenii,* p. 378, for Lysenko brushing aside his denial of intraspecific competition as irrelevant to the issues of the meeting.

92. The originator of this sort of work was Lysenko's junior colleague, V. K. Karapetian. See his "Izmenenie prirody tverdykh pshenits v miagkie," *Agrobiologiia,* 1948, No. 4, pp. 5–21. The major article on the wheat-rye transformation was by a genuine wheat specialist who should have known better. See M. M. Iakubtsiner, "Materialy k voprosu o nakhozhdenii zeren rzhi v kolos'iakh pshenitsy," *ibid.,* 1952, No. 1.

93. The work with viruses and bacteria was done by G. M. Bosh'ian. See his *O prirode virusov i mikrobov.* He made a big enough splash to be reported seriously in *BSE* (2nd ed.), VIII (1951), 158, and to be the object of an official investigation and refutation in 1954. For the transformers of plant into animal tissue and chicken into rabbit, see the critical report by V. P. Efroimson, "O roli eksperimenta i tsifr v sel'skokhoziaistvennoi biologii," *BMOIP, otedel biol.,* 1956, No. 5, p. 84.

94. See, e.g., Lysenko and Nuzhdin, *Za materializm.* Nuzhdin's chapters purvey the sophisticated new views; Lysenko's are as crude as ever, though he blandly denies that he ever tried to outlaw such techniques as inbreeding and the use of radiation to induce mutations. For a massive presentation of sophisticated Lysenkoism, see Feiginson, *Korpuskuliarnaia genetika.* See also the journal *Agrobiologiia* during its last ten years for

a steady stream of translations and summaries of Western articles, with editorials purporting to show that Western scientists are willy nilly coming around to the Lysenkoite outlook.

95. A. N. Studitskii is the man. See his "K diskussii o vidoobrazovanii, *Uspekhi sovremennoi biologii,* 1954, No. 2 (5). After Lysenko's fall he continued to be not only an important editor but also head of a laboratory at the Academy of Sciences. See Medvedev, p. 239.

96. See, e.g., *Selektsiia i semenovodstvo,* 1965, No. 3, pp. 40–2; 1966, No. 6, pp. 19–21; 1967, No. 6, pp. 65–70. A rare Lysenkoite item also makes its appearance. See the book favorably reviewed in 1967, No. 1, pp. 71–2.

97. An approach to such a textbook was made by Turbin, *Genetika s osnovami selektsii,* but he defected in 1952, and by 1954 the Lysenkoites were retreating, if only tacitly, from some of the extreme positions that he took in this textbook. For another, less-than-authoritative textbook, see Vorob'ev, *Osnovy michurinskoi genetiki,* which had only two editions, 1950 and 1953. From the establishment of Lysenkoite dominance over biological education in the late 1930s until the fall of Lysenko in 1964 the most popular textbook of genetics and breeding — such a linkage was mandatory — was Iur'ev et al., *Obshchaia selektsiia.* It is a chimera, that is a hodgepodge of Lysenkoite and scientific passages. During the quarrel over biological education in the late 1930s Lysenko said that an anthology of Michurin's writings would be a suitable textbook of genetics (*PZM,* 1939, No. 11, p. 160). When he fell in 1964 his school was no closer to an authoritative textbook.

98. See Lévi-Strauss, *The Savage Mind.*

99. S. S. Chetverikov, "O nekotorykh momentakh evoliutsionnogo protsessa s tochki zreniia sovremennoi genetiki," *Zhurnal eksperimental'noi biologii, seriia* A, 1927, No. 1, pp. 4–5. There is an English version of this pioneering article in *Proceedings of the American Philosophical Society,* 105:167–195 (April, 1961).

100. See the delightful quotation adduced by P. A. Baranov, "O vidoobrazovanii," *BZ,* 1953, No. 5, p. 674.

101. Lysenko, *Novoe v nauke,* pp. 27–8.

102. The search was done by Iakubtsiner, as cited in note 92. For a critique of his methods, see E. V. Bobko, "K voprosu o metodike izucheniia obrazovannia novykh vidov," *BZ,* 1953, No. 3, pp. 401–3.

103. See Lysenko's letter defending Karapetian, in *BZ,* 1953, No. 6, p. 891. The proof of fraud immediately precedes Lysenko's letter. See also *ibid.,* 1954, No. 6, pp. 882–9. For the exposé of the pine-spruce transformation, see *ibid.,* 1953, No. 3, pp. 418–21. Conscious fraud was not as common among Lysenkoites as many observers have assumed; "far more often the conclusions of [Lysenkoite] authors are disproved by the very data that they adduce in support of their sensational conclusions." Efroimson, as cited in note 93, p. 84.

104. Lysenko himself made a few gestures in this direction. See, for example, his "Teoreticheskie osnovy napravlennogo izmeneniia nasledstvennosti s.-kh-ykh rastenii," in Akademiia Nauk, Institut Genetiki, *Genetika — sel'skomu khoziaistvu,* p. 11. For his followers' extensive efforts along these lines, see works cited in note 94.

105. Lysenko, "Rezul'taty opytnykh i proizvodstvennykh posevov lesnykh polos gnezdovym sposobom," SZ, 1952, April 3.

106. Lysenko, "O biologicheskom vide i vidoobrazovanii," *Agrobiologiia,* 1956, No. 7. p. 18.

107. Lysenko, as cited in note 104, pp. 14–15.

108. The resemblance was noted, for example, between the speculations of the Lysenkoites and the frankly anti-Darwinian writings of S. I. Korzhinskii, a prerevolutionary botanist. The response of the Lysenkoites was usually to deny kinship with him, and we should take them at their word. Chances are they had never heard of him until their critics raised his name. But then one Lysenkoite read Korzhinskii and decided that he did not deserve the epithet vitalist. See B. G. Ioganzen's defense in *BZ,* 1962, No. 6, pp. 879–885. For an accurate analysis of Korzhinskii's views, see Timiriazev, VII, 229–31 *et passim;* also VI, *passim.*

109. See, e.g., G. B. Medvedeva, "Oplodotvorenie rastenii i praktika rastenievodstva," in Akademiia Nauk, Institut Genetiki, *Genetika — sel'skomu khoziaistvu,* pp. 71–81. For the cruder Lysenkoite writings on the subject, see the earlier citations in her bibliography. For analogous work in animal fertilization, see M. M. Lebedev, "Oplodotvorenie u sel'skokhoziaistvennykh zhivotnykh," *ibid.,* pp. 82–93. For a more sophisticated Lysenkoite treatment, which ignores the early writings and shows considerable knowledge of genuinely scientific studies of plant fertilization, see I. M. Poliakov, "Sovremennoe sostoianie problemy izbiratel'nosti oplodotvoreniia u rastenii," in Akademiia Nauk, Institut Genetiki, *Nasledstvennost' i izmenchivost'* I, 106–120. For a scientific review of selective fertilization, see O. K. Miriuta, "Problema selektivnosti oplodotvoreniia u rastenii," *Genetika,* 1967, No. 5, pp. 148–161.

110. See Lysenko's account in *PZM,* 1939, No. 11, pp. 149–151.

111. Lysenko and Prezent, writing to the Commissariat of Agriculture, as quoted in *Iarovizatsiia,* 1940, No. 5, p. 4. Cf. Lysenko's explanation cited in the previous note, which excepts certain plants from the operation of this rule.

112. M. M. Zavadovskii, "Protiv zagibov v napadkakh na genetiku," *SRSKh,* 1936, No. 8, p. 93. Cf. C. H. Waddington, "Talking to Russian Biologists," *The Listener,* 1963, Jan. 17, pp. 119–121, for a scientist trying to take the Lysenkoites seriously, but breaking into giggles when he comes to marriage for love.

113. See above, p. 103.

114. M. M. Zavadovskii, "Genetika, ee dostizheniia i bluzhdaniia," in *Sbornik diskussionnykh statei,* p. 76.

115. Elsewhere M. M. Zavadovskii tried to be specific. See *Spornye voprosy,* pp. 402–3, where he credited Lysenko with great enthusiasm to help agriculture. When someone broke into this speech with the question, "Is there anything besides enthusiasm?" Zavadovskii backed off from an answer.

116. A. S. Serebrovskii put it plainly, with his usual audacity: "Some people who have been connected with Serebrovskii have begun to wonder whether they shouldn't run away from Serebrovskii before it's too late. There are such fainthearted comrades. I myself am not fainthearted, and I don't place my wager on the fainthearted." *Spornye voprosy,* p. 448. It should be borne in mind that Serebrosvkii was a dangerous man not only because he was exceptionally outspoken, but also because, as chief planner of stock breeding, he was blamed for its terrible shortcomings. Yet he would not yield an inch to the savage attacks on his science and

himself. When he died, in June 1948, his sympathizers turned his burial into a public demonstration, at the Novodevichii Cemetery in Moscow. See Glushchenko's indignation, *VAN*, 1948, No. 9, p. 61.

117. N. Vavilov, "Blizhaishie zadachi sovetskoi agronomicheskoi nauki," *Front nauki i tekhniki,* 1936, No. 2, p. 63.

118. *Spornye voprosy,* p. 228. The defecting protégé was G. N. Shlykov.

119. See D. Joravsky "The Vavilov Brothers," *Slavic Review,* 24:381–394 (Sept. 1965).

120. In the letters of N. K. Kol'tsov to L. C. Dunn, 1928–1930, a supercilious attitude toward Zhebrak is clearly expressed. (Professor Dunn kindly allowed me to examine his papers.) Not from Dunn, but from other scientists who admired Zhebrak's courage, I have heard sneers at his scientific abilities. Very likely he heard them too.

121. A. R. Zhebrak, "O nekotorykh itogakh chetvertoi sessi VASKhNILa," *Izvestiia Akademii Nauk SSSR, seriia biol.,* 1937, No. 3, pp. 686–695.

122. The appeal is *ibid.,* pp. 671–702. The attack on Zhebrak is included in a report of turmoil at the Timiriazev Academy, in *SZ,* 1937, April 8.

123. See, e.g., *SZ,* 1939, June 14, for Zhebrak's defense of genetics courses. At the conference of October 1939, his speech was so militant that the reporter in *PZM,* 1939, No. 11, p. 98, would not repeat his argument. Another of his militant speeches was quoted above, p. 136.

124. He gave a strong speech at the August Session (*O polozhenii,* pp. 393–401), and did not join those who apologized as soon as Lysenko revealed that the Central Committee had approved his speech. Zhebrak's letter is in *Pravda,* 1948, Aug. 15.

125. Meister, *Kriticheskii ocherk.*

126. *Spornye voprosy,* p. 476. The words were spoken by A. I. Muralov, president of the Lenin Academy. The wide reprinting of Meister's speech showed that Muralov was not speaking for himself alone. Meister's speech may be found in *ibid.,* pp. 406–432.

127. Of the 77 plant breeders who spoke, 37 evaded the controversy, 11 attempted a compromise, 15 took a Lysenkoite stand, and 14 were anti-Lysenkoite. The proceedings of the conference were printed in *Spornye voprosy,* the record of the avowedly controversial sessions, and in *Dostizheniia sovetskoi selektsii,* the record of the supposedly noncontroversial sessions. The reader will note however that 19 speakers in the second volume took a stand in the controversy.

128. The anti-Lysenkoite breeder was P. M. Zhukovskii, the distinguished Lysenkoite was P. P. Luk'ianenko. See *O polozhenii,* pp. 347–351 and 383–393.

129. See *ibid.,* pp. 292, 321, 474, 505, 519.

130. V. Ia. Iur'ev was a striking example of an outstanding plant breeder who spoke for Lysenkoism already in the 1930s (see, e.g., *Iarovizatsiia,* 1938, No. 1–2, pp. 147–151, and 1939, No. 2, pp. 117–122), yet did his work in such a way that the geneticists were still claiming him at the August Session. To the end of his ambiguous life in 1962 he continued to work both sides of the street. He could, for example, publish an article in *SZ,* 1954, Aug. 6, which was hailed by Lysenko's opponents for its implicit criticism of Lysenkoite methods (including one that Iur'ev had

endorsed in the 1930s), and he could also appear as a feature speaker at a major Lysenkoite conference, using the occasion not to endorse Lysenkoism but to criticize shortcomings in plant breeding. See V. Ia. Iur'ev, "Voprosy selektsii pshenitsy," in Akademiia Nauk, Institut Genetiki, *Nasledstvennost' i izmenchivost'*, I, 65–9.

A list of plant breeders who paid tribute to Lysenkoism would be very long. A few, like V. S. Pustovoit, a distinguished improver of the country's chief oil-bearing crop (sunflower), seem to have been genuine sympathizers. Others, like the cotton breeder S. S. Kanash, were probably hypocritical in their repeated, emphatic endorsements of Lysenkoism. Most, however, seem to have been pliable men of principle like the potato breeder A. G. Lorkh or the corn breeder B. P. Sokolov, willing to give only the minimum tribute that seemed necessary to keep their work going. These are intuitive generalizations of much evidence.

131. The original staff is listed and described in Ia. Ia. Lus et al., "Vazhneishie rezul'taty raboty Instituta genetiki AN SSSR," *Izvestiia Akademii Nauk SSSR, seriia biol.*, 1937, No. 5, pp. 1469–92. The staff members who held their jobs past 1940 can be discovered by consulting Akademiia Nauk, Institut Genetiki, *Trudy*, Nos. 13 *et seq*.

132. Within the Institute of Genetics in the 1930s S. Ia. Kraevoi and N. S. Butarin were such people. S. M. Gershenzon was a would-be conciliator, who persistently played into the hand of the Lysenkoites. (See, e.g., *PZM*, 1939, No. 11, pp. 118 and 209; also Akademiia Nauk Ukrainskoi SSR, *Filosofskie voprosy*, pp. 362–6 and 397.) M. S. Navashin, after losing his laboratory because of his vigorous opposition to Lysenkoism, published an incredible endorsement of Lysenkoism: "O roli nekletochnogo zhivogo veshchestva v protsesse vosproizvedeniia u rastenii," *Izvestiia Akademii Nauk SSSR, seriia biol.*, 1952, No. 5, pp. 8–32. I do not wish to besmirch the reputation of these people; I only wish to show that transparent hypocrisy marked the few instances of Lysenkoite phrasemongering by geneticists. Cf. the case of Lobashev, reported on p. 121.

133. See his article on "wrecking," in *Moskovskii universitet*, 1938, May 9. See his article on the textbook of Delone and Grishko, *ibid.*, Sept. 16.

134. See Alikhanian's brief report in *PZM*, 1939, No. 11, pp. 94–5, and 104. See Mitin's praise, *ibid.*, No. 10, pp. 146, 155.

135. *O polozhenii*, pp. 358–370.

136. *Ibid.*, pp. 525–6.

137. That was Kh. S. Koshtoiants. See *VAN*, 1948, No. 9, p. 67.

138. *O polozhenii*, pp. 525–6.

139. S. I. Alikhanian and L. N. Borisova, "Vegetativnaia gibridizatsiia gribov iz roda Penicillium," *Izvestiia Akademii Nauk SSSR, seriia biol.*, 1956, No. 2. Nonsense concerning vegetative hybridization in microorganisms was begun by N. F. Gamaleia. See his "K voprosu ob izmenchivosti mikrobov," *Agrobiologiia*, 1946, No. 3.

140. Alikhanian, *Teoreticheskie osnovy*.

141. The only genuine exception was V. S. Kirpichnikov, who published, with the biochemist Zh. A. Medvedev, "Perspektivy," *Neva*, 1963, No. 3. See also the negative review of a Lysenkoite book, in *BMOIP, otdel biol.*, 1961, No. 3, pp. 153–158. The ecologist A. A. Nasimovich was the reviewer. This sort of restrained criticism on restricted technical issues

continued. See, e.g., L. I. Lipaeva, "O prirode geterozisa u rastenii," *BMOIP, otdel biol.*, 1961, No. 5, pp. 108–11.

142. Oleg Pisarzhevskii, "Pust' uchenye sporiat," *Literaturnaia gazeta*, 1964, Nov. 17. For a list of some Varangians, see *ibid.*, 1964, Nov. 24. For another list, which includes one man who does not deserve the honor (B. M. Kedrov), see *Genetika*, 1967, No. 10, p. 13.

143. See *VAN*, 1948, No. 9. The especially virulent denunciations occur on pp. 59–66 (Glushchenko's speech), 79–84 (Kushner's), 132 (Avakian's), and 155 (Grashchenkov's).

144. *Ibid.*

145. *Ibid.*, pp. 85–91. Perhaps Khrushchov was especially frightened because he had been a colleague of Levit's in the 1930s. But it was hardly necessary for him to heap abuse on other people, or to give repeated endorsements in the next few years to Lepeshinskaia's anticytology.

146. The neurophysiologist was N. I. Grashchenkov (*ibid.*, pp. 148–155). The two philosophers were M. B. Mitin (112–19), and G. F. Aleksandrov (135–45).

147. *Ibid.*, p. 53.

148. See Glushchenko's autobiography in Lipshits, II. Note especially the first eight items in his bibliography.

149. For the Council of Commissars forcing Vavilov to start Lysenkoite work in the Institute, see above, pp. 106–7. I can imagine no other reason for Glushchenko's move into the Institute in 1939.

150. Glushchenko's progress in sophistry can be measured by comparing his *Vegetativnaia gibridizatsiia* with his *Nasledstvennost'*. Kushner and Nuzhdin, who knew what they were doing from their first adherence to the Lysenkoite movement, simply wrote what the times called for, crude or sophisticated as the occasion required.

151. See Glushchenko, *U zarubezhnykh druzei*, and his *Strany, vstrechi, uchenye*.

152. See, e.g., Haldane's editorial comment in *Evolution*, VII, p. xi. (The volume contains typical essays by Waddington and Hinshelwood.) Scattered through Haldane's *Biochemistry of Genetics* are similar comments on "the Soviet school." Haldane assumed that untranslated works which he had not seen would show the Lysenkoites to be reasonable critics of prevailing views in genetics. He could not bring himself to face the facts of Soviet intellectual life, for they seemed to reinforce the smug superciliousness of the average Western philistine. Cf. Donald Michie, "The Moscow Institute of Genetics," *Discovery*, 1957, No. 10, pp. 432–4. See also Michie, "The Third Stage in Genetics," in Barnett, ed., *A Century of Darwinism*, pp. 56–84.

153. In the middle of 1965 Glushchenko quit the editorial board of Lysenko's journal, *Agrobiologiia*, and, following the dissolution of Lysenko's Institute of Genetics, went to work in Kiev with a leading opportunist of plant physiology, P. A. Vlasiuk. For refernces to his publications since then, see *Sel'skokhoziaistvennaia literatura SSSR*, 1965 to the present. Note in particular I. E. Glushchenko, "Izmenenie nasledstvennykh priznakov u tomatnykh rastenii pod vozdeistviem x-luchei," *Vestnik sel'skokhoziaistvennoi nauki*, 1966, No. 2, pp. 13–18. Kushner and Nuzhdin were already moving back to science before Lysenko's Institute was dissolved.

154. The highest bosses, unwilling to discourse on this subject in pub-

lic, have passed the word through their subordinates. See, e.g., V. D. Pannikov, "XXIII s'ezd KPSS i zadachi sel'skokhoziaistvennoi nauki," *ibid.*, 1966, No. 5, pp. 1–10. This was a speech at a meeting of the Lenin Academy, given by the assistant director of the Central Committee's Division of Agriculture. It is worth noting that he, Pannikov, was another opportunist, in this case a soil scientist who had been advocating Lysenko's fertilizer composts in the last year of Lysenko's power. Such types dominate the administration of agricultural science, presumably because they fully share the highest bosses' antipathy toward recrimination and critical retrospection. See, for more examples, most of the other speeches at the April, 1966, meeting of the Lenin Academy, *ibid.*, 1966, No. 8, pp. 131–46. For Pannikov's earlier Lysenkoism, see his *Pochvy*.

8. ACADEMIC ISSUES: MARXISM

1. For notable examples, see Zirkle, *Evolution*; and Huxley, especially pp. 173 ff. where he answers the question "Why?" Huxley (p. 24) declares that Hudson and Richens, *The New Genetics in the Soviet Union,* "resolves itself into an enumeration of the influence on the formulation of Michurinism exerted by dialectical materialism in general, and by the authority of various historical figures in particular." Actually Hudson and Richens only intermittently exaggerate the role of dialectical materialism in the rise of Lysenkoism.

2. See Joravsky, p. 6 *et passim.*

3. See especially Engels, *Dialectics of Nature,* and *Anti-Dühring.*

4. Lichtheim suggests that Plekhanov invented the name and the *Weltanschauung,* or at least that Plekhanov gave the name to Engels' denatured, positivistic version of Marxism. To sustain his argument that there was a sharp philosophical difference between Marx and Engels, Lichtheim brushes aside Marx' approval of Engels' philosophical writing as "a factual circumstance which need not concern us, though it may be of interest to his biographers" (p. 245). This "factual circumstance" is usually regarded as evidence that Marx *and* Engels moved from a metaphysical philosophy toward a form of positivism, while insisting that they had transcended both. The one serious flaw in Lichtheim's superb book is his refusal to face the question squarely, Did Marx really transcend the choice between metaphysics and positivism?

5. See below, pp. 256–7.

6. Plekhanov, *Izbrannye,* IV, 279 ff.

7. See Lenin, *Polnoe sobranie,* LIV, 300–1. See also above, Chapter 2, note 13.

8. See references in Joravsky, p. 322, note 96.

9. See Stalin, I, 301, 303, 309. The article was written in 1906.

10. K. Kautsky, "Ein Brief über Marx und Mach," *Der Kampf,* 1909, No. 10, p. 452.

11. There were a few prerevolutionary natural scientists who showed a strong interest in Marxism, for example, Marx's friend Karl Schorlemmer, and the Bolshevik astronomer P. K. Shternberg. On the whole issue of the Bolshevik Revolution and the scientific intelligentsia, see Joravsky, Chapter 4.

12. *Ibid.,* Chapter 19.

13. See, e.g., Timiriazev, VII, *passim;* and C. D. Darlington, "Pur-

pose and Particles in the Study of Heredity," in *Science, Medicine, and History,* II, 472–81.

14. See, e.g., Beckner, *The Biological Way of Thought.*

15. Danilevskii, p. 529. This polemic was long remembered. See, e.g., Taliev, p. 50.

16. Whittaker Chambers, p. 16.

17. S. S. Chetverikov, "On Certain Aspects of the Evolutionary Process from the Standpoint of Modern Genetics," *Proceedings of the American Philosophical Society,* 105:193 (April, 1961). The translation is Malina Barker's, edited by I. Michael Lerner.

18. R. A. Fisher shows a pronounced interest in the sociological implications of population genetics, but very little in the philosophical problem discussed above. For a list of Sewall Wright's pioneering papers, see Rieger and Michaelis, p. 647.

19. See, e.g., von Bertalanffy; and Matson; and Koestler.

20. For a neat statement of the conclusions which most Soviet Marxist biologists had reached by the end of the 1920s, see Agol, "Problema organicheskoi tselesoobraznosti," *Estestvoznanie i marksizm,* 1930, No. 1, pp. 3–20. For a detailed account of the discussions that led to these conclusions, see Joravsky, Chapter 19.

21. Agol's paper is cited in note 20. A. S. Serebrovskii's appeared in the same journal. Cf. also Serebrovskii, "Problema svedeniia v evoliutsionnom uchenii," *Nauchnoe slovo, 1930,* no. 9, pp. 29–48. For an account of the conference, including a synopsis of Prezent's paper, see *Priroda,* 1930, No. 9, pp. 925–31.

22. Iu. M. Vermel', for example, who had been vigorously defending Lamarckism until 1929, abstained from such publications for a while thereafter. See his comment at a March 1931 meeting that criticized Agol and Serebrovskii: Kommunisticheskaia Akademiia, *Protiv mekhanisticheskogo materializma,* pp. 45–47. At that time Vermel' published a long article on directed evolution. (He was subsequently destroyed by the terror.) See also the comment of the Lamarckist E. S. Smirnov, *ibid.,* pp. 41–3. He survived to endorse Lysenkoism, as did the notorious S. S. Perov.

23. See Joravsky, Chapters 16–17.

24. Kommunisticheskaia Akademiia, *Protiv mekhanisticheskogo materializma,* p. 78. At greater length, see Tokin, *Voprosy biologii,* which consists of lectures given in 1932.

25. *Marks, Engel's, Lenin o biologii,* p. 175.

26. Bukharin, "Darvinizm i marksizm," *Sorena,* 1932, No. 5, p. 22.

27. *Ibid.,* pp. 31–3.

28. See Il'in (3rd ed., 1936), and various translations, including English.

29. See the reports of Director B. P. Tokin, "Raboty biologicheskogo instituta komakademii im. K. A. Timiriazeva," *VKA,* 1932, No. 9–10, pp. 161–173, and "V Biologicheskom Institute im. K. A. Timiriazeva," *PZM,* 1934, No. 4, pp. 184–9. See also O. Krasovskaia, "Biologicheskii Institut im. K. A. Timiriazeva," *FNIT,* 1934, No. 4, pp. 82–6.

30. Tokin gave her views a contemptuous brush-off in Kommunisticheskaia Akademiia, Biologicheskii Institut, *Probleme der theoretischen Biologie,* p. 269. Lepeshinskaia countered in *PZM,* 1936, No. 5, pp. 206–8. Tokin replied, *ibid.,* No. 8, pp. 166–9.

31. For a list of its personnel and their positions in the Lysenko controversy, see Chapter 5, note 15. For a characteristic sample of its work, see the symposium cited in this chapter, note 30. Kommunisticheskaia Akademiia, *Pamiati K. A. Timiriazeva,* its last publication, is marred by uncharacteristic contributions from outsiders, such as A. A. Maksimov, but most of the papers are typical of Soviet Marxist biology in the mid-1930s.

32. A. S. Sereiskii, "Problema individual'nogo razvitiia v osveshchenii sovremennoi fiziologii rastenii," *Za marksistsko-leninskoe estestvoznanie,* 1932, No. 5–6, pp. 22–55.

33. *PZM,* 1932, No. 11–12, pp. 230–6. The textbook under criticism was Natali. The critic was N. Veselov.

34. I. I. Prezent, "Predislovie," in Filipchenko, *Eksperimental'naia zoologiia;* and Prezent, *Teoriia Darvina.*

35. See above, p. 84, for Commissar Iakovlev's endorsement of Lysenko. Prezent recalled the event in a conversation with me on May 25, 1962, in Moscow, at the *Dom druzhby.*

36. Prezent, *Klassovaia bor'ba,* p. 45.

37. The last pre-Lysenkoite publication of Prezent's was his "Uchenie Lenina o krizise estestvoznaniia i krizis burzhuaznoi biologicheskoi nauki," in Ral'tsevich, pp. 234–280. Though published in 1935, this essay was prepared long before. The first fruit of his collaboration with Lysenko was their *Selektsiia i teooriia stadiinogo razvitiia rastenii,* which was sent to the press in 1934.

38. In December 1935, at the *Soveshchanie peredovikov urozhainosti,* p. 191, Lysenko gave Prezent credit for revealing the relevance of Darwin. At the conference of December 1936 Lysenko still spoke of Prezent as his most important collaborator. See *Spornye voprosy,* pp. 39–71.

39. Vil'iams, *Sobranie sochinenii,* X, 126.

40. There were a few other exceptions. Most notable was A. A. Maksimov. See his "Filosofiia i estestvoznanie za piat' let," *PZM,* 1936, No. 1, pp. 63–5, for an endorsement of Lysenko and Prezent.

41. *PZM,* 1939, No. 10, p. 156.

42. *Ibid.,* pp. 156–7. Cf. No. 11, p. 188.

43. In a letter to Vil'iams, inviting him to the conference, Mitin made clear the philosophers' intention of avoiding a complete condemnation of genetics. See Vil'iams annoyed response, in his *Sobranie sochinenii,* XI, 32–3.

44. *PZM,* 1939, No. 11, p. 159.

45. T. D. Lysenko, "Engel's i nekotorye voprosy darvinizma," *VAN,* 1941, No. 1, pp. 1–11. Variously reprinted, including Lysenko, *Agrobiologiia,* various editions. The reader should note how few and vague are the texts from Engels.

46. *PZM,* 1939, No. 10, p. 161.

47. Mitin, "Za rastsvet sovetskoi agrobiologicheskoi nauki," *Literaturnaia gazeta,* 1947, Dec. 27. Notice that he avoids condemnation of genetics, while gently disagreeing with Lysenko on the issue of intraspecific competition.

48. The conscious elevation of a scientific controversy to the level of theoretical ideology was one of the recurrent themes in the speeches at the August Session (see *O polozhenii*), and in the follow-up conference at the Academy of Sciences (see *VAN,* 1948, No. 9). These comments led many outside observers to conclude that theoretical ideology was the

cause of the elevation, which is equivalent to the belief that wars are caused by declarations of war.

49. *O polozhenii,* p. 506.

50. See *VAN*, 1948, No. 9, pp. 112–119 and 135–145, for the speeches of Mitin and G. F. Aleksandrov at the follow-up meeting in the Academy of Sciences.

51. *Ibid.,* p. 144.

52. *Ibid.,* p. 139.

53. Stalin, XIII, 38–9.

54. *Plenum TsK* (1961), pp. 342–3.

55. Gaidukov, pp. 229 *et passim,* uses this phrase to describe the general, historical relationship between science and "practice." No Soviet author has tried to spell out precisely what this means for the relationship between Soviet scientists and the bosses of Soviet practice.

56. To be precise, none had ever won the prize by the time that the philosopher, P. F. Iudin, in *Izvestiia,* 1964, March 8, blamed the prize commission.

57. L. F. Il'ichev, "Metodologicheskie problemy estestvennykh i obshchestvennykh nauk," in Akademiia Nauk, SSSR, *Metodologicheskie problemy nauki,* p. 39.

58. See the review article by A. P. Ogurtsov, in *VF,* 1967, No. 7. There is an English translation, "Perspectives on Practice as a Philosophical Category," *Soviet Studies in Philosophy,* 7:26–45 (Summer 1968).

59. See N. K. Kol'tsov, "Struktura khromosom i obmen veshchestv v nikh," *Biologicheskii zhurnal,* 1938, No. 1, pp. 3–46, for a review of the articles he had published on the subject beginning in 1928. See also Kol'tsov, *Les molécules héréditaires.*

60. See the typical editorial, "Edinstvo teorii i praktiki," *VF,* 1954, No. 2, p. 8, which included gentle criticism of G. V. Platonov, one of the chief Lysenkoite philosophers. But then Platonov himself was given space for a polemic against the anti-Lysenkoites, "Nekotorye filosofskie voprosy diskussii o vide i vidoobrazovanii," *ibid.,* No. 6, pp. 116–132, in the course of which he acknowledged the justice of a little mild criticism of his side. This sort of article appeared in the journal from time to time for the next ten years.

The editors of *VF* rarely gave space to forthright defenders of genetics. See N. P. Dubinin, "Metody fiziki, khimii i matematiki v izuchenii problemy nasledstvennosti," *VF,* 1957, No. 6, and N. N. Semenov, "O sootnoshenii khimii i biologii," *ibid.,* 1959, No. 10. For a complete bibliography of the journal's publications on biology, see *Bibliographie der sowjetischen Philosophie.* The period covered is 1947–1960.

61. S. A. Iakushev, "O glavnykh zadachakh biologicheskoi nauki i putiakh ikh resheniia," *Agrobiologiia,* 1962, No. 4, p. 489.

62. The chief philosophical defenders of Lysenkoism were Prezent, G. V. Platonov, V. M. Kaganov, and I. I. Novinskii. For a list of their works, see the bibliography cited in note 60. For characteristic samples see Platonov, ed., *Ocherk,* Platonov, *Dialekticheskii materializm,* and Akademiia Nauk SSSR, Institut Filosofii, *Problema prichinnosti.*

63. Adapted from Mishnev, pp. 127–9.

64. *VF,* 1962, No. 2, p. 147.

65. For notable examples, see the articles by Dubinin and Semenov, cited in note 60.

66. See *Filosofskie problemy,* pp. 12–31 for Mitin's paper, *et passim* for comments by other philosophers, who talked about physics and physiology, cybernetics and epistemology.
67. *Ibid.,* p. 193.
68. *Ibid.,* p. 479. The speaker was K. Iu. Kostriukova.
69. *Ibid.,* pp. 251–2. The authors were S. L. Sobolev and A. A. Liapunov.
70. For a good example of such analysis, see Beckner.
71. The speaker was N. M. Rutkevich, in *Filosofskie problemy,* p. 431. For his previous enthusiastic endorsement of Lysenkoism, see Rutkevich, *Praktika,* pp. 241–2. In his textbook, *Dialekticheskii materializm,* Rutkevich took a slightly restrained Lysenkoite viewpoint. See, e.g., pp. 460–1 and 489–92.
72. *Filosofskie problemy,* p. 552. The speaker was I. Marculescu-Hurduc.
73. See, e.g., some of the essays in Akademiia Nauk Belorusskoi SSR, Institut Filosofii i Prava, *Dialekticheskii materializm kak metodologiia;* and, by the same Institute, *Rol' dialektiki.* See also *VF,* 1965, No. 7, *et seq.*
74. There is a tiny report of a comment that Prezent offered at a meeting of the Lenin Academy in 1966. See *Vestnik sel'skokhoziaistvennoi nauki,* 1966, No. 8, p. 143. Otherwise he went unpublished until his death in 1969. For Platonov's defense of Lysenkoism, see his "Dogmy starye i dogmy novye," *Oktiabr',* 1965, No. 8, pp. 149–62; and, with M. A. Lisavenko, "I. V. Michurin i sovremennaia biologiia," *Vestnik s.-kh. nauki,* 1967, No. 4, pp. 118–23. The reader will note that Platonov has retreated very far from the position he held while Lysenkoism enjoyed political support.
75. Akademiia Nauk SSSR, Institut Filosofii, *Marksistsko-leninskaia filosofiia i sotsiologiia,* p. 174.
76. *Ibid.,* p. 257.
77. I. T. Frolov, "Materialisticheskaia dialektika i sovremennaia biologiia," *Kommunist,* 1966, No. 2, pp. 61–70. Cf. his book, *Ocherki metodologii;* its concluding chapter, "The Role of Practice in Biological Research," manages to avoid comment on the role of practice in the Lysenko affair.
78. For other examples, see Fedorenko, ed., *Filosofskie voprosy meditsiny i biologii,* which includes a Lysenkoite article as well as the obfuscating type; and Pilipenko, pp. 87 ff.
79. See the poignant expression of regret by Bochenski, "Thomism and Marxism-Leninism," *Studies in Soviet Thought,* 7:154–168 (June, 1967), especially sections VII–X.
80. *New York Review of Books,* 1968, Feb. 29, p. 19. Waddington seems to have been misled by Bernal, *The Origin of Life.*
81. For a firsthand report of Oparin's intellectual development, see his *Fermenty* (1923), where his interest in enzymes leads him to an interest in colloid chemistry, and so to his *Proiskhozhdenie zhizni* (1924), which he prefaces with a little history of theories of the origin of life and sprinkles with references to his respected predecessors, chiefly Edward Pflüger. Oparin's underlying philosophy is plain old mechanistic reductionism. There is not a breath of Marxism in this pamphlet, conscious or subconscious. In the late 1920s Oparin lectured to the Communist Academy on "Khimicheskaia teoriia proiskhozhdeniia zhizni," *VKA,* 1927, *kniga* 21,

pp. 229–43. Once again he gave no evidence of Marxist inspiration. It would have been most surprising if he had, for the consciously Marxist biologists of the 1920s did not think of the origin of life as a topic that distinguished them from ordinary, non-Marxist mechanists. See, e.g., Azimov, *Proiskhozhdenie zhizni* (1925), and B. M. Zavadovskii, *O korobke konservov* (1926).

82. The process began in Oparin, *Proiskhozhdenie zhizni* (1935), and the expanded *Vozniknovenie zhizni* (1936), which is the basis of the English translation: Oparin, *The Origin of Life.* In untranslated publications of the period from the late 1930s to the mid-1950s Oparin's distortion was far worse. See the items listed in his bibliography, *Aleksandr Ivanovich Oparin.*

83. See above, Chapter 6, note 46.

84. See, e.g., Oparin, "Nesostoiatel'nost' predstavlenii mendelistov po voprosu o proiskhozhdeniia zhizni," in Akademiia Nauk, Institut Genetiki, *Protiv reaktsionnogo,* pp. 47–70; and see the discussion of the origin of life, in *VAN,* 1962, No. 10, pp. 105–8, at which Oparin stood alone with Prezent against the biochemical approach. It must be borne in mind that Oparin talked this way only when his Lysenkoite commitment required it. In other contexts he dispensed with nonsense about the "purely biological" approach. See, e.g., his "Life, Origin Of," in Gray, pp. 561–3. Note too the complete absence of references to Marxism.

85. See Engels, *Anti-Dühring,* and especially his posthumous *Dialectics of Nature.* In the latter Engels does take a position on scientific aspects of the problem or rather, he endorses the position that he found in Haeckel and other favorite biologists.

86. Emily Dickinson, p. 249.

87. Pisarev, "Podvigi evropeiskikh avtoritetov," in his *Sochineniia,* V (1894), pp. 123–42. The essay is omitted from the Soviet edition of Pisarev's *Sochineniia.*

88. Lepeshinskaia, p. 8. Even A. A. Maksimov showed better judgment. See his comment on Pisarev's *gaffe,* in *PZM,* 1924, No. 6–7, p. 119. Since the 1920s Soviet admirers of Pisarev have preferred to ignore the clash between his materialism and science.

89. See relevant passages in Akademiia Nauk, Institut Filosofii, *Osnovy marksistskoi filosofii,* and in the various editions of Rozental' and Iudin.

90. See Loren R. Graham, "Cybernetics," in Fischer, pp. 83–106. For a characteristic example of the technical emphasis in Soviet writings on this subject, see Akademiia Nauk Ukrainskoi SSR, *Modelirovanie v biologii i meditsine.*

91. Blake, p. 469.

92. Frost, pp. 473–4.

93. See, e.g., E. Nagel, "The Meaning of Reduction in the Natural Sciences," in Stauffer, pp. 99–135.

94. E. W. Caspari and R. E. Marshak, "The Rise and Fall of Lysenko," *Science,* 149:275–278 (July 16, 1965). For the most extended argument along these lines, see Zirkle, *Evolution.* The fable crops up in the most unexpected places, e.g., Harris, *The Rise of Anthropological Theory,* p. 237.

95. Marx-Engels, *Correspondence,* p. 126, for the first quotation. Marx-Engels, *Werke,* XXX, 131, 578 for the original of both. It is worth noting that Engels called Marx's attention to the *Origin* in December 1859, as

soon as the book appeared, pointing out the "destruction of teleology" and the historical interpretation of nature as its great merits. *Ibid.*, XXXIX, 524.

96. See, e.g., *ibid.*, XXXII, 685–6, translated in Marx-Engels, *Correspondence*, p. 301. Cf. also Engels' letter to Lavrov, *Werke*, XXXIV, 169–172.

97. *Ibid.*, XIX, 335 et seq.

98. See M. O. Kosven, "Problema doklassovogo obshchestva v epokhu Marksa-Engel'sa," *Sovetskaia etnografiia*, 1933, No. 2, pp. 1–38. Kosven condemned those who doubt Engels' originality in this matter, but he was an honest, thorough scholar, and actually proved what he set out to condemn in this fascinating article.

99. See Kautsky, *Der Einfluss* (1880), and *Vermehrung* (1910). The latter version was immediately translated into Russian and was subsequently included in Kautsky, *Sochineniia*, XII (1923), with a valuable bibliographical preface by Riazanov. The reader should also take note of Kautsky's introduction, in which he sketches his intellectual development from social Darwinism to Marxism. Cf. Engels' inconclusive comment to Kautsky on the first version of the book, in Marx-Engels, *Werke*, XXXV, 431–2.

100. K. A. Timiriazev, "Evgenika," *Entsiklopedicheskii slovar' Granata*, XIX (1914?), reprinted in his *Sochineniia*, VIII, 469–473. Cf. *ibid.*, pp. 406–9, for his tribute to Francis Galton.

101. See the anthology, Ravich-Cherkasskii, ed., *Darvinizm i marksizm* (1923), and especially the expanded second edition of 1925, which contains a bibliography of the subject compiled by Ia. Rozanov. It should be noted that the anthology includes other than orthodox views on the connection between Darwinism and Marxism. For example, there is a selection from Ludwig Woltmann, who was regularly denounced by Soviet writers of the 1920s as a social Darwinist and racist. It would be difficult to compile a substantial anthology of orthodox Marxists, for they have tended to skirt the subject.

102. See Kautsky, *Ethics*. (There are several Russian editions, including Soviet ones.) See also his *Obshchestvennye instinkty*, and his *Vozniknovenie braka*, and his *Nauka, zhizn' i etika*.

103. Kautsky, *Materialistische Geschichtsauffassung*, I, 199 *et passim*.

104. The words are Riazanov's, the leading editor of the Marxist classics in the 1920s. See Kautsky, *Sochineniia*, XII, p. vi. Kautsky's magnum opus, cited in note 103, aroused only hostile criticism in Soviet Marxist publications.

105. Conveniently printed in Engels, *Dialectics of Nature*. In Russian it was first published in 1923. It is the only Marxist selection in the anthology edited by Gremiatskii, *Evoliutsiia cheloveka* (M., 1925), published by Sverdlov Communist University.

106. Lenin, *Sochineniia*, XIV, 315.

107. See Plekhanov, I, 494, 607–10, 612 ff., 762–3; II, 90–3, 153–4, 162–3, 179, 206, 244–5, 677–8; IV, 279 ff.; V, 288 ff.

108. *Ibid.*, I, 690–1.

109. *Ibid.*, II, 151.

110. Quoted in Joravsky, p. 63.

111. Semashko, *Nauka o zdorov'e* (1922), pp. 29–30.

112. *Ibid.*, pp. 30–31.

113. *Ibid.*, pp. 53–4. Cf. remarks by S. Kaplun, in B. Khaies (B. Chajes), especially pp. 291–2.

114. N. K. Kol'tsov, "Uluchshenie chelovecheskoi porody," *Russkii evgenicheskii zhurnal,* I, pp. 14–15.

115. For one of the few explicit criticisms of Kol'tsov on the ultimate ideal, see B. M. Zavadovskii, "Darvinizm i lamarkizm i problema nasledovannia priobretennykh priznakov," *PZM,* 1925, No. 10–11, p. 106. Evidence of the Bolshevik commitment to the ideal of universal man abounds — in pedagogical publications, for example.

116. For Filipchenko's work, see Akademiia Nauk SSSR, *Izvestiia biuro po evgenike,* Nos. 1–3 (1921–1925); Filipchenko, *Chto takoe evgenika?*; and Filipchenko, *Puti uluchsheniia.* For Kol'tsov's work, see *Russkii evgenicheskii zhurnal,* 1922–1929 (1928, No. 2–3, contains a thorough bibliography of all Russian publications on the subject); and Kol'tsov, *Uluchshenie chelovecheskoi porody.*

117. Articles on the genealogies of famous people, with a repeated stress on noble origins, abound in *Russkii evgenicheskii zhurnal.* For a particularly funny passage, see Kol'tsov's solemn analysis of Darwin's descent from Yaroslav the Wise (1922, No. 1, p. 69). On the Decembrists, see 1927, No. 1; and 1928, No. 4.

118. *Sotsial'naia gigiena,* 1924, No. 3–4, pp. 170–2. For another eugenicist arguing along similar lines, see Kanaev, the chapter on human heredity.

119. N. K. Kol'tsov, "Rodoslovnye nashikh vydvizhentsev," *Russkii evgenicheskii zhurnal,* 1926, No. 3–4, pp. 103–143. He repeatedly wrote of the *gen aktivnosti.* See, for example, his notorious explanation of the genetic effects of revolutions and civil wars: While they have the eugenic effect of opening careers to talents, they have the dysgenic effect of killing, on both sides of the barricades, "bearers of one and the same gene of activism." Kol'tsov, *Uluchshenie,* p. 56.

120. Slepkov, pp. 164–5.

121. See *VKA,* 1927, No. 20, pp. 212–254, the record of a discussion where all the participants took this line. None agreed with the anthropologist, S. A. Vaisenberg (Weissenberg), whose survey, "Theoretische und praktische Eugenik in Sowjetrussland," *Archiv für Rassen- und Gesellschaftsbiologie,* 18:69–83 (1926, No. 1), declared Marxism and eugenics to be quite incompatible.

122. See Akademiia Nauk SSSR, *Izvestiia Biuro po Genetike i Evgenike,* IV (1926), for the change in Filipchenko's research center. At the time they still talked of continuing eugenics work, but successive issues make clear the abandonment of it. See especially Filipchenko's obituary in IX (1932), p. 5, for a forthright statement. Kol'tsov's Institute of Experimental Biology quietly dropped its work in eugenics about the same time, and the Russian Eugenics Society expired at the end of the 1920s.

123. See especially G. A. Batkis, "Sotsial'nye osnovy evgeniki," *Sotsial'naia gigiena,* 1927, No. 2 (10), pp. 7–27, which takes the argument of his earlier speech ("Sovremennye evgenicheskie techeniia v svete sotsial'noi gigieny," *ibid.*, No. 1 [9], pp. 97–98) to the verge of a complete repudiation of eugenics. See also Batkis, "Evgenika," *BSE* (1st ed.), XXIII (1931), 812–19, which ignores the possibility of a socialist eugenics and restricts applied genetics to plant and animal breeding.

124. *VKA,* 1927, No. 19, pp. 231–2. The speaker was M. M. Mestergazi.

125. Semashko, *Vvedenie*, pp. 83–4.

126. See Volotskoi (1923 and 1926), for the only Soviet advocacy of enforced sterilization that I have been able to find. See *Pechat' i revoliutsiia*, 1924, No. 1, for a dissenting review by Filipchenko. As it happens, Volotskoi was a Marxist eugenicist. The other eugenicists agreed that "negative" eugenics could have only a slight effect, if any; "positive" eugenics — i.e., the encouragement of desirable breeding — was their great hope.

127. See, e.g., the treatment of eugenics in the curriculum for social hygiene in medical schools, *Sotsial'naia gigiena*, 1928, No. 2–3 (12–13), p. 200. See also A. Fisher (Alphonse Fischer), pp. 253–282, where the Soviet editors of this translation disapprove of the author's Lamarckist view of heredity but offer no criticism of his eugenic proposals. See also Safonov, pp. 72–6, for a thoroughly anti-Lamarckist plea for a Marxist eugenics; and Keller, *Genetika*, pp. 88–9. It is amusing to note that Safonov and Keller subsequently became Lysenkoites.

128. *VKA*, 1927, No. 19, pp. 226 *et passim*.

129. See Slepkov, pp. 155–6; and the discussion of eugenics at the Communist Academy in 1926, reported in *VKA*, 1927, No. 20. Five Lamarckists defended their faith against the attacks of five Mendelians.

130. N. Sh. Melik-Pashaev, "Chelovek budushchego," in *Zhizn' i tekhnika budushchego*, pp. 412–31.

131. *Ibid.*, pp. 385–8. Ilya Ehrenburg, *Trest D. E.*, is conveniently reprinted in his *Sobranie sochinenii*, I. Ehrenburg seems to have got the idea from remarks that Bertrand Russell made in a debate with Haldane.

132. See the convenient biography of A. D. Speranskii (1888–1961), with brief bibliography, in *Biograficheskii slovar'*, and cf. *Razvitie biologii v SSSR*, p. 512, for evidence that he still enjoys some influence. But note that this version of his doctrine is very restrained by comparison with Speranskii, *Elementy*.

133. See especially B. M. Zavadovskii, as cited in note 115, who tells how he was convinced of this by an argument between the Marxist eugenecist Volotskoi and the "bourgeois" eugenicist Filipchenko.

134. See Haldane, *Heredity*. Muller caused a sensation by identifying Lysenkoism with Lamarckism and presenting this analysis to the December 1936 conference. See above, pp. 117–18. Cf. also his "Lenin's Doctrines in Relation to Genetics," in *Pamiati V. I. Lenina*, pp. 563–92.

135. See especially Dobzhansky, *Mankind Evolving*; and Dunn, *Heredity and Evolution*.

136. See Moody, for a sober compilation of what is known on this subject.

137. UNESCO, *The Race Concept*, p. 19. The geneticist who submitted this comment was Walter Landauer.

138. A. S. Serebrovskii, "Antropogenetika," *Mediko-biologicheskii zhurnal*, 1929, No. 5, pp. 3–19. For more details see Joravsky, pp. 305–7.

139. *Ibid.* For "the insult to Soviet womanhood," see *Spornye voprosy*, pp. 305–7.

140. See the brief report of a discussion — "Does Marxism Recognize Eugenics?" — in *Pod markso-leninskim znamenem*, 1934, No. 2, p. 153. H. J. Muller was one of the few Marxists in the Soviet Union who were willing in the 1930s to go on record in favor of a socialist program of eugenics. See his "Evgenika na sluzhbe u natsional-sotsialistov," *Priroda*,

1934, No. 1, pp. 100–106. So was Rokitskii (1934), pp. 245–6, and (1935), pp. 227–8. And so was Tomilin, pp. 189–200. Semashko's last friendly association with eugenics came in 1929, when the medical encyclopedia of which he was editor published T. I. Iudin, "Evgenika," *BME*, IX, 663–670. In 1936 Semashko participated in the condemnation of Levit for studying human heredity.

141. See G. Frizen, "Genetika i fashizm," *PZM*, 1935, No. 3; N. N. Cheboksarov, "Iz istorii svetlykh rasovykh tipov Evrazii," *Antropologicheskii zhurnal*, 1936, No. 4; Julian Huxley and A. C. Haddon, "Rasovyi vopros — teoriia i fakty," *Uspekhi sovremennoi biologii*, 1936, No. 6; and S. G. Levit, "Rasizm pered sudom mezhdunarodnoi nauki," *PZM*, 1936, No. 7, with citation of other articles in the Soviet press by Muller, Haldane, and Hogben.

Actually the German émigré Max Levin and the Austrian émigré Julius Schaxel were the earliest in the USSR to sound the alarm against Nazi misuse of biological science. See Levin, "Stimmen aus dem deutschen Urwalde," *Unter den Banner des Marxismus*, 1928, No. 2; and Schaxel (Shaksel'), "Germanskaia biologiia gosudarstva," *PZM*, 1934, No. 4. It should also be noted that Soviet biologists and anthropologists in the 1930s began to suppress the casual expressions of racism that were common before Nazism made everyone conscious of them. Usually they had taken the form of comments that placed Negroes and Australian aborigines lower than whites in the evolutionary scale. (Engels' comment on the Bushmen is of this type.) Compare, e.g., Anuchin (1927), p. 109, with Vishnevskii (1935), p. 139. Both have the same sketch of facial profiles arranged in ascending order from ape through fossil humanoids to contemporary European. In the 1927 book a Negro profile is just below the culminating Caucasoid; in the 1935 book the Negro profile is omitted.

142. *Komsomolskaia pravda*, 1936, Nov. 15, reports the meeting, which was called by the Science Division of the Moscow Party organization, with Ernst Kol'man giving the main denunciation.

143. See Karlik's attack on Levit's Institute, *PZM*, 1936, No. 12, pp. 178–86. The Central Committee Decree of July 4 is cited on p. 180. Cf. E. Kol'man, "Chernosotennyi bred fashizma i nasha mediko-biologicheskaia nauka," *ibid.*, No. 11, pp. 64–72.

144. See above, pp. 171 *et seq*. It must be noted that the charge of racism was demagogically hurled at Ukrainians, as part of a charge of nationalist deviations, before it was hurled at the students of human genetics. See, e.g., the symposium, *Rasovaia teoriia na sluzhbe fashizma*, especially I. M. Poliakov's speech.

145. SZ, 1937, April 12. A. A. Nurinov was coauthor with Prezent.

146. *VAN*, 1948, No. 9, pp. 55, 117–18, 150–3. The sponsor was Orbeli. Note that Mitin, who denounced Davidenkov in 1948, in 1939 had granted the propriety of his studies. See *PZM*, 1939, No. 10, p. 164. Davidenkov appeared at the 1939 conference to defend medical genetics (*ibid.*, No. 11, pp. 116–17), and that same year was allowed to write the article on human heredity for *BSE* (1st ed.), XLI. But the man's utter isolation was evident in his remarks at the 1939 conference. The only other person who seems to have been doing similar research in those years — as part of his psychiatric investigations — was T. I. Iudin. See his biography in *BME* (2nd ed.), XXXV (1964), which neglects to point out that he was a leading eugenicist in the 1920s.

147. Belatedly reported in *Literaturnaia gazeta,* 1964, Nov. 24. The book at issue was Efroimson, *Vvedenie.* The translation was Neel and Schull, *Nasledstvennost' cheloveka.* Note the introduction, which criticized the chapter on eugenics, from which the sections on sterilization and differential fertility have been excised.

148. Pavlenko, p. 43. Note that his bibliography does not contain a single "Michurinist" item on the subject.

149. Lysenko, *Raboty v dni,* p. 7.

150. See, e.g., Batkis, p. 5, for the simple declaration that the population problem disappears with the advent of socialism. It was not until the post-Stalin period that Soviet demography began to revive discussion of the problem.

151. See, e.g., S. P. Tolstov, "V. I. Lenin i aktual'nye problemy etnografii," *Sovetskaia etnografiia,* 1949, No. 1; and V. P. Iakimov, "Rannie stadii antropogeneza," in Akademiia Nauk SSSR, Institut Etnografii, *Trudy, novaia seriia,* XVI (1951), pp. 7–88. Iakimov defends the common-sense view that the transformation of ape to man did not occur the moment that the first hominid fashioned the first crude tool, but he introduces his argument with a frightfully scholastic interpretation of a fleeting comment in Lenin.

152. See, e.g., the contribution of Ia. Ia. Roginskii to Bunak *et al.,* pp. 270–3. For an apology, for this and other anthropological publications that used the data of genetics, see the editorial in *Sovetskaia entografiia,* 1948, No. 4, pp. 9–22. Another serious sinner was A. A. Malinovskii, "Biologicheskie i sotsial'nye faktory v proiskhozhdenii rasovykh razlichii u cheloveka," *Priroda,* 1947, No. 7, pp. 40–48.

153. See the editorial cited in note 152, p. 12. Cf. Iakimov, as cited in note 151, pp. 24–5.

154. For a convenient review of the discussion, see Semenov, pp. 18–25, with a good bibliography on pp. 525–554. The items that Bunak published in 1951 were among the earliest in the revival after Stalin's articles on linguistics.

155. Semenov, p. 25.

156. M. B. Mitin said this. *PZM,* 1939, No. 10, p. 161.

157. See, for example, Akademiia Nauk SSSR, Institut Etnografii, *Trudy,* novaia seriia, XCII, especially the concluding essay by Bunak.

158. See A. R. Wallace, *On Miracles*; and A. R. Wallace, *Studies,* II, chapter 21.

159. Nesturkh, in Bunak *et al.,* p. 131. The reference to A. R. Wallace is on p. 122.

160. M. F. Nesturkh, "Darvin i sovremennye problemy antropogeneza," *Voprosy antropologii,* 1960, No. 2, p. 14.

9. CRITERION OF PRACTICE

1. See G. B. Orlob, "The Concepts of Etiology in the History of Plant Pathology," *Pflanzenschutz — Nachrichten "Bayer,"* 17:185–268 (1964, No. 4); Salaman, *The History and Social Influence of the Potato*; and R. N. Salaman, "Outline of the History of Plant Virus Research," in *Agriculture in the Twentieth Century,* pp. 261–289.

2. This was D. I. Ivanovskii. See his biography in Lipshits, III. Actually, it is an oversimplification to say that he discovered plant viruses. See Orlob, as cited in note 1.

3. See Ryzhkov, pp. 22–5, and V. L. Ryzhkov, "Tridsat' let izuchenii virusnykh boleznei rastenii v SSSR," *Mikrobiologiia,* 1947, No. 5, p. 375.
4. The American practice is described in Stuart, pp. 98–106. The Ukrainian and French experience is referred to in Favorov and Kotov, pp. 168–170; and in G. N. Linnik, "Semenovodstvo kartofelia," in Iur'ev, (1940), pp. 443–4.
5. *Pravda,* 1936, Sept. 2.
6. See *Narodnoe khoziaistvo SSSR v 1960 godu,* pp. 405, 417; and *Sel'skoe khoziaistvo SSSR,* p. 201; and *Comparisons of U.S. and Soviet Economies,* I, pp. 211, 229, 237.
7. See, e.g., S. N. Evstigneev, "Ser'eznyi ekzamen," *Plodoovoshchnoe khoziaistvo,* 1934, No. 3.
8. *Pravda,* 1931, Aug. 3.
9. See Salaman, pp. 164–5; and Bukasov, *Revoliutsiia.*
10. See, e.g., *Plodoovoshchnoe khoziaistvo* 1933, No. 5–6, pp. 14–15, 17–18; 1935, No. 1, pp. 18–21; and Veselovskii, *Kartofel' semenami;* and *Gossortoset', sbornik* No. 6 (1934), pp. 42–3.
11. *SZ,* 1935, March 26.
12. See Lysenko, "Kartofel' na iuge," *Izvestiia,* 1935, Sept. 15, for a brief reference to such an "experiment." Cf. A. M. Favorov, "Voprosy kul'tury i selektsii kartofelia v usloviiakh iuzhnoi chasti USSR," *Iarovizatsiia,* 1935, No. 2, pp. 23–35, for a detailed report by Lysenko's chief potato expert, which makes no reference at all to such an "experiment." Much later Favorov reported that it had been performed in 1933. See Favorov and Kotov, pp. 46–8.
13. *Ibid.*
14. See, e.g., Lysenko, *Agrobiology* (1954), pp. 201–4, 325–339. The Russian originals are cited in the bibliography.
15. Lysenko and Babak, *Bor'ba s vyrozhdeniem,* pp. 17–18; *SZ,* 1935, March 26; *Izvestiia,* 1935, Sept. 15. Note that the three accounts do not agree with each other. The closest to a complete report — and that is not very close at all — was given almost twenty years after the experiments, in Favorov and Kotov, pp. 55–7.
16. *SZ,* 1935, March 26.
17. *SZ,* 1935, Sept. 16.
18. Reported with variations by Favorov, as cited in note 12, p. 34; Lysenko, "Teoriia razvitiia rastenii i bor'ba s vyrozhdeniem kartofelia na iuge," *Iarovizatsiia,* 1935, No. 2, pp. 15–16; *Izvestiia,* 1935, Sept. 15; and Favorov and Kotov, pp. 58–9, 77.
19. Lysenko and Babak, *Bor'ba,* p. 22, for the plan; Lysenko and Favorov, *Letnie posadki,* and Favorov and Kotov, for highly selective reports of performance.
20. Target and achievement are reported with some variation in *Izvestiia,* 1937, Feb. 3; Lysenko and Favorov, *Letnie posadki;* BSE (1st ed.), XXXI, 659; and other works cited in previous notes.
21. *Izvestiia,* 1938, March 21. Cf. March 15 for Molotov giving the news to a conference of agricultural officials. For further figures, see works cited in previous notes.
22. Favorov and Kotov.
23. *Ibid.*
24. Lysenko and Babak, *Bor'ba.*
25. Favorov and Kotov, p. 81.

26. *Izvestiia,* 1937, Feb. 3.

27. Lysenko and Babak, *Bor'ba,* p. 26.

28. The Seventeenth Party Congress in January, 1934 was especially important in setting this theme. See Stalin, XIII, pp. 317 ff., for the keynote.

29. See above, p. 83.

30. See Konstantinov et al., in *SRSKh,* 1936, No. 11, pp. 123–4. For a representative sample of the characteristically timid reaction of potato specialists and plant pathologists, see VASKhNIL, *Trudy, vypusk* XII (1936).

31. See *ibid.,* p. 51, for Lysenko interrupting a specialist to shout, "I do not deny viruses!" Yet the whole tenor of his, and his followers', comments strongly suggested the opposite. See also his reply to Konstantinov et al., in *SRSKh,* 1936, No. 11, pp. 134–6.

32. Dunin, ed., *Virusnye bolezni,* p. 4. Cf. Dunin's article in *SRSKh,* 1936, No. 10.

33. See, e.g., Ryzhkov, for the record of a 1940 conference on virus diseases of plants, at which degenerative diseases of potatoes were almost ignored. Note pp. 29–34, for Dunin trying to discuss these diseases without offending the Lysenkoites. See also Bukasov, *Istoriia kartofelia,* pp. 76–77, for one of the country's leading potato specialists giving tribute to Lysenkoism.

34. The most extreme were L. V. Rozhalin and O. D. Belova, "K voprosu o gotike ili veretenovidnosti kartofelia," *Agrobiologiia,* 1948, No. 6. For a critique, which attempts however to mollify the Lysenkoites, see M. K. Fomiuk, "Vliianie raznykh temperatur i vlazhnosti pochvy na bolezni vyrozhdeniia kartofelia," in Akademiia Nauk Ukrainskoi SSR, Institut Entomologii i Fitopatologii, *Trudy,* IV (1953).

35. See Kruzhilin, pp. 26–7; and A. M. Puchkov et al., "Kak poluchit' vysokii urozhai nevyrozhdennogo kartofelia na iuge," *Plodoovoshchnoe khoziaistvo,* 1936, No. 6, pp. 58–60; and Moscow, Nauchno-issledovatel'skii Institut Kartofel'nogo Khoziaistva, *Semenovodstvo i aprobatsiia* (1946), *passim.*

36. See, e.g., Bukasov, *Kul'tura kartofelia, passim.*

37. Favorov and Kotov, *Letniaia posadka.* See G. N. Linnik, "O prichinakh vyrozhdeniia kartofelia," *BZ,* 1955, No. 4, p. 529, for evidence that a special commission was established in 1951 to check on the quality of seed potatoes.

38. See especially the translation of Burton, *Kartofel'* (1952). The preface by A. G. Lorkh praises the Lysenkoite program, but the text undermines it.

39. *Kartofel',* 1956, No. 5, p. 39. Cf. *Sel'skoe khoziaistvo* SSSR (1960), p. 184.

40. Linnik, as cited in note 37.

41. See the brief report in *Kartofel',* 1958, No. 3, pp. 60–3.

42. Bukasov and Kameraz, *Osnovy,* pp. 284–5 *et passim.* Note the continuing elements of kowtowing in this massive compendium. For other evidences of the cautious return to scientific methods of potato growing, see various articles in *Kartofel',* 1958. For a typically inconsistent mixture of science and Lysenkoism, see Chesnokov, *Bolezni* (1961).

43. See, e.g., Bukasov, *Kartofel'* (L., 1965), and various articles in *Kartofel' i ovoshchi,* 1965 to the present.

44. There are many works on the history of maize. See e.g., Mangelsdorf, *The Origin of Corn*; and Weatherwax, *Indian Corn*. For a brilliant insight into the Indian's thought processes, see Lévi-Strauss, pp. 73–4. Lévi-Strauss shies away from talk of genuine understanding, but I find some such epistemology implicit in his placing of savage thought "between the basic absurdity Frazer attributed to primitive practices and beliefs and the specious validation of them in terms of a supposed common sense invoked by Malinowski."

45. The story of the new hybrids has been frequently told. For one of the pioneering works, see E. M. East, "The Distinction between Development and Heredity in Inbreeding," *American Naturalist*, 43:173–181 (1909). For a retrospective account by another of the pioneers, see George H. Shull's paper in Gowen.

46. See various essays in Gowen.

47. Wallace, *Corn and Corn-Growing* (1923), pp. 18 and 184–5.

48. *Ibid.*, pp. 38–9.

49. For a striking example of discord between traditional ideology and the social requirements of hybrid corn, see R. W. Jugenheimer, "Hybrid Maize Emigrates to the Old World," *Foreign Agriculture*, 18:63–66 (1954, No. 4).

50. N. N. Kuleshov, " 'Inbriding' v selektsii kukuruzy," *Semenovodstvo*, 1931, No. 9–10, p. 6. He makes the same point in his introduction to the translation of Richey, *Selektsiia kukuruzy* (1931), pp. 3–4. Sometimes it is difficult to distinguish between genuine utopianism on the part of the specialist and deference to the utopianism of his political bosses. See, e.g., Talanov's articles in VASKhNIL, *Rastenievodstvo* (1933), tom I, chast' 2, pp. 230–47, and chast' 1, pp. 71–3.

51. On 16 March and 13 April 1930 STO (Council of Labor and Defense) decreed an enormous increase in the goal that SNK had set in 1929. Corn sowings were to reach 15 million hectares by 1932. (The average in the preceding decade had been 1.5 to 2.5 million, and SNK at the end of 1929 had ordered a 15 percent increase in 1930.) The decrees are constantly referred to in the outpouring of literature on corn during the early 1930s. See, e.g., A. M. Chekotillo, "Kukuruza v bor'be za sotsialisticheskoe sel'skoe khoziaistvo," *SRSKh*, 1930, No. 6, pp. 70–88. For an especially extravagant statement of the utopianism that inspired this campaign, see Krzhizhanovskii's introduction to Chekotillo and Kagan, *Kukuruza* (1930). G. M. Krzhizhanovskii, an electrical engineer, may be considered the founder of science fiction as an operative element in Soviet policy; in 1920, at one of the lowest points in Bolshevik fortunes, he raised his comrades' spirits with a wildly ambitious program of electrification.

52. He was conscious of the fact. See his reference to the earlier campaign, in Khrushchev, II, 79.

53. The sharply fluctuating fortunes of corn in Russia are described in many works. For a guide to them, see Emel'ianov.

54. *SRSKh*, 1930, No. 6, p. 78.

55. Chekotillo and Kagan (1932), p. 8.

56. Dnepropetrovsk, Institut Kukuruzy, *Biulleten'*, 1930, No. 5–6, p. 60.

57. For samples of the blustering, see Chekotillo and Kagan (1932).

58. For delicate acknowledgment of these facts by Soviet corn specialists, see Talanov et al., *Kukuruza*, pp. 177–8, and 205–7; V. I. Pisarev,

"Intsukht," in Vavilov, ed., *Teoreticheskie osnovy,* I, p. 613. Cf. M. I. Khadzhinov, "Selektsiia kukuruzy," in *ibid.,* II, p. 429, for a repetition of the boast — socialist agriculture is fully capable of utilizing the new hybrids — with a peculiarly ambiguous twist.

59. For a description of the Mexican program, see E. J. Wellhausen's article in Gowen.

60. B. P. Sokolov, "Gibridy kukuruzy, ikh poluchenie i ispol'zovanie," *Selektsiia i semenovodstvo,* 1946, No. 1–2, pp. 36–7.

61. See above, Chapter 4.

62. See *Izvestiia,* 1935, July 15, for a report of Lysenko's first attack on inbreeding. Cf. Lysenko, "O perestroike semenovodstva," *Iarovizatsiia,* 1935, No. 1, p. 45, for "the impoverishment of the hereditary foundation." Conveniently reprinted in Lysenko, *Agrobiologiia* (1949), pp. 138–9. This paper was given at a crucial meeting of VASKhNIL in Odessa, in June, 1935, at which the discussion of the wheat seed business was briefly enlarged to touch on corn breeding.

63. See especially the articles by Plachek, Krasniuk, and Vakar, in *SRSKh,* 1936, No. 12, following Sizov's Lysenkoite attack on inbreeding. The reader will note that all parties to this debate were concerned with breeding such cross-pollinating plants as rye, sunflower, and sugar beet, far more than they were with maize.

64. A. B. Salamov, "Nekotorye voprosy selektsii kukuruzy metodom samoopyleniia linii (intsukht)," *Selektsiia i semenovodstvo,* 1936, No. 10, pp. 37–8. Later on Salamov capitulated to Lysenkoite views on the essential harm of inbreeding. See, e.g., his article in *ibid.,* 1940, No. 3.

65. Vavilov to F. D. Richey, 1935, Nov. 5. U.S. National Archives.

66. Richey to Vavilov, 1935, Nov. 25. That Vavilov already knew what Richey wrote him is evidenced in the volume he sent to Richey along with the letter of Nov. 5, and in the second volume of the same work, which went to press on the same day. See the essays by Khadzhinov and by Pisarev in Vavilov, ed., *Teoreticheskie osnovy,* I and II. Note in II, p. 429, Khadzhinov's unsupported assertion that Soviet agriculture was fully capable of utilizing double-cross hybrids. Vavilov's knowledge that this was not so apparently prompted his effort to discover if the American breeders had discovered a simpler use of inbred lines.

67. *Spornye voprosy,* pp. 33 and 52–3. Meister, who gave the official summation of this conference, cited the American experience as evidence against Lysenko's condemnation of inbreeding (*ibid.,* pp. 424–5), but he refrained from comment on its applicability in the Soviet Union. Three other speakers simply took it for granted that widespread use of first-generation double-cross hybrids was out of the question. See *Dostizheniia sovetskoi selektsii,* pp. 75–8 and 80–83. All three of these corn specialists, A. G. Shapoval, A. B. Salamov, and B. P. Sokolov, also took it for granted that such first-generation seed was far more productive than any other.

After the conference of December 1936, when Vavilov recognized the impossibility of placating the Lysenkoites, he boldly cited the American success and blamed the Lysenkoites for the failure of Soviet farms to benefit from the new hybrids. See especially *PZM,* 1939, No. 11, pp. 129–30. Of course there was an element of truth in the accusation, but Vavilov refrained from stating the larger reason why Soviet farms were not using the new hybrids.

68. A. S. Musiiko, "Novoe orudie povysheniia urozhaev kukuruzy," SZ, 1936, Oct. 30.

69. *Ibid.*, 1938, Oct. 20. See also Musiiko's article in *ibid.*, June 17.

70. See A. S. Musiiko, "Iskusstvennoe opylenie perekrestnoopyliaiushchikhsia kul'turnykh rastenii," *Iarovizatsiia,* 1939, No. 1, pp. 76–82, with introduction by Lysenko. See *ibid.*, 1940, No. 5, p. 4, for a letter that Lysenko and Prezent wrote to the Commissariat of Agriculture in December 1938 urging a change in the seed rules for cross-pollinating plants. In January 1939 the Commissariat arranged a conference that began the change. See Salamov's approving report in *Selektsiia i semenovodstvo,* 1940, No. 3, pp. 25–7.

71. See the report in *Iarovizatsiia,* 1939, No. 5–6 (26–27), pp. 226–9.

72. M. A. Popovskii, "Selektsionery," *Novyi mir,* 1961, No. 8. Cf. Tsitsin, *Sorta* (1944), pp. 217–220, for the hybrids certified.

73. *KPSS v rezoliutsiiakh,* II, 1052.

74. Lysenko, *Agrobiologiia* (1949), pp. 592–3.

75. The conference, held in June, 1949, is reported in *Selektsiia i semenovodstvo,* 1949, No. 9, pp. 72–9.

76. For an early example of B. P. Sokolov's effort to curry favor with Lysenko while continuing standard corn breeding, see his "Gibridy kukuruzy Dnepropetrovskoi gos. sel. st.," *ibid.*, 1940, No. 3, pp. 27–30. For the direct order he received, in September 1948, to stop work with inbred lines, see Popovskii, as cited in note 72, p. 204.

77. *Ibid.* Cf. Dudintsev, *Not By Bread Alone.* Note that Popovskii ignores Sokolov's effort to curry favor with Lysenko; his heroes, like Dudintsev's, are purely heroic.

78. Popovskii, as cited in note 72, p. 204, implies that 1951 was the last year of the ban. Cf. P. A. Baranov et al., "Problema gibridnoi kukuruzy," *BZ,* 1955, No. 4, p. 490, for an indication that interlinear hybrids were certified for farm use in 1952. Cf. also *Kukuruza,* 1956, No. 3, pp. 1–5; and *Mezhdunarodnyi sel'skokhoziaistvennyi zhurnal,* 1957, No. 2, p. 126.

79. For an illuminating review of Khrushchev's corn campaign, see J. Anderson, "A Historical-Geographical Perspective on Khrushchev's Corn Program," in Karcz, *Soviet and East European Agriculture,* pp. 104–134; and Jasny, *Khrushchev's Crop Policy.*

80. Shevchenko, pp. 16 ff.

81. *Plenum TsK* (1958), p. 420.

82. See, e.g., *Vsesoiuznoe soveshchanie po proizvodstvu kukuruzy. Krasnodar', 9–13 fev. 1960 g., passim.* By contrast, see Luk'ianiuk, for a 1956 conference, at which only two papers were concerned with the economics of corn growing.

83. See the first item cited in note 82, pp. 40 *et passim.*

84. Letter of C. O. Erlanson to D. Joravsky, Sept. 3, 1964. Dr. Erlanson was also kind enough to send me a copy of his mimeographed "Trip Report," in which he explains his assessment of Soviet hybrid corn, pp. 13 and 18.

85. See, e.g., I. E. Glushchenko, "Osnovnye printsipy i pervye rezul'taty selektsii kukuruzy v In-te genetiki AN SSSR," *Izvestiia Akademii Nauk SSSR, seriia biologicheskaia,* 1956, No. 3, pp. 31–49; Musiiko's paper in *Vsesoiuznoe soveshchanie po proizvodstvu gibridnykh semian kukuruzy, Dnepropetrovsk, 28–30 marta 1956 g.*, pp. 73–8; and Sokolov's criticism

of Lysenkoite corn breeding, at the 1960 conference cited in note 82, pp. 197–9.

86. See, e.g., Baranov, as cited in note 78, pp. 503–4; and the 1960 conference cited in note 82, *passim.*

87. For an American evaluation of the economics of this technique, as compared with detasselling, see Sprague (1955), pp. 391–3. For a contrasting Soviet evaluation, see Khadzhinov's paper at the 1960 conference cited in note 82, pp. 185–96. Dr. Sprague has informed me that American seed companies have changed their minds about cytoplasmic male sterility since his book was published.

88. See the data in *Vestnik statistiki,* 1963, No. 9, p. 83. Compare the sober assessment of Anderson, as cited in note 79.

89. From 1948 to 1955 the percentage of corn sowings under first-generation hybrids fell from 10 percent to 1.4 percent. See the 1956 conference cited in note 85, p. 394. Unfortunately the speaker does not specify whether the figures are for interlinear or varietal hybrids. Chances are they include both, the bulk of which were probably open-pollinated varietal hybrids.

90. For the post-Khrushchev fortunes of corn breeding and corn growing, see M. I. Khadzhinov and G. S. Galeev, "Sostoianie i perspektivy selektsii kukuruzy," *Genetika,* 1966, No. 10, pp. 56–66; cf. G. Lisichkin, "Spustia dva goda," *Novyi mir,* 1967, No. 2, pp. 160–85.

91. See Prianishnikov, *Azot v zhizni;* Slicher van Bath; Chambers and Mingay; and, for a convenient summary, Singer, IV, Chapter 1.

92. *Ibid.*, V, pp. 5–7.

93. See Shevchenko, for the impression Garst made on Khrushchev.

94. See Vakar, pp. 208–213.

95. See above, pp. 23–4.

96. See Vil'iams, *Sobranie sochinenii,* or the more convenient *Izbrannye.*

97. See Kononova, pp. 108–110.

98. See W. H. Allaway, "Cropping Systems and Soil," in USDA *Yearbook* (1957), pp. 389 *et passim.*

99. See *Pochvovedenie,* 1902, No. 4, pp. 434–6, for a brief expression of courteous amazement at Williams' views. The only other notice of him, in this chief journal of Russian soil scientists, was Sukachev's devastating critique, *ibid.*, 1916, No. 2.

100. See above, pp. 21, 24–5.

101. *Akademik V. R. Vil'iams,* pp. 24–5. See above, p. 28.

102. See *Izvestiia,* 1928, No. 159, for the celebration of the event.

103. See e.g., *Puti sel'skogo khoziaistva,* 1926, *passim,* for intensive discussion of *travopol'e* in such sparsely settled areas as Siberia. For the early 1930s, see above, pp. 65–7.

104. See above, p. 65.

105. At the Sixteenth Party Congress in June 1930 Stalin endorsed regional specialization. See Stalin, XII, 325–6. For the plan worked out by the specialists, see Cherdantsev. Cf. W. A. D. Jackson, "The Problem of Soviet Agricultural Regionalization," *Slavic Review,* 20:656-678 (Dec. 1961); R. G. Jensen, "The Soviet Concept of Agricultural Regionalization and Its Development," in Karcz, ed., *Soviet and East European Agriculture,* pp. 77–103; and *Bibliografiia po raionirovaniiu.*

106. See above, pp. 65–7.

107. See B. Blomkvist, "S sevooborotami neblagopoluchno," *SZ*, 1937, April 8; and M. Chernov, "O vvedenii pravil'nykh sevooborotov," *ibid.*, 1937, July 14.

108. See I. Kantyshev, "Za marksistsko-leninskuiu teoriiu i praktiku v agronomii," *Na agrarnom fronte*, 1934, No. 2–3 (Feb.–March), pp. 36–67, and No. 5–6, pp. 54–67. This heated argument for *travopol'e* carried the tag, "For discussion," but the celebration of Williams in *Pravda*, Nov. 18, 1934, endorsed his theories without that disclaimer of official authority. Between spring and fall of 1934 the public record reveals the following sequence: In June 1934 a plenary meeting of the Party's Central Committee issued an order for rapid increase in all sorts of fodder crops, with a major emphasis on grass. In July the Council of People's Commissars adopted an unpublished resolution "On the Experimental Verification of the *travopol'naia* System of Agriculture." That was probably the official evaluation of the experiment called for in the resolutions of the October 1931 Conference on Drought Control. That unpublished resolution of July 1934 was the first to be rescinded twenty-eight years later, when the Soviet government turned against *travopol'e*. (See *Sobranie postanovlenii pravitel'stva SSSR*, 1962, No. 3, p. 47.) On November 11, 1934, the Council of People's Commissars ordered a celebration of the fiftieth anniversary of Williams' scientific career, adding five years to the career in order to do so. (In 1914 Williams' students celebrated his twenty-fifth anniversary.) The article in *Pravda*, Nov. 18, 1934 followed.

109. Dmitriev, p. 54. For a representative comment on *plodosmen* by Williams himself, see his *Izbrannye sochineniia*, II, 707–8. Consistency was not a characteristic of Williams and his followers. Before nativism reached its peak, in the period 1948–1953, they sometimes defended *travopol'e* with the argument that it was widely used in Western Europe, in effect identifying *travopol'e* with *plodosmen*. At the same time they condemned the techniques of capitalist agriculture, calling them inapplicable in socialist agriculture.

110. T. D. Lysenko, "Ob agronomicheskom uchenii V. R. Vil'iamsa," *Pravda*, 1950, July 15. This article was widely reprinted. Lysenko seems to have been spelling out a decree of the Council of Ministers and the Party's Central Committee.

111. The last decree to be rescinded in 1962 had been issued in 1954. See *Sobranie postanovlenii pravitel'stva SSSR*, 1962, No. 3, p. 59.

112. See Khrushchev, VI, 56 *et passim*.

113. See, for example, S. P. Anikeev, "Sel'skokhoziaistvennoe proizvodstvo i nauka o pochve," *Voprosy filosofii*, 1954, No. 1, pp. 128–142.

114. See, for example, V. I. Anisimov, "Sel'skoe khoziaistvo," *Entsiklopedicheskii slovar' Granat*, XXXVIII (1916?), 110–23.

115. See above, p. 65.

116. See, for example, V. R. Vil'iams, "Travopol'naia sistema zemledeliia i 'mineral'naia agrokhimiia,'" *Khimizatsiia sotsialisticheskogo zemledeliia*, 1937, No. 10, pp. 3–18. In another article of 1937 Williams set the figure a bit higher. See his *Izbrannye sochineniia*, II, p. 114.

117. D. N. Prianishnikov, "I plodosmen i travopol'e," *SZ*, 1937, Sept. 20. Note that Prianishnikov objects to the sloppy use of the term *travopol'e* to mean any kind of rotation that includes some grass.

118. Lysenko, as cited in note 110.

119. *Sel'skoe khoziaistvo SSSR* (1960), pp. 134–5.

120. Prianishnikov, as cited in note 117. For an extended discussion see his *Azot v zhizni.*

121. See G. Pisarev, "Nazrevshaia popravka," *SZ,* 1937, June 22, for an attack on the Commissariat of Agriculture for leaving crop rotation to local authorities; the result, the author claims, has been little or no effort to improve rotations. Twenty-five years later I. Strakhov, "Ob ekonomicheskoi nesostoiatel'nosti travopol'noi sistemy zemledeliia," *Voprosy ekonomiki,* 1962, No. 2, p. 31, attributed the necessity of administrative orders and fines to "the theoretical groundlessness of the *travopol'naia* system."

122. See Zh. Medveder, p. 94, for the beginning of cooperation in 1936. After a little friction in 1937 (see "U zakrytykh dverei Timiriazevskoi Akademii," *SZ,* 1937, April 18), Lysenko and Williams intermittently praised each other in public. In 1938 Williams sponsored Lysenko's candidacy for election to the Academy of Sciences, and in 1939 Lysenko spoke at Williams' funeral. But the major Lysenkoite effort on behalf of Williams' school came after the war. See especially Lysenko's speech at the August Session in 1948, which made Williams and Michurin twin gods of Lysenkoite agrobiology. Note especially the speech by V. S. Dmitriev, Head of Gosplan's Division of Agricultural Planning, in *O polozhenii,* pp. 259–269. See also *VAN,* 1948, No. 9, pp. 75–7 and 119–129, for evidence of the campaign against standard soil science in the Academy's Institute of Soil Science. For evidence of the low state to which the All-Union Institute of Fertilizer and Agricultural Soil Science (*VIUA*) was brought, see *Partiinaia zhizn',* 1962, No. 13, pp. 40–4. See also the report of a team of U.S. soil scientists: USDA, *Soil and Water Use in the Soviet Union.*

123. See T. D. Lysenko, "Neotlozhnye zadachi sel'skokhoziaistvennoi nauki," *SZ,* 1961, December 15. Note that Lysenko fastened on a passage in one of Khrushchev's anti-*travopol'e* speeches where Williams was praised for his theory of soil formation. Cf. note 127.

124. V. R. Vil'iams, "Rukovodstvo k deistviiu," *Pravda,* 1937, March 7.

125. See, for example, V. Egorov and B. Dospekhov, "Opytnoe delo — nauchnaia osnova zemledeliia," *SZ,* 1965, March 21.

126. *Sobranie postanovlenii pravitel'stva SSSR,* 1962, No. 3, pp. 47–59.

127. Khrushchev, VI, 283–5, 383, *et passim.* Note that Khrushchev confined his remarks almost entirely to *travopol'e* as a system of crop rotation. He seemed quite neutral on problems of soil science. At one point, p. 176, he paid tribute to Williams for his theory of soil formation; but at another point, p. 182, he said that agricultural experience had "overturned many of Williams' theoretical assumptions." As a result, Lysenko and other followers of Williams were able to continue advocacy of Williams' pseudoscience, while genuine soil scientists were able to revive their discipline. See, for example, the discussion of the American system of soil classification in *Pochvovedenie,* 1964, No. 6, pp. 14–48.

128. *Izvestiia,* 1964, October 25.

129. A. Basov, "Rossiia zavershaet sev," *Sovetskaia Rossiia,* 1965, June 1.

130. Anatolii Agranovskii, "Nauka na veru nichego ne prinimaet," *Literaturnaia gazeta,* 1965, January 23.

10. IDEOLOGIES AND REALITIES

1. See Sakharov. One cannot help noting that he is a student of Tamm's, who publicly challenged the administrators of terror in 1937. See above, p. 124 and references in note 76 of Chapter 5.

2. V. V. Matskevich, "Nashe sel'skoe khoviaistvo vchera, segodnia, i zavtra," *Nauka i zhizn'*, 1965, No. 7, pp. 2–8.

3. Trapeznikov, *Leninizm,* II, 172.

4. Quoted in Joravsky, p. 241.

5. See *VAN,* 1966, No. 6, pp. 4–7, and No. 8, pp. 3 *et seq.*, for part of the story. See Sakharov, pp. 56–57 for the rest.

6. Maiakovskii, *The Bedbug,* pp. 220 ff. I have altered George Reavey's translation; his is somewhat euphemistic ("crap" for *govno*) and metrical, while mine is literal and prosaic.

7. Pasternak, p. 65.

8. Lysenko, *Raboty v dni,* p. 174. The original appeared in *Pravda,* 1941, Oct. 3.

9. See above, p. 31. The metaphor derives from Timiriazev and from Klebs.

Index

Index does not necessarily include names listed alphabetically in Appendices A and B.

Index

Index

Index

Index